Psychological Perspectives on Reality, Consciousness and Paranormal Experience

Tony Jinks

Psychological Perspectives on Reality, Consciousness and Paranormal Experience

palgrave
macmillan

Tony Jinks
School of Social Sciences and Psychology
University of Western Sydney
Penrith, NSW, Australia

ISBN 978-3-030-28901-0 ISBN 978-3-030-28902-7 (eBook)
https://doi.org/10.1007/978-3-030-28902-7

This Palgrave Macmillan imprint is published by the registered company Springer Nature Switzerland AG
The registered company address is: Gewerbestrasse 11, 6330 Cham, Switzerland

Introduction

Tom is 31 years old and works for a large multinational company as an internal IT consultant. That means he helps maintain the company's intranet and fixes all the problems employees encounter, or most times, create. To get an impression of Tom, imagine a Caucasian male of average height and weight, clean-shaven with mouse-brown hair and pale skin who wears thin-rimmed glasses and expresses a serious disposition (I can describe him this way because I guarantee he'll never read this book).

In mid-2015 Tom's company held a conference for employees at a hotel and function centre on the outskirts of a city, not far from the company's head office. The building was of late nineteenth-century sandstone construction, two stories high and in the shape of a square "U", with meeting spaces and dining facilities in the main (horizontal) section facing a busy arterial road, and guest accommodation in each of the two wings projecting away from the traffic. As the conference was a two-day affair, Tom and his colleagues stayed overnight. Tom's room was on the top floor of the southern wing, second door from the end of a long and well-lit hallway.

Following the completion of the first day's session, Tom made his way upstairs to his room to freshen up before dinner. Arriving at the door, he dumped a bundle of conference paperwork from under one arm onto the carpeted floor and extracted the room's key-card from his pocket.[1] Tom inserted the card into the slot, but as he did so he felt an urge to look up and to his left. From his vantage point, he noticed the remaining corridor didn't end abruptly in a wall or a window but instead was intersected by *another* corridor running at right-angles to the main hallway he was standing in (something he hadn't really considered to that point in time). Tom's analytical mind imagined the floor plan at this location to resemble a "T" shape, with him standing almost where the two lines joined. He also concluded the horizontal arms could be nothing more than short stubs. That made sense, since the building's wing appeared to be only as wide as his corridor and the two rooms on each side. Yet for all Tom knew, there was a whole *other* section of the hotel that used this right angle corridor as a connection.

However these musings came later. When Tom looked up he was entirely focussed on what he *saw*. To his surprise, it was a man walking swiftly across the gap, travelling from left to right. At least Tom *assumed* the figure was male because the posture and walking style appeared masculine, and he seemed to be wearing dark trousers and a dark jacket. The face in profile was pasty and clean-shaven, and the hair short and black. The man was in view for only a few seconds, so that was the best Tom can remember on re-telling the story. The overall impression he got was that the man was walking to *be somewhere*, with purpose.

Initially, Tom wasn't the least bit disturbed. The corridor was empty and no-one else *seemed* to be upstairs, but if a man was roaming the far end of the building then it meant there were more corridors and rooms beyond. The man might have been a worker or guest with a perfectly good reason to be on the second floor, crossing from one part of the hotel to another. Curious about the man's identity, Tom left his room door half-open and walked the few steps to the corridor's junction. Sticking his head into the gap and looking back-and-forth, he expected to see a flight of stairs or an open door…or the man just standing

[1]There's some licence included here, for the purpose of storytelling. However the description of events is entirely accurate.

there. Instead he was disturbed to find this cross-corridor was indeed "stumpy", and held nothing more than a small window to his right overlooking the hotel's courtyard, and a door leading an external set of fire stairs to his left.

Tom thought the man must have been hiding in the recess next to the fire escape to begin with. After all, had the mystery figure recently entered the corridor from the outside staircase an alarm would have activated the moment the fire-door was opened. While troubling, that wasn't Tom's real concern. What irked him most was where the man had *gone*. Recall the man had walked left to right across the corridor's gap. That is, from the fire-door in one alcove to the window in the other alcove. However in the alcove there was no man, no hiding place, and no open window (which was securely locked). Needless to say Tom made a hasty retreat back to his room, trying to comprehend what he'd just witnessed.

After reading this tale you would assume that Tom kept his story quiet, for fear of ridicule. He might admit that it happened, but many months later and only when he was trusting enough that the events he described wouldn't be dismissed as nonsense. Surprisingly, that's not the case. Tom is happy to say that he went straight down to pre-dinner drinks and treated his fellow workers to a detailed commentary of his sighting. Furthermore, his colleagues were equally happy to accept what he told them on face-value, and they spent the rest of the night musing on the event in the corridor (and other similar encounters the team had heard about, or even had themselves). No-one argued that Tom had *not* seen what he *said* he saw.

Who was the pale man in the dark clothing? Tom and his colleagues decided that if the man couldn't be found in the nook of the tiny corridor, and hadn't entered via the fire-door or exited via the window, then he must have been a *ghost*. Reflecting on the story, the acceptance of a ghost "explanation" is understandable. I associate with a lot of "average" people and I'd be hard-pressed to find someone who *wouldn't* agree that Tom had seen a ghost. It's a pervasive conclusion.

So what *is* a ghost? In advance won't be providing a definitive answer to this question. That's something I believe will *never* be achieved, by anyone. Instead the focus of this book is to summarise three distinctly different perspectives on what ghosts (and other strange things) might be, and why people report seeing them. Presented in this way, you can make up your own mind. After all, the emphasis is on the journey, rather than the destination—and it's a wild ride!

Contents

Part I

Paranormal Experience

1
What Is a Ghost?

I've chosen Tom's ghost story as the centrepiece *because* it is so humdrum. There are plenty of more interesting (although much rarer) examples I could have used that reside on the other end of the ghost-story spectrum, such as this one from a 56-year-old man named Dan;

> In December 2002, just a few days before Christmas, I woke up about 5:30 or 6 a.m. and walked over to the bedroom window. I parted the vertical blinds and cursed under my breath because it was still raining like it had been for days. As I turned away from the window, my body chilled from head to toe. Standing right in the middle of our bedroom doorway was this huge ghost dressed in a hooded brown robe. The guy was at least seven feet tall and filled the whole doorway. Where his face should have been in the big hood was just like a swirling mass of energy, and you also couldn't see any hands or feet – but he totally filled the whole doorway. As I looked at him he extended his arms forward like he was offering me this short staff or rod that he was holding – it was maybe 24 inches long. Then I turned my head away for a moment, and when I looked back he was gone.[1]

[1] Belanger (2007, pp. 201–202).

T. Jinks, *Psychological Perspectives on Reality, Consciousness and Paranormal Experience*,
https://doi.org/10.1007/978-3-030-28902-7_1

This unsettling tale is *too* dramatic for me to use, as it distracts from the book's primary aim to summarise the origins of "everyday" unusual witnessed events that are far more commonly reported, like Tom's man-in-a-suit.

At this stage it is worth taking a brief pause, and recognise that attributing Tom's (or even Dan's) story to a "ghost" is unhelpful. It begins at the wrong end. Instead of *starting* with this label, there should be a slow and methodical examination of the details of Tom's sighting without any reference to what a "ghost" might be. I'll be laborious and detailed and overly systematic at times, but I think that's necessary as this is not a method commonly employed—after all most books on "ghosts" (or anything else strange) set out to argue a particular case from the beginning. I'm not doing that, for I have no particular perspective to favour, and they are all of equal significance.

So back to ghosts.

While Tom's audience relied on this particular word to discuss the sighting, they might instead have used "apparition" to characterise the experience as something visual—that *appeared*. Apparitions can be in the solid human form of Tom's entity, known prior to the nineteenth century as the "living dead" because the ghost looked just like a real flesh-and-blood person until it disappeared. Alternatively it might seem substantial but be of the much rarer "living cadaver" type if its physical appearance showed signs of decomposition.[2] Luckily for Tom's state of mind he didn't see a rotting corpse. Then again, an apparition might be as elusive and insubstantial as a cloud of steam,[3] the typical representation being the wraithlike and shrouded "phantom".

Other words available to describe the ghost were "spirit" and "presence", and yet all these *terms in the manner in which they are used* imply the same thing. In the words of Welsh philosopher Henry Price,[4] rightly or wrongly Tom's friends are referring to the pre-scientific idea of a dead human being in a duplicate form to the human body still lingering on earth.[5] That

---

[2]Harvey (2004).

[3]Green (1976, p. 34).

[4]Tyrell (1953, p. 8).

[5]Of course, Dan's spectre doesn't give the impression of *ever* having been human, but it's a very unusual story.

idea is expressed in various ways, with degrees of additional information. For example, in the words of long-time ghost hunter Andrew Green, a ghost specifically refers to the earth-bound essence of a deceased person that for whatever reason continues to haunt a particular location, and will do so until they are persuaded to "pass on".[6] Indeed, by Gary Schwartz' definition a ghost is a combination of a spirit *and* a presence;

> .... spirit is used to refer to the hypothesized continued existence of the consciousness and information (and associated energy) after physical death.... [whereas] presence is used to refer the potential hypothesized localization of the consciousness, information, and energy/spirit of the deceased person.[7]

Ideas like "passing on" (alongside "apparition", "spirit" and "energy") are intuitively understood, even if the mechanisms underlying these states, and how they could transpire, remain a mystery. A listener might even fill in the missing gaps in Tom's story, concluding that the ghost (or whatever label they use) was indeed someone who had died on the hotel's second floor at some undisclosed time in the past, perhaps tragically. Feasibly, he committed suicide in one of the end rooms because of a business decision or an affair gone-wrong, and has never "passed on" to the afterlife. That's why the ghost was seen by Tom—he needs assistance in the *passing on* process!

Taking this idea further, a ghost traipsing the corridor as a *man*, albeit in a diminished form, is a sentient *active spirit presence*.[8] This is arguably the most popular contemporary conception of what a ghost is. On the other hand, maybe we're giving too much clout to Tom's ghost? The ghost-concept has been intensely studied for well over 100 years. Indeed the best work was done prior to (and just after) the turn of the twentieth century by illustrious names associated with the *Society for Psychical Research* (SPR); Barrett, Sidgwick, Myers, Gurney, Lang, Crookes, Rayleigh and Lodge. They each provided alternative (and creative) theories about what ghosts are, that don't require them to be "active". For instance, the seventh President of the Society, Frederick Myers, suggested a ghost need not be

[6] Green (1976, p. 33).
[7] Schwartz (2010, p. 166).
[8] Schwartz (2010).

sentient (or even partially sentient). Rather it is a *veridical after-image*, meaning the entity doesn't force its perception onto the witness. In Myer's words;

> Instead of describing a "ghost" as a dead person permitted to communicate with the living, let us define it as a manifestation of persistent personal energy, or as an indication that some kind of force is being exercised after death which is in some way connected with a person previously known on earth.[9]

And further;

> ...there is strong evidence for the recurrence of the same....figures in the same localities, but weak evidence to indicate any purpose in most of these figures, or any connection with bygone individuals, or with such tragedies as are popularly supposed to start a ghost on its career. In some of these cases of frequent, meaningless recurrence of a figure in a given spot, we are driven to wonder whether it can be some deceased person's past frequentation of that spot, rather than any fresh action of his after death—which has generated what I have termed the veridical after-image—veridical in the sense that it communicates information, previously unknown to the percipient, as to a former inhabitant of the haunted locality.[10]

Applied to Tom's encounter, his ghost is merely *residual.* It is the impressions of a former guest or worker imprinted onto the built environment through some habitual and emotion-charged activity they undertook when they were alive,[11] and that can be visually "played back" to a living witness until the imprint of this energy has dried up. An alternative name for this notion is the *stone-tape* effect (or stone tape theory[12]), although in this case "wood-tape" or "plaster-tape" might be more appropriate in deference to the composition of the structure that the remnant is perceived in.

---

[9] Myers (1903/ 1978, p. 703).
[10] Ibid.
[11] Deveraux (2001, p. 187).
[12] Brown (2003).

But there's more to ghosts than active spirits or veridical afterimages. From the nineteenth century onwards, investigators identified *apparitions of the dying*. Here is an example from the writings of engineer and parapsychologist George Tyrrell;

> A case is given in which a man came out of an upstairs room, in which he had been alone, on to the landing, noting the time he did so. He saw the apparition of an old lady, dressed like his mother, come out of another room on the same floor and descend the stairs. As he watched the figure going down, his wife came up the stairs, and passed so close to the descending figure that she appeared to him to brush against it. But she had neither seen nor felt it. At precisely the same time the percipient's mother, dressed in outdoor clothes as the figure had been, died in another town.[13]

Tyrrell also identified a much subtler style of ghost, the *experimental apparition*. To create this form of apparition, an individual purposefully projects an image of themselves to a remote percipient. The resulting apparition possesses all the characteristics of a ghost, but isn't a *real* ghost (in the sense that it represents the spirit of a deceased human). In addition there is the *crisis apparition*, which is not a deliberate projection from the agent but occurs—as suggested in the title—when the individual is suffering from significant personal crisis. For psychoanalyst and ghost investigator Nandor Fodor, the "driving motive" for the crisis apparition is a need to inform the unsuspecting witness of vital information. That could be significant anxiety, danger, severe illness or impending death.[14] Of course these categories are flexible, since there are examples of living apparitions that are not the result of a crisis, but are also not strictly deliberate either.[15] They more closely resemble a *spontaneous apparition*. Spontaneity is represented in the following case from an early issue of the SPR journal, where the "motivation" is as routine as a worrying dream experienced by Miss L. Bourne of her father Canon Bourne;

---

[13]Tyrell (1953/1963).

[14]Fodor (1933, pp. 60–66) (From "Cases Supplied to the Literary Committee", *Journal of the Society for Psychical Research*, No. CIII, Vol. VI, p. 129, October 1893). While death is certainly a usual outcome, as the name implies it is not a *necessary* condition that the agent of the crisis apparition die shortly after they are seen in spectral form.

[15]Irwin and Watt (2007, p. 197).

> On February 5th, 1887, my father, sister, and I went out hunting. About the middle of the day my sister and I decided to return home with the coachman, while my father went on. Somebody came and spoke to us, and delayed us for a few moments. As we were turning to go home, we distinctly saw my father, waving his hat to us and signing us to follow him. He was on the side of a small hill, and there was a dip between him and us. My sister, the coachman and myself all recognized my father, and also the horse. The horse looked so dirty and shaken that the coachman remarked he thought there had been a nasty accident. As my father waved his hat I clearly saw the Lincoln and Bennett mark inside, though from the distance we were apart it ought to have been utterly impossible for me to have seen it. At the time I mentioned seeing the mark in the hat, though the strangeness of seeing it did not strike me till afterwards. Fearing an accident, we hurried down the hill. From the nature of the ground we had to lose sight of my father, but it took us very few seconds to reach the place where we had seen him. When we got there, there was no sign of him anywhere, nor could we see anyone in sight at all. We rode about for some time looking for him, but could not see or hear anything of him. We all reached home within a quarter of an hour of each other. My father then told us he had never been in the field, nor near the field, in which we thought we saw him, the whole of that day. He had never waved to us, and had met with no accident. My father was riding the only white horse that was out that day.[16]

When reviewing the case, Frederick Myers believed that Canon Bourne was himself dreaming of (or imagining) having had a fall, and projected this dream information into the waking state of another person; his daughters and the coachman. Myers labelled the particular talent possessed by Canon Bourne *psychorrhagic diathesis*. This psychic ingredient is able to disconnect from an individual's body and become a *phantasmogenetic centre* that influences the physical space surrounding other people as a ghostly impression. The mechanisms behind the process were unknown to Myers, although he believed it was not a material (physical) or optical effect. As best explained, it was a form of mental intrusion that *immaterially* and non-*optically* created the phantasm of Canon Bourne (and the horse) to be apprehended by the three witnesses.

[16] Fodor (1933, p. 64).

Without even positing a subconscious motivation of fear or dread, a spontaneous apparition might be no more than an accidental *apparition of the living*, epitomised by this brief tale of the bilocation of Mrs Milman in 1890s London;

> The other day a friend took leave of me in my work-room…Scarcely had he stepped out of the door, when he found me again on the landing of the stairway. Dumbfounded, he shrank aside to let me pass.[17]

Mrs Milman felt she was located in her work room, not in the landing, so she was unaware who the other "her" might be. Although this case might represent the appearance of a *doppelgänger* (a body double), that term better refers to cases where people confront *themselves* when going about their everyday business. In the case of Mrs Milman, she never actually saw her double face-to-face, but it seems she was aware that she possessed a double, and had heard it making noises (like opening and closing doors) at various locations around her house.[18] She was dumbfounded as to why her living apparition was being seen by other people.

Returning to Tom's ghost, the inference of all those who heard his story (and Tom's own opinion) was that it was a modern-style, active spirit. Had he spoken with more erudite people, he might have been introduced to the unusual observation that it was a veridical afterimage, crisis apparition or even his doppelgänger. Tom has never suggested what he saw was indeed *himself*, or someone he knew many miles away in the throes of dying at the very moment he was opening his hotel door in order to take a shower. And without the typical maudlin glances or burial shrouds of a nineteenth century tale to inform us of what category of apparition we are dealing with, there's nothing to suggest Tom's sighting was *any* one of these examples.

For this reason, rather than assuredly labelling Tom's sighting a ghost, it would be more beneficial to place the encounter (at this stage at least) within the broader label of a *paranormal experience*.

---

[17] Inglis (1985/1986, p. 186)

[18] Ibid.

References

Belanger, J. (2007). *The Ghost Files: Paranormal Encounters, Discussion, and Research*. Franklin Lakes, NJ: Career Press.

Brown, A. (2003, October 26). Thoughts That Keep Going Bump in the Night. *The Sunday Times*, p. 3.

Deveraux, P. (2001). *Haunted Land: Investigations into Ancient Mysteries and Modern Day Phenomena*. London: Piatkus Ltd.

Fodor, N. (1933). *Encyclopedia of Psychic Science*. London: Arthurs Press.

Green, A. (1973/1976). *Ghost Hunting: A Practical Guide*. St Albans: Mayflower Books.

Harvey, J. (2004). The Photographic Medium: Representation, Reconstitution, Consciousness, and Collaboration in Early-Twentieth-Century Spiritualism. *Technoetic Arts: An International Journal of Speculative Research, 2*(2), 109–124.

Inglis, B. (1985/1986). *The Paranormal: An Encyclopedia of Psychic Phenomena*. London: Paladin.

Irwin, H. J., & Watt, C. A. (2007). *An Introduction to Parapsychology* (5th ed.). Jefferson, NC: McFarland Inc.

Myers, F. (1903). Phantasms of the Dead. In M. Ebon (Ed.) (1978), *The Signet Handbook of Parapsychology* (pp. 273–285). New York: Lombard Associates.

Schwartz, G. E. (2010). Possible Application of Silicon Photomultiplier Technology to Detect the Presence of Spirit and Intention: Three Proof-of-Concept Experiments. *Explore, 6*(3), 166–171.

Tyrell, G. N. M. (1953/1963). *Apparitions*. New York: Collier Books.

2

What Is the Paranormal?

Of *course* Tom had a paranormal experience. That goes without saying—even if you don't believe he truly saw something extraordinary. But what do these two words actually mean? To begin to understand Tom's sighting there needs to be separate appreciation of what the paranormal *is*, what paranormal *phenomena* are, and what it means to have a paranormal *experience* (without necessarily concluding that what Tom saw was indeed paranormal—whatever that might mean). This is not an original idea, but it is certainly worth pursuing because Tom's sighting cannot be resolved (if that is even possible) without an understanding of these very different concepts.

Beginning with *paranormal*, many definitions have been provided over the last 100 years but arguments still occur over what the word actually means. The problem is its ambiguous etymology. "Para" is a prefix of Greek origin, and can take on various meanings depending on the word it's attached to. That could be "beside" or "beyond" to "resembling" or even "auxiliary to" (in the manner of words like paralegal, paramedic or paramilitary). While "besides" and "auxiliary to" best represent what paralegals and other "paras" might do in their work practice, the application

T. Jinks, *Psychological Perspectives on Reality, Consciousness and Paranormal Experience*,
https://doi.org/10.1007/978-3-030-28902-7_2

most appropriate to the paranormal is *beyond* normal. That's certainly the case with Tom's sighting. It wasn't *next* to normal, or resembling normal. His extraordinary sighting, and its investigative aftermath, was something beyond what Tom would consider typical of his everyday life.

The word *paranormal* appears to have been in irregular use before 1883, but by that year it explicitly referred to phenomena or conditions that diverged from normality, although not necessarily of a mystical or psychical nature. That is, the events it described were instead merely "abnormal, unusual, or irregular".[1] During that time *supernormal* was a more popular term used by the likes of Myers to describe events that occur beyond expectation, or more specifically beyond the known laws of nature. The physician Joseph Maxwell acquired the word *paranormal* from French in 1905, although he used it mainly to describe premonitions of death.[2] It was not until the 1930s that *paranormal* finally overtook *supernormal* as the preferred expression for all those "beyond normal" events people were interested in at the time, in particular the extraordinary powers of the mind.

Another expression available to describe Tom's encounter is *supernatural*, although similar semantic problems exist on account of the word's origins. Supernatural evidently means *super to nature*, but since this was a religious concept introduced towards the end of the Middle Ages the word actually refers to things that belong *higher* than nature.[3] In other words, God, Heaven and Angels. On account of this overtly religious association, nineteenth century spiritualists (and the SPR) tended to avoid using *supernatural*, leaving *supernormal* and *paranormal* the preferred terms. Nevertheless, the pursuit of things *super* to the *natural* does seem very similar to examining things *beyond* the *normal*, and in the modern age the two terms are often used interchangeably.

Debating the meaning of *the paranormal* is only of academic interest and rarely side-track authors of popular books with "paranormal" in their title. These authors tend to assume readers are familiar with what the label

[1] Penman (2015, p. 34).
[2] Penman (2015).
[3] Lindeman and Svedholm (2012).

represents. As a result, "the paranormal" becomes a function of the topics that are discussed in the book, and not the other way around.

Hence from an A to Z perspective, the author might begin with *abductions by aliens.* Just about everyone is familiar with the notion of alien abductions, starting from the idea that aliens are extraterrestrial lifeforms from some other planet flying to earth in nuts-and-bolts spacecraft. Apparently they are interested enough in people (for whatever hidden reason) to want to kidnap them from cars or bedrooms in order to physically inspect them, or discuss environmental concerns with them (or both) before conducting a tour of the ship and returning the victim to where they came from. In this sense the notion of aliens as paranormal is a little shaky, because as physical entities they would be expected to conform to the laws of the universe like everyone else. They are a slightly more *advanced* version of normal, rather than being entirely *beyond normal.*

Then again, aliens often do things that are weird enough to be beyond normal. Here's a snippet of an abduction story that emphasises the seemingly paranormal nature of the phenomenon. One night in 1976 four men aged in their early twenties were (allegedly) taken from their canoe on the Allagash River in Maine by a beam of light attached to a flying saucer. They were miraculously transported inside the craft and examined by aliens (of the typical "gray" variety later pervasive in popular culture). Following the physical examination;

> … the abductees were made to dress and enter another room, which had a round portal in one of its walls. They were lined up and made to walk into the portal. Strange sensations surged through their bodies as they found themselves floating down the hollow beam of light into their canoe, which was now floating in shallow water and their campsite. The tube-like beam of light seemed to hold the canoe steady as each person was placed in it, in the same seating position that they were in prior to their abduction.[4]

Floating up and down beams of light is strange enough to warrant a paranormal attribution, even if exotic theories from the fringes of physics (a field labelled *paraphysics* by SPR President Bernard Carr in a 2002

[4]Fowler (2001, p. 57).

address[5]) could be offered to bring a comfortable scientific perspective to the mechanism behind the beam's action. This type of alien activity is not unusual, so it's understandable why abductions would be the first entry of the typical paranormal encyclopaedia.

After a few hundred pages of the encyclopaedia the last entry might be *zombies*,[6] or if the author is savvy with the latest terminology, *zooforms*—a neologism created by Jon Downes[7] to differentiate the various types of peculiar animals people report seeing in different places around the world. Downes proposes three categories of peculiar animal; *cryptids*, *pseudo-cryptids* and *zooforms*. Cryptids are animals that haven't been discovered yet (or potentially rediscovered, in the case of the Thylacine—Tasmanian Tiger—which *might* still roam the Tasmanian wilderness). Bigfoot and Yeti, if they are actual species of undiscovered hominid, are more extreme examples of cryptids. Pseudo-cryptids are simply out-of-place animals that might have been let go in the wild, such as a panther roaming in Britain or a wallaby hopping around in the mid-west United States. These categories are not paranormal.

Zooforms, the third type of peculiar animal, are entirely unexpected. They are entities that possess a wide variety of animal physiognomies and yet (to use the prefix *para-* a second time) possess a *paraphysical* nature. That means their existence is not grounded in the physical world like a "typical" Thylacine, Bigfoot, Yeti, panther or wallaby. *Paraphysical* originally described the physical aspects of psychical events generated by mediums and their spirit guides in the séance room.[8] When applied to zooforms, the implication is that these entities are not just beside (or resembling, or auxiliary to) normal animals, but *beyond physical* animals. They therefore display typical apparitional qualities, as shown in this nineteenth century example of a spectral dog from an early reference work of unusual occurrences in nature, the *Book of Dates*, published in 1866;

[5]Carr (2008).

[6]E.g., Gordon (1992/2000) (in Gordon's book, Zooforms is the *second* last entry; Zoastrianism comes last).

[7]Coleman (2013).

[8]Besterman and Gatty (1934).

> I was returning home late at night in a gig with the person who was driving. When we came near the spot where a portion of the gibbet had lately stood, we saw on the bank of the roadside a flame of fire as large as a man's hat. 'What's that?' I exclaimed. 'Hush!' said my companion, and suddenly pulling in his horse, came to a dead stop. I then saw an immense black dog just in front of our horse, the strangest looking creature I ever beheld. he was as big as a Newfoundland, but very gaunt, shaggy, with long ears and tail, eyes like balls of fire, and large, long teeth, for he opened his mouth and seemed to grin at us. In a few minutes the dog disappeared, seeming to vanish like a shadow, or to sink into the earth, and we drove on over the spot where he had lain.[9]

That briefly summarises the first and last topics of my hypothetical A–Z encyclopaedia of the paranormal; and somewhere in the middle will be a section dedicated to ghosts, with accompanying stories resembling Tom's (or Dan's). Generating an A–Z list of the paranormal in this fashion is perfectly reasonable. Whether you accept these topics as legitimate or not, there's no escaping that as the stories are told, encounters with aliens or zooforms, and everything in between, do appear to be *beyond normal*.

A serious encyclopaedia might further apply a more discerning set of sub-categories to its ghostly subject matter, using the *para-* prefix for a third occasion to identify particular themes. For instance, *parapsychology* is a term introduced in the late nineteenth century by philosopher-psychologist Max Dessoir to describe phenomena that fall outside (*beside* or *beyond*) normal *psychological* events.[10] In this context the subject matter of parapsychology is undoubtedly paranormal in nature, rather than something unusual but grounded in accepted psychological theory.

Traditionally, *life-after-death* (LAD) is the foremost concern of parapsychology, and implies a person's psychological essence—their mind, or soul, or spirit—survives after the demise of the physical body. Interest in the afterlife became the bread-and-butter of paranormal investigation by the turn of the twentieth century. Indeed, parapsychology *was* the paranormal (or based on the years in question, the supranormal). However, there's more to LAD than investigation of apparitions. Mediumship,

---

[9] Keel (1975/1976, p. 27).

[10] Ebon (1978, p. 18).

defined as supposed communication with the dead,[11] was essentially the research expression of LAD, with early investigators exploring the range of disparate practices employed by mediums in a séance to facilitate the contact with the spirit world (which assumes a *survivalist* interpretation, rather than fraud or telepathy).

A typical investigative method is *mental mediumship*, where the medium gains essential information about a deceased individual in the absence of any earthly clues. In other words, the spirit mentally communicates to the medium information about things that had happened during the time they were alive, or potentially events that are happening to living sitters in the séance room.[12] Independent of, or concurrent to the display of mental mediumship may be the spectacle of *physical mediumship*. This would involve the materialization, levitation or apport (appearance)[13] of objects in the séance room, created by the deceased spirit or through the medium's spirit guide.[14]

Further expressions of LAD include *past life recall*, which assumes retention of an individual's personal essence during cycles of birth, death and rebirth (i.e., reincarnation)[15] and also *near death experiences* (NDEs). Regarding the latter, it's not surprising that NDEs have an association with LAD as they both have "death" in their title. Of course, a person undergoing an NDE doesn't actually *die* despite being close to the point of death (or demonstrably dead for a period of time). They need to be able to report on what happened to them! There's also no necessity to be even close to dying, since NDE-*like* reports emerge individuals undergoing physical or emotional peril,[16] and even meditation, fainting and falling asleep.[17] The experiential elements from these antecedents are so compatible with "true" NDEs that the syndrome refers to the range of symptoms reported, rather than any putative cause.

---

[11] Luke (2011).

[12] Sudduth (2009).

[13] Ibid.

[14] Alvarado (2014a).

[15] Jones (2017).

[16] Cassol et al. (2018).

[17] Cassol et al. (2018).

The NDE achieved popular recognition as a consequence of Raymond Moody's popular 1975 book *Life After Life*[18] which brought the topic to a wide audience. They're also considered common enough, at 4% of the adult population, to warrant impartial study (and its own scientific journal).[19] In terms of progression, successive researchers have attempted to map the stages of an NDE, which usually start with the feeling of peace, calm and wellbeing followed by the sense of being divorced from the physical body,[20] or indeed *any* physical grounding. This can be accompanied by *autoscopy*—a term describing the personal view of one's own body from an external perspective.[21] However this is a special type of autoscopy, since the individual no longer identifies with the body they perceive from a distance.[22] Instead, they are most connected with their new non-physical state. Such a statement is predicated on an understanding of "normal" embodiment, which is grounded in a first-person sense of self, coupled with knowledge of the where the body is positioned in time and space.[23] In this story, the sense of *dis*embodiment is narrated by a patient in a life threatening situation;

> I knew something was going to happen…and then I went unconscious … and I was looking down and could see myself going into convulsions and I was starting to fall out of bed. And the girl in the next bed screaming for the nurses… The nurse caught me and put me back and by then there were two other nurses there and one came back almost immediately with a tongue depressor on my tongue. And they got the rides up on the bed and they called the doctor.. ..It was a feeling of height, great distance, a light feeling, like being up in a balcony looking down and watching all this and feeling very detached as though I was watching someone else, like you might watch a movie.… It was a very calm, relaxed feeling, a feeling of well-being if anything.. .everything was clearly seen like watching television.[24]

18 Moody (1975/2001).
19 Hanes (2012).
20 Cassol et al. (2018).
21 Arnette (2012).
22 Rabeyron and Caussie (2016).
23 Blanke and Metzinger (2008).
24 Sabom and Kreutziger (1977, p. 199).

This represents the first stage of an NDE, the *locational* component.[25] They are now primed to enter the second stage of the NDE, the *noetic* component where they gain knowledge not ordinarily accessible to them, and where their state of awareness operates very differently to normal.[26] The process begins with the voyager transitioning to a "dark tunnel" before being surrounded by (and merging with) a fabled tunnel of light that is variously described as "brilliant", "warm" and "loving".[27] Although these symptoms define the NDE, they are not always encountered in this order, and components can be missed completely. Here's one interesting variant;

> …and then not only did I see the colored lights, it looked like it was coming towards us. And this time instead of having a sensation of just hanging in space, I had the sensation of going up like I was lifting up...there was a big white light, but it was not like one man, but maybe two in light. It wasn't bright, blinding light at all. It was a white light that looked like two people but I didn't actually see who it was …. They were coming toward me…I had a complete relaxed feeling.. contentment.[28]

The quintessential NDE event—the *end* of the journey—involves contact with deceased relatives and friends coupled with a life review[29] and perhaps a "life-changing" communication with a divine "Being of Light".[30] This could be a religious figure or Angel, but may just as likely be one of those dead friends or relatives. There are even reports of the end journey involving a meeting with unknown people, and even deceased pets (if the voyager is a child).[31] Proponents argue that the purpose of these encounters is to ensure the voyager's disconnected soul is sustained after bodily death, which is where the NDE becomes an expression of LAD. Of course, the voyager needs to return to their living body, which might

[25] Tart (1998).
[26] Ibid.
[27] Serdahely (1990).
[28] Sabom and Kreutziger (1977, p. 201).
[29] Mobbs and Watt (2011).
[30] Stafford-Betty (2004).
[31] Serdahely (1990).

occur automatically, or through instruction from the guide who they meet in this alternative place.

If NDEs imply the continuance of the human spirit or soul following death, then it's reasonable to argue this putative spirit or soul can indulge in disembodiment when the owner is still perfectly alive. After all, we've learned how there are plenty of reports of apparitions of people who are, or end up, completely healthy. This is best conveyed through another paranormal event; the spontaneous and benign *out-of-body experience* (OBE). In this tale, the traveller was doing nothing more than gardening;

> [While working in the garden] I got happier and happier. All I was thinking about was I was doing something worthwhile. I was not thinking about anything else. Just doing the garden. I was completely carried away. I was so happy after about one and half hours (worked it out after) with digging that I felt myself rising into my head like not connected to my body anymore, floated there for a while, think only a few moments, then suddenly I was outside my body, looking down on it. A distance away, I didn't think this was strange, I knew exactly who I was and what I was and I was 'me'. The real me. I knew that my body was part of me but was not the real me. My body was my puppet and I loved it very much. As I was very concerned with what it was doing. I felt extremely contented and fulfilled. My body meanwhile was actually carrying on by itself digging. This is what confused me. A thought came into my head: "How could I possibly be up there?" and I was just back into my body in a flash. I was slightly annoyed at having come back.[32]

Contemporary interest tends to focus on the OBE (rather than the NDE) as it is a broader term for human *separation experiences.* They're also reported much more than NDEs—as many as 14% of the general population report having had them (compared with 4% for NDEs).[33] OBEs are also an older subject matter than NDEs for laboratory (or medical-style) research. Parapsychologist Carlos Avarado has traced the term back

[32] Alvarado (2016, p. 6).

[33] Rabeyron and Caussie (2016).

to at least 1911, when they were known as an "out of *the* body" experience.[34] And by the mid-1950s psychologist (and parapsychologist) Charles Tart, in collaboration with Robert Munroe (who popularised self-induced OBEs in his 1971 book *Journeys Out of the Body*[35]), performed controlled experiments with OBE travellers.[36]

In terms of characteristics, OBEs resemble the early stage of the NDE. The traveller reports a temporary spatial disconnect with their physical body[37] with an emphasis on the autoscopic feature often found with an NDE. Indeed, disembodiment is of *more* relevance to OBEs than NDEs, because the outcome of an NDE—reaching the afterlife—is the defining feature, rather than autoscopy. Here Robert Monroe describes one of his early OBEs in 1958, with the accompanying spatial disconnect;

> .... four weeks later, when the vibrations came again, I was duly cautious about attempting to move an arm or leg. It was late at night, and I was lying in bed before sleep. My wife had fallen asleep beside me. There was a surge that seemed to be in my head, and quickly the condition spread through my body. It all seemed the same. As I lay there trying to decide how to analyze the thing in another way, I just happened to think how nice it would be to take a glider up and fly the next afternoon (my hobby at that time). Without considering any consequences—not knowing there would be any—I thought of the pleasure it would bring. After a moment, I became aware of something pressing against my shoulder. Half-curious, I reached back and up to feel what it was. My hand encountered a smooth wall. I moved my hand along the wall the length of my arm and it continued smooth and unbroken. My senses fully alert, I tried to see in the dim light. It was a wall, and I was lying against it with my shoulder. I immediately reasoned that I had gone to sleep and fallen out of bed. (I had never done so before, but all sorts of strange things were happening, and falling out of bed was quite possible). Then I looked again. Something was wrong. This wall had no windows, no furniture against it, no doors. It was not a wall in my bedroom. Yet somehow it was familiar. Identification came instantly. It wasn't a wall, it was the ceiling. I was floating against the ceiling, bouncing

34Alvarado (2014b).
35Inglis (1985/1986).
36Tart (1998).
37Rabeyron and Caussie (2016).

> gently with any movement I made. I rolled in the air, startled, and looked down. There, in the dim light below me, was the bed. There were two figures lying in the bed. To the right was my wife. Beside her was someone else. Both seemed asleep.[38]

OBEs can involve more than a feeling of "mere" disembodiment. Following a brief loss of consciousness there is a common report that one's own body shape has been distorted—perhaps in the form of a sphere or a cloud. This leads not so much the impression that the voyager has left their body (i.e., has become *disembodied*), but rather they've been *differently* embodied.[39]

To use the para-prefix yet again, this impression of "disembodied embodiment"[40] is known as *parasomatic*, meaning beyond or beside the *body*, rather than the mind.[41] And where an actual disconnect between the physical body and the mental state is postulated, the connection is said to be through an *astral cord*.[42] That's why the parasomatic shape has also been labelled the *astral body*. Once disconnected from the sleeping physical form, this body can potentially travel beyond the bedroom ceiling (as in Monroe's example) to the next room, or somewhere else in the local vicinity or even somewhere completely unexpected and unknown.[43] Here is a report of *astral travel* reported from a woman named Nadia;

> After going to bed, she woke up outside her body, near the bedroom ceiling. She then felt her body as "a circular thing", and felt as if she was drawn upwards, with the impression of going through the ceiling and travelling at dizzying speed in "dark surroundings". She suddenly stopped "in the middle of the cosmos", with stars all around her, before being seized with great anxiety, accompanied by a certainty: if I go further I will never be able to return to my body. She then felt herself drawn downwards and "reintegrated" her sleeping body a few seconds later.[44]

---

[38] Tart (1998, p. 83).
[39] Rabeyron and Caussie (2016).
[40] Ibid.
[41] Irwin and Watt (2007, p. 178).
[42] Ibid.
[43] Dane (1975–1976).
[44] Rabeyron and Caussie (2016, p. 61).

In the opinion of some practitioners, what Nadia is reporting here is in fact *astral projection*. More specifically, astral travel is where the traveller's thoughts are projected onto a location leading to an impression they are there, whereas in astral projection something called an *astral counterpart* actually leaves the physical body and *does* relocate to that other, alien place.[45]

The astral journey is frequently identified as a category of *mystical experience*. These possess features separate from, or additional to, the disembodiment or altered body perception[46] of the OBE and astral journey. In a mystical experience there is a particular emphasis on the affective sensations of joy and sacredness. The traveller no longer relies on their conventional senses, hence personal identity is replaced by a feeling of oneness or unity with existence, of standing outside of space and time. This in turn leads to a belief in an understanding of the nature of *everything*.[47] On account of these emotional and noetic qualities an explicitly religious interpretation might be applied, with a religious figure, practice or tome incorporated into the journey.

Summarising these different states, it is apparent that NDEs, OBEs, astral journeys and mystical experiences share many similarities, but also show plenty of variations. The early stages might involve autoscopy, alternative embodiment and partaking of a journey, but they terminate at different destinations and potentially have different causes. Most importantly, this differentiation is not a necessary function of the process of dying, hence none of the claims provide necessary evidence for LAD. Charles Tart makes clear that nearness to physiological death might only encourage an NDE, but it is not necessary for OBEs and that's presumably the case for astral journeys and mystical experiences too.[48] However,

[45]Anonymous http://www.forumforastral.com/techniques-and-help-to-astral-projection/difference-between-astral-travel-astral-projection/); The notion of a true disconnect between the traveller's thoughts and their earth-bound body is too controversial for "regular" OBE researchers, since its origin can be traced to the aetheric self of Theosophy. Consequently expressions such as astral projection are avoided so as to evade association with the Theosophic theory of separation (e.g., Inglis, p. 56).

[46]Blackmore (1992) and Tart (1994).

[47]E.g., Lynn and Evans (2017).

[48]Tart (1998).

all four fall under the parapsychological umbrella—hence they are often discussed as related phenomena.

So what does this have to do with Tom's "ghost" sighting? The suggestion he witnessed the apparition while in the process of an NDE, an OBE or astral journey, seems ridiculous; none of the symptoms he reports involves a personal sense of disembodiment or a local or distant journey to *somewhere*. And he certainly wasn't dying. What he says he felt and "saw" occurred in an entirely natural manner, and in the corridor he was standing—not in some Heavenly tunnel or another cosmos. However the (alleged) mechanisms, processes and language of disembodiment might still be relevant if there are difficulties with clarifying how someone can actually see a ghost, which will be made clear later in the book.

Instead of travelling, Tom ghost might be attributed to another field of parapsychology; *psi*. Psi a term devised during the early 1940s by physiologist Bertold Wiesner and psychologist Robert Thouless to encompass things not necessarily associated with LAD (including forms of "soul" and disembodied journeying), but which have a definite parapsychological nature. These are the particular functions of the mind that are beyond acceptable, mainstream psychology. The two most significant categories of psi are *extrasensory perception* (ESP) and *psychokinesis* (PK).[49] ESP refers to the gleaning of information about physical or mental states, from the past, present or future, without the use of standard senses[50] and includes the potentially interrelated concepts of telepathy, clairvoyance and retrocognition/precognition.[51]

Alternately, PK implies a peculiar mental *exertion* rather than the reception of information—the ability to affect matter without direct physical contact. Most celebrated is the movement, lifting or bending of objects, or at its most subtle the influence of the mind over the output of digital

[49]Thalbourne (2009) and Neppe and close (2015).

[50]Thalbourne (2009).

[51]E.g., von Lucadou and Wald (2014). It should be noted that Etzel Cardeña (2018) argues that the "ESP" is not an appropriate expression because it emphasizes *perception* as the process behind the alleged capability, and there's no necessary requirement for this to be the case, as I'll shortly discuss.

equipment,[52] although it can also refer to the healing or "hexing" of people through psi action.[53] These are respectfully referred to as *receptive psi* and *active psi*.[54]

On face value, Tom's ghost is a representation of LAD because he and his colleagues accepted the sighting as a traditional ghost. It's hard *not* to assume the figure was a deceased former resident of the top floor of the south wing of the hotel. However, the sighting might equally be identified as a parapsychological example of psi—an *apparition of the living*. Tyrrell pointed out the strong connection between ghost sightings and psi using the framework of telepathic "transmission" by a psychic agent to a receiver. This idea has been popular since the mid-nineteenth century when telegraphy was invented, giving rise to the concept of "invisible" communication between different communication stations.[55] This analogy was cemented with the discovery of X-rays which possessed high frequency and strong penetrating ability compared with ordinary light. Other rays of even higher frequency were also speculated as the transmitting medium for telepathy, perhaps created by brain cell oscillations in one person producing high frequency energetic pulse in another person's receptive brain.[56]

The telegraph analogy has been in use for well over a century, although little is known about how this energetic but subtle signal transmission allegedly takes place, and what receptive parts of the brain receive and process this information.[57] In the case of Tom's ghost, there's also no indication of who the transmitter might be. For this reason, alternative psi solutions are just as applicable, for instance *retro-cognitive* or *precognitive* perception of a man who once was—or will one day be—in the hotel's stubby corridor. From a PK perspective, perhaps there *was* an external agent (or even Tom himself) who was able to affect the matter in the corridor and the components of the atmosphere, mould these ingredients into a short-lived physical shape that can be perceived through normal sensory channels.

[52] Roll and Joines (2013).
[53] Luke (2011).
[54] Roney-Dougal et al. (2013).
[55] Harvey (2004).
[56] Noakes (2008).
[57] Shiah (2012).

Assertions of entity creation through concentration or other magical rites using the "subtle matter" of an unobservable psychic space (as Theosophist Annie Besant suggested)[58] have a long tradition in the spiritual and paranormal literature. They're commonly referred to as *tulpas*, which link the practice to a Tibetan tradition—although these self-created entities have more in common with the Theosophical thought-form referred to by Besant and her colleagues, rather than the traditional Buddhist concept of *sprul pa*. However the tulpa name was "rediscovered" by modern paranormalists and applied to human-derived paranormal phenomenon. For instance, journalist and paranormal author John Keel refers to tulpas as actual physical creations of the mind that might be mistakenly identified as ghosts or even extraterrestrial aliens.[59]

To recap, parapsychology—incorporating LAD and psi—is the original pursuit of things "beyond normal" in a psychological context, and anything "discovered" since then (like aliens or zooforms) fit into the newer and broader field of paranormal phenomena which *incorporates* parapsychological things. Having briefly covered these topics, it's apparent that Tom's ghost might be attributed to *any number* of parapsychological or paranormal possibilities. Rather than a discarnate entity or tulpa, perhaps the dark-suited man in the corridor was truly substantial but transient in this place and timeframe—a time traveller from the future or a scientist currently testing a teleportation project for a black government program. Or maybe it wasn't a man, but an alien with human physiognomy who beamed in and out of the corridor by accident? After all, there are plenty of stories of alleged extra-terrestrial aliens not only doing paranormal things (like the Allagash abduction mentioned previously), but acting like apparitions—as this famous story attests (considered a classic "cross-over" example);

> On September 19, 1963, four children playing behind a school in Saskatoon, Saskatchewan, Canada, reportedly saw an oval-shaped light descent from the sky after circling overhead. A 'box' seemed to fall from it and hover a few feet away. The children cautiously approached it and were startled

[58] Mikles and Laycock (2015).

[59] Mikles and Laycock (2015).

> when a 'man' stood up, appearing suddenly. 'After that we didn't see and box', Brian Whitehead, eleven, later told investigators. The children estimated that the 'man' was about 10 feet tall, dressed in clothes 'like a monk's' which were 'white like a crayon.' 'Sometimes I could see right through him' Brian said. The 'man' made a moaning sound, held out his hands, [and] floated towards the youngsters. They ran in terror. One of the girls in the group became so hysterical that she actually had to be hospitalised. Their parents and the local police were impressed with their genuine fright and took their story seriously. UFO-type lights were seen in the district.[60]

Accepting the colossal range of "explanations" for Tom's ghost, it's really not possible for him (or anyone else) to determine which paranormal label is the best fit. Indeed, discerning folk might apply the more academically fashionable expression *anomalous* to his sighting.[61] It's a word that refers to unusual events or circumstances, but doesn't assume it was actually paranormal.[62]

Therefore, it's not just "typical" paranormal topics like LAD or psi that are identified as anomalous, but also abilities like *synaesthesia*. This is where two distinctly different sensory modalities, like vision and hearing, are seamlessly linked in a highly unusual way.[63] In its most common form, numbers, letters and words provoke the sensation of colours when they are seen (termed "colour–graphemic" synaesthesia) or when they are heard (termed "colour–phonemic" synaesthesia).[64]

To take an example involving word-related synaesthesia, one "sufferer" reports that;

> Tuesday is yellow. I don't 'see' it anywhere in particular; rather, I have a general awareness of yellowness in relation to the word.[65]

Alternately, but far less often, colours can be evoked by musical tones or everyday sounds, such as the barking of a dog or the blast of a car horn.

---

[60]Keel (1975/1976, pp. 150–151)

[61]Polito et al. (2010).

[62]Drinkwater et al. (2013).

[63]Reznik et al. (2018).

[64]Rich and Mattingly (2002).

[65]Ibid., p. 44.

Even specific smells or tastes can elicit a synaesthetic response, and vice versa. For some synaesthetes, the consequence is not a seen colour but an olfactory or gustatory sensation following a sound;

> I avoid middle C when I play the piano, because it has an earthy, musky smell I don't like.[66]

Regarding the anomalous nature of synaesthesia, it's not so much that the cause is mysterious—there's plenty of evidence for a neurophysiological origin—but that what synaesthetes *sense* is not normal, and that's why it is a phenomena of medical interest.

But now I'm getting off track. Returning to Tom's ghost, to say it was paranormal implies that what he saw was authentic with a *beyond normal* explanation (if it's even possible to explain something beyond normal). However, to deem the event anomalous acknowledges he might *believe* he saw a ghost, and this is a highly unusual event, but that doesn't necessarily mean he has come to an accurate conclusion. As with synaesthesia, only the person's *report* is anomalous, not necessarily the event itself. There's a very strong possibility the cause of Tom's ghost is able to be determined rationally, however strange that explanation might be.

So that's where things stand. Conveniently ignoring the chance Tom is lying, we're left with two basic possibilities; either (1) *he saw a paranormal ghost* (whatever this label potentially entails) or (2) *his anomalous encounter is able to be explained without recourse to any word possessing a para- prefix.* Where to now?

References

Alvarado, C. S. (2014a). The Term 'Out-of-Body Experience'. *Journal of the Society for Psychical Research, 78*(914), 61.

Alvarado, C. S. (2014b). On W. J. Crawford's Studies of Physical Mediumship. *Journal of Scientific Exploration, 28*(2), 351–357.

[66] Ibid., p. 44.

Alvarado, C. S. (2016). Out-of-Body-Experiences During Physical Activity: Report of Four New Cases. *Journal of the Society for Psychical Research, 80*(922), 1–12.

Arnette, J. K. (2012). Recollections of Death: A Medical Investigation. *Journal of Scientific Exploration, 26*(1), 207–211.

Besterman, T., & Gatty, O. (1934). Investigation of Paraphysical Phenomena. *Nature, 133*(3363), 569–570.

Blackmore, S. J. (1992). *Beyond the Body: An Investigation of Out-of-the-Body Experiences.* Chicago: Academy Chicago Publishers.

Blanke, O., & Metzinger, T. (2008). Full-Body Illusions and Minimal Phenomenal Selfhood. *Trends in Cognitive Sciences, 13*(1), 7–13.

Cardeña, E. (2018). The Experimental Evidence for Parapsychological Phenomena: A Review. *American Psychologist, 73*(5), 663–677.

Carr, B. (2008). Can Psychical Research Bridge the Gulf Between Matter and Mind? *Proceedings of the Society for Psychical Research, 59*(221), 1–96.

Cassol, H., Pétré, B., Degrange, S., Martial, C., Charland-Verville, V., Lallier, F., et al. (2018). Qualitative Thematic Analysis of the Phenomenology of Near-Death Experiences. *PLoS ONE, 13*(2), 1–14.

Coleman, L. (2013). Are Zooforms Cryptids? *Cryptozoonews.* Retrieved July 17, 2018, from http://www.cryptozoonews.com/zooforms-cryptids/.

Dane, L. (1975–1976). Astral Travel: A Psychological Overview. *Journal of Altered States of Consciousness, 2*(3), 249–258.

Drinkwater, K., Dagnall, N., & Bate, L. (2013). Into the Unknown: Using Interpretative Phenomenological Analysis to Explore Personal Accounts of Paranormal Experiences. *Journal of Parapsychology, 77*, 281–294.

Ebon, M. (1978). History of Parapsychology. In M. Ebon (Ed.), *The Signet Handbook of Parapsychology* (pp. 17–40). New York: Lombard Associates.

Fowler, R. E. (2001). Allagash Abductions. In R. D. Story (Ed.), *The Mammoth Encyclopedia of Extraterrestrial Encounters* (pp. 54–59). London: Constable and Robinson.

Gordon, R. S. (1992/2000). *The Paranormal: An Illustrated Encyclopedia.* London: Caxton.

Hanes, K. (2012). Unusual Phenomena Associated with a Transcendent Human Experience: A Case Study. *Journal of Transpersonal Psychology, 44*(1), 26–47.

Harvey, J. (2004). The Photographic Medium: Representation, Reconstitution, Consciousness, and Collaboration in Early-Twentieth-Century Spiritualism. *Technoetic Arts: An International Journal of Speculative Research, 2*(2), 109–124.

Inglis, B. (1985/1986). *The Paranormal: An Encyclopedia of Psychic Phenomena.* London: Paladin.

Irwin, H. J., & Watt, C. A. (2007). *An Introduction to Parapsychology* (5th ed.). Jefferson, NC: McFarland.

Jones, H. A. (2017). Reincarnation. *Journal for Spiritual and Consciousness Studies, 40*(2), 140–149.

Keel, J. (1975/1976). *Strange Creatures from Time and Space*. London: Sphere Books.

Lindeman, M., & Svedholm, A. M. (2012). What's in a Term? Paranormal, Superstitious, Magical and Supernatural Beliefs by Any Other Name Would Mean the Same. *Review of General Psychology, 16*(3), 241–255.

Luke, D. (2011). Anomalous Phenomena, Psi, and Altered Consciousness. In E. Cardeña & M. Winkelman (Eds.), *Altering Consciousness: Multidisciplinary Perspectives: History, Culture, and the Humanities; Biological and Psychological Perspectives* (Vol. 2, pp. 355–374). Santa Barbara: Praeger.

Lynn, S. J., & Evans, J. (2017). Hypnotic Suggestion Produces Mystical-Type Experiences in the Laboratory: A Demonstration Proof. *Psychology of Consciousness: Theory, Research, and Practice, 4*(1), 23–37.

Mikles, N. L., & Laycock, J. P. (2015). Tracking the Tulpa: Exploring the 'Tibetan' Origins of a Contemporary Paranormal Idea. *Nova Religio: The Journal of Alternative and Emergent Religions, 19*(1), 87–97.

Mobbs, D., & Watt, C. (2011). There Is Nothing Paranormal About Near-Death Experiences: How Neuroscience Can Explain Seeing Bright Lights, Meeting the Dead, or Being Convinced You Are One of Them. *Trends in Cognitive Sciences, 15*(10), 447–449.

Moody, R. A. (2001). *Life After Life: The Investigation of a Phenomenon—Survival of Bodily Death*. New York: HarperOne (Original work published 1975).

Neppe, V. M., & Close, E. R. (2015). The Concept of Relative Non-locality: Theoretical Implications in Consciousness Research. *Explore: The Journal of Science and Healing, 11*(2), 102–108.

Noakes, R. (2008). The 'World of the Infinitely Little': Connecting Physical and Psychical Realities Circa 1900. *Studies in History and Philosophy of Science Part A, 39*(3), 323–334.

Penman, L. T. I. (2015). The History of the Word Paranormal. *Notes and Queries, 62*(260), 31–34.

Polito, V., Langdon, R., & Brown, J. (2010). The Experience of Altered States of Consciousness in Shamanic Ritual: The Role of Pre-existing Beliefs and Affective Factors. *Consciousness and Cognition: An International Journal, 19*(4), 918–925.

Rabeyron, T., & Caussie, S. (2016). Clinical Aspects of Out-of-Body Experiences: Trauma, Reflexivity and Symbolisation. *L'Évolution Psychiatrique, 81*(4), e53–e71.
Reznik, D., Gertner-Saad, L., Even-Furst, H., Henik, A., Ben Mair, E., Shechter-Amir, D., et al. (2018). Oneiric Synesthesia: Preliminary Evidence for the Occurrence of Synesthetic-Like Experiences During Sleep-Inertia. *Psychology of Consciousness: Theory, Research, and Practice, 5*(4), 374–383.
Rich, A. N., & Mattingly, J. B. (2002). Anomalous Perception in Synaethesia: A Cognitive Neuroscience Perspective. *Nature Reviews—Neuroscience, 3*(1), 43–52.
Roll, W. G., & Joines, W. T. (2013). RSPK and Consciousness. *Journal of Parapsychology, 77*(2), 192–211.
Roney-Dougal, S. M., Ryan, A., & Luke, D. (2013). The Relationship Between Local Geomagnetic Activity, Meditation and Psi Part 1: Literature Review and Theoretical Model. *Journal of the Society for Psychical Research, 77*(911), 72–88.
Sabom, M. B., & Kreutziger, S. (1977). The Experience of Near Death. *Death Education, 1*(2), 195–203.
Serdahely, W. J. (1990). Pediatric Near-Death Experiences. *Journal of Near-Death Studies, 9*(1), 33–39.
Shiah, Y.-J. (2012). A Possible Mechanism for ESP at the Initial Perceptual Stage. *Journal of Parapsychology, 76*(1), 147–159.
Stafford Betty, L. (2004). Mind, Paranormal Experience, and the Inadequacy of Materialism. *International Philosophical Quarterly, 44*(3), 373–392.
Sudduth, M. (2009). Super-Psi and the Survivalist Interpretation of Mediumship. *Journal of Scientific Exploration, 23*(2), 167–193.
Tart, C. T. (1994). *Living the Mindful Life*. Boston, MA: Shambhala.
Tart, C. T. (1998). Six Studies of Out-of-Body Experiences. *Journal of Near-Death Studies, 17*(2), 73–99.
Thalbourne, M. A. (2009). Reports of Paranormal Experiences: Can Transliminality Tell Us Anything About Them. *Archive for the Psychology of Religion, 31*(3), 375–386.
von Lucadou, W., & Wald, F. (2014). Extraordinary Experiences in Its Cultural and Theoretical Context. *International Review of Psychiatry, 26*(3), 324–334.

3

What Is *Experience*?

Acknowledging these two distinct possibilities, let's start by examine Tom's ghost story without any presuppositions. In other words, avoid consulting the fluctuating A–Z list of paranormal topics other people have formulated in an attempt to match the best solution to the evidence. Was the figure of the man an alien? Probably not—he looked human, and aliens are supposed to be small and grey with dark elliptical eyes (or perhaps *tall* and grey with long noses, or reptilian, or "Nordic", or dwarflike…). They're unlikely to wear dark suits. Was it a zooform? That's even less likely, since the figure was actually *wearing* clothes. Bigfoot is hairy. What about a tulpa? They're supposed to be created through acts of concentration, and have neem known to hang around the creator causing acts of mischief. The problem is, Tom doesn't *know* what a tulpa is, but perhaps he was unaware he'd created one? Alternately was it someone else's tulpa? This is getting confusing. Was the figure a time-traveller? Now that's a possibility, although the clothes seemed contemporary and not futuristic. What about a haunting ghost? Yes, that sounds better, but I really can't differentiate that possibility from a telepathic thought-form beamed in by the agent from somewhere else, or a precognitive or retro-cognitive expression of

T. Jinks, *Psychological Perspectives on Reality, Consciousness and Paranormal Experience*,
https://doi.org/10.1007/978-3-030-28902-7_3

Tom's own psi ability, or even an accidental doppelgänger. The approach is truly hit-and-miss. We're at the whim of the encyclopedia's content, and there are too many overlapping "solutions" to choose from.

It is best to avoid picking an off-the-shelf paranormal category for his encounter, and while acknowledging it as anomalous, use a "bottom-up" approach which examines the raw events without any causal attribution. By necessity this removes a subdivision of the paranormal (or better still, anomalous) concerned with topics that *aren't* self-reports, but are still ostensibly *beyond normal*. Examples are the unusual structures built in antiquity popularly attributed to "ancient astronaut" technology or forgotten human expertise, such as the three giant hewn stones at Baalbek in Lebanon or Göbekli Tepe in Anatolia.[1] There are also crop circles (if not created by people for a lark) and controversial "out of place objects"—modern human footprints or recognisably manufactured objects entombed in ancient samples of gypsum rock, or flint or coal (to give a few possibilities).[2]

These are paranormal themes that people *read* about in literature or *watch* on media platforms of dubious integrity.[3] Learning about such strange things is not the kind of direct first-hand encounter we imagine when contemplating *anomalous experiences* like Tom seeing a ghost.

The first step in appreciating anomalous reports from an experiential (rather than a topic) perspective is to consider Tom's story once again—but not so much in terms of its content but what it means for someone to believe they've had an "experience" of anything *at all*. A succinct definition regards an experience as an *event*—something that happens or takes place—that has been *observed*, *lived through*, or *participated in* or *perceived* by a person or persons, which in turn leads to the gaining of knowledge.[4]

That assumes *perception*, which is the capacity to see, hear, feel, taste or smell stimuli through the conventional five senses. Whatever the event might be (ordinary or anomalous), it is happening "out there" somewhere *beyond the senses*. The living through it, or participation or perception of

---

[1] Hubert (1977, p. 308).

[2] Rickard and Michell (2007, pp. 214–216).

[3] It's very rare to actually see a crop circle being formed.

[4] E.g., *The Concise Oxford Dictionary* (p. 365).

it (or all of these together), is a representation of the observer's interaction with their external environment.

The first thing to recognise that such a definition of "experience" places no necessary requirement for Tom to have only *seen* his ghost as an apparition. Playing along with the idea that Tom encountered something that was taking place in the external world as a lived experience—rather than *seeing* the ghost he instead might have *heard* the ghost's footsteps. Perhaps the man's spirit (if it was a spirit) wasn't able to muster enough energy to be visualised, or whatever spiritual residue was left in the corridor could only harness enough power to generate a few sharp creaks of the floorboards. Alternately, Tom might have smelled the faint aroma of the spirit's aftershave or a lingering pungent cigar smoke (I've yet to hear of a gustatory ghostly encounter, but I'm sure there's a story out there somewhere of a spectral taste).

And at its most subtle manifestation the ghost's perception would be *somatosensory*: Tom might merely have indulged in an involuntary shiver while unlocking his hotel room door as he was lightly touched by an invisible presence. These more intangible ghosts are by definition still experiences, although their sensory effects might not have even impinged on Tom's awareness, and the pre-dinner drinks conversation would have been devoid of any ghostly discussion.

This leads to the awkward possibility that if ghosts are authentic, then they can be experienced *without* gaining the percipient's attention. By that logic, non-visual ghostly encounters could be very frequent but never realised—we might be ignoring one even now! Luckily the working definitions of an experience include reference to *awareness* of the sensory knowledge, which conveniently circumvents the awkward notion of experiencing something that hasn't been noticed.

Regarding the requirement for awareness, the majority of paranormal experiences (in the literature, at least) are solely visual because vision is a human's primary sense—the chief source of environmental information is in the form of light.[5] Hence Tom *saw* a ghost, the aliens were *seen* by the children in Saskatchewan, and OBEs, NDEs and even the astral journey are *visual* narratives. An exception to this "visual" rule of paranormal

[5]González-Casillas et al. (2018).

experience might be a *poltergeist* event, where the responsible agent (whatever its might be) is not initially seen but heard—poltergeist translated from the German means "noisy spirit". Nevertheless, in plenty of cases the effects of the poltergeist are eventually *seen*, as in the subsequent levitation or apporting of objects, fires breaking out or even demonstrable bite marks on residents of the location.[6]

What about telepathy, clairvoyance and precognition? Can these be considered true experiences if they don't involve any of the five conventional senses?—after all, they represent the *sixth sense*! That said, the "sixth sense" simply refers to the anomalous reception of information that is still recognised by the percipient in the form of something visual, or auditory (or even something felt or smelt). This might be *called extra sensory* perception, but there's no requirement for an additional sensory channel. Consider Tom's story as a *cameo* expression of Tyrrell's psychic ghost (since Tom doesn't remember ever having experienced anything remotely "psychic" in his life). For some unique reason, Tom was able to unexpectedly and directly receive information *in visual form* of a normally invisible figure—bypassing the retina and optic nerve. Yet this solution is indistinguishable from sensing a ghost through the visual system as a corporeal figure. It's not possible to determine which option is more likely unless there is sophisticated equipment available then-and-there in that hotel corridor to determine the actual mechanism of perception.

A similar situation exists for the olfactory ghost. You fleetingly smell the scent of perfume or cigar smoke as you step into a room that no-one has been in for days. Are there actual molecules of odorant floating in the atmosphere that react with olfactory receptor neurons situated in your olfactory mucosa? If so, where do they come from—are they spontaneously created from surrounding matter before returning to their insubstantial form? Or is the odour impression generated directly in the olfactory mechanisms of the brain, with these patterns "mistakenly" recognised as an odorant?

For this reason, it is more appropriate to differentiate *normal* sensations (from the five senses) from *mirrored* sensations (the sixth sense)—the latter

[6]Rogo (1979, pp. 39–40, 190); Again, we're not assessing the authenticity of the case yet, just how they are perceived/experienced.

having all the qualities of the former but achieved through some alternative and unknown mechanism. They are two methods of comprehending environmental information, and the outcome is identical. Therefore, if an experience refers to gaining knowledge of an event, then both normal sensations and mirrored sensations are legitimate experiences, although true perception is a function of the five senses only. The sixth sense is an example of *alternative perception.*

However it is too limiting to expect that only the five senses can create normal or mirrored experience. We constantly receive non-sensory impressions, and these may be interpreted as mundane or anomalous. If mundane, they are *secondary perceptions* because a preceding energetic interaction has already taken place between the individual and their surrounds via the conventional senses, which then translates into a non-sensory reaction. If the event is labelled paranormal (a more common usage than anomalous, which I will use solely from now on), then *alternative perception* has occurred, since the impression is presumably captured by the actions of those mysterious extra-sensory channels mentioned previously.

A good illustration of a non-sensory experience is the unexpected emotion a person attributes to a particular location. As a secondary perception, the feeling of sadness might be triggered by a feature of the corridor, such as the distinctive aroma of musty carpet reminding Tom of his childhood home. As an alternative perception, one might envision Tom *not* seeing his ghost, but instead being overcome by a fleeting melancholy when unlocking his door—as though whatever loitered in the corridor only had the power to actively trigger this singular emotion. Alternately, Tom may have made connection with an unthinking, residual baseline energy that elicited this same affective reaction.

A more sophisticated expression of a paranormal non-sensory experience, albeit initiated by sensory perception, is *déjà vu* (meaning "already seen"). The characteristic of déjà vu is *familiarity* distinct from psychic impressions involving flashbacks of sensory data in the form of visual imagery.[7] In the words of the philosopher and early psychologist William James;

[7]Moulin (2018).

> ...there is a curious experience which everyone seems to have had – the feeling that the present moment in its completeness has been experienced before – we were saying just this thing, in just this place, to just these people.[8]

From a paranormal or mystical perspective, déjà vu signifies the gleaning of hermetic information through alternative (psychic) perception, or is the *feeling* component of a reincarnation memory surfacing to awareness. Of course, less dramatic explanations are available, which I will explore later in the book (for now, I am simply describing the varieties of possible experience).

A third example of a non-sensory impression is the *sense of presence.* Here the individual feels that there is someone (or a sentient some*thing*) is standing or hovering in their vicinity; behind them, or just out of sight. And yet this presence is not actively perceptible.[9] The person simply *knows* that something is there, and this is frequently ascribed to an invisible guardian.

Spontaneous emotion, déjà vu and the sense of presence by no means exhaust the list of non-sensory paranormal experiences. The following story is particularly difficult to classify, since words like *emotion, familiarity* and *presence* don't apply. Nor do these events conform to a typical "sighting" of a paranormal entity. The location was a long-abandoned hospital complex built in the early 1900s, which has a long reputation for paranormal activity.

On a late afternoon visit to the old buildings in August 2016, Grahame entered the southern-most ward, which consisted of a very large empty and derelict timber room. The ceilings were 15 feet high, and the many windows were tall and rectangular. It was still light outside and the room didn't give a "spooky" impression, so Grahame felt confident enough to sit on the dirty and splintered floor and start talking out loud, to no-one in particular, about the experiences of other investigators at this location. For instance, in the early hours of the morning three years earlier Grahame had witnessed a male investigator partake in terse "communication" with an

[8] Brown (2003, p. 395).

[9] Comparelli et al. (2000).

alleged ghostly nurse in the same building. Shortly after, the same investigator began taunting something else that appeared to be lurking *behind* this invisible nurse. This exchange triggered the man to become highly aggressive, and it took four colleagues to drag him out of the building. Prior to that, another investigator had interacted with this same entity, and in a fit of rage had broken his hand while punching a wall during a similar "possession".

Back in the sun-lit ward, Grahame recounted these earlier tales and was immediately overcome with a very strange sense of being "rushed" by something that wasn't visible, although his eyes were fully open. The only way he could describe the event was being overwhelmed by a void of intense *blackness*, although the word didn't really fit the encounter. Grahame felt drawn to meet (mentally rather than physically) whatever this *blackness* was—and the exceptional nature of the encounter did not match anything he would consider normal human interaction (living or deceased). Apparitions like Tom's possess human characteristic, but Grahame's encounter was with something unspeakable—not in the dramatic sense of being evil but rather something that could not be defined using regular language. His use of the visual word for "blackness" was all he had available to convey the essence of this *non-visual* experience.

Alongside the very strange sensation of being engulfed, Grahame's mood state changed from surprise to anger when he became dimly aware that a colleague (the one who'd broken his hand during a similar encounter) was entering the room to drag him out. Grahame remembers being prepared to fight the friend to prevent contact being broken with the *blackness*, which was quite out-of-character. This unexpected anger surged from out of no-where. Luckily for Grahame he was quickly bundled out of the room and soon recovered (and to his relief didn't report any further effects of the encounter).

A paranormal experience is therefore a lived and informative event that is unusual enough to be considered *beyond* the types of things is normally encountered in the environment, or different enough to what the percipient accepts as a normal psychological state. This *lived information* can be garnered through the five senses directly, although there is scope for arguing that an indistinguishable experience might be achieved if or when the sensory channels are bypassed in some exceptional manner. Whatever the source, visual impressions are most frequently reported, however the encounter is occasionally supplemented by information from other senses (e.g., seeing *and* hearing the unusual event), or sometimes by those senses alone. In the latter case, the veracity of the event may be dismissed by the percipient since secondary channels alone do not possess the authenticity of a visual impression.

Occasionally, however, non-sensory paranormal experiences are reported in the form of cognitive or emotional awareness (or both), or complex impressions that are not able to be described easily with language, as alluded to with Grahame's hospital encounter with *blackness.*

From this analysis, we can conclude that Tom's alleged paranormal experience was quite simple; brief and entirely visual, with no other sensory or non-sensory components. The next step is to try and determine *what* he actually saw, or even if he saw anything at all?

References

Brown, A. (2003, October 26). Thoughts That Keep Going Bump in the Night. *The Sunday Times*, 3.

Comparelli, A., De Carolis, A., Kotzalidis, G. D., Masillo, A., Ferracuti, S., & Conesa, J. (2000). Geomagnetic, Cross-Cultural and Occupational Faces of Sleep Paralysis: An Ecological Perspective. *Sleep and Hypnosis, 2*(3), 105–111.

González-Casillas, A., Parra, L., Martin, L., Avila-Contreras, C., Ramirez-Pedraza, R., Vargas, N., et al. (2018). Towards a Model of Visual Recognition Based on Neurosciences. *Biologically Inspired Cognitive Architectures, 25*(1), 119–129.

Hubert, E. (1977). Gazetteer of Mysterious Sites Around the World. In *The World's Last Mysteries* (pp. 297–318). London and New York: Reader's Digest Association.

Moulin, C. J. A. (2018). *The Cognitive Neuropsychology of Déjà Vu.* London: Routledge.

Rickard, B., & Michell, J. (2007). *The Rough Guide to Unexplained Phenomena* (2nd ed.). London: Rough Guides Ltd.

Rogo, D. S. (1979). *The Poltergeist Experience: Investigations into Ghostly Phenomena.* New York: Penguin.

Part II

The First-Path

4
The 5W1H Method

This is the first stage in assessing Tom's sighting in a forensic, "bottom up" manner. To assist with the second stage, Rudyard Kipling's well-known verse is helpful, as it lists the six questions that can be asked about any event or incident that's experienced—even paranormal ones;

> *I have six honest serving men*
> *They taught me all I knew*
> *I call them What and Where and When*
> *And How and Why and Who*

The clever ideas contained in his rhyme were known long before Kipling wrote of them—they were deliberated in classical times.[1] However Kipling's four lines pithily sum up the approach to take when analysing human experience, and is widely used in many investigative fields (journalism is a prime example) using the contemporary label *5W1H*. On balance, the 5W1H approach is an effective way to analyse Tom's story, and requires six pieces of information;

[1] Sloan (2010).

T. Jinks, *Psychological Perspectives on Reality, Consciousness and Paranormal Experience*,
https://doi.org/10.1007/978-3-030-28902-7_4

When his experience took place
Where his experience took place
What happened during his experience
How his experience happened
Why his experience happened
and finally.....
Who caused his experience

It's easy to document *where* Tom's story happened (the hotel corridor) and *when* it took place (mid-2015). And yet without the simplicity of an off-the-shelf A–Z paranormal explanation like a "ghost" (providing a ready-made *what, how, why* and *who*), a significant amount of extra work is required to account for these additional four steps. Using the bottom-up approach, the *what* is most conveniently associated with Tom's description of *what* he saw—a clean-shaven man with very pale skin, short black hair and wearing a in a dark suit, walking across his field of vision for a matter of seconds.

This is the *phenomenological* description of the event, and refers to Tom's subjective experience of phenomena without necessary interpretation. Such an austere definition provides no extra information about what Tom might have actually witnessed, regardless of whether his subsequent investigation of the corridor pointed towards a paranormal or mundane solution. All that can be discussed are the subjective elements of Tom's experience—the vision of an ephemeral man.

It's reasonable to argue that Tom's audience—even if they are paranormal sceptics—would be happy to acknowledge not only the *when* and *where* of his sighting, but also the *what* in its phenomenological form. Similar to their acceptance of his ghost sighting being "anomalous", one would hope most sceptics would grudgingly accept Tom's personal description of what happened (discounting the hard-core few who might reject his story and consider it nothing more than a fabrication to amuse his work colleagues).

The acceptance of a storyteller's phenomenological *what* is something philosopher Henry Price contemplated in the first half of the twentieth century. He wrote that a person's reaction to a ghost story isn't so much about whether they believe in ghosts, but whether they believe that people can *experience apparitions.*[2] To Price the answer is "yes" because the statement doesn't necessarily imply that ghosts *exist* or that apparitions are genuine. Tom lived through an event and gained knowledge from it—indeed he captured enough detail to regale his work colleagues with an eerie story. It's perfectly appropriate to accept he did experience something that might be called an apparition, which just happened to assume the substantial form of a human.

Beyond the *where, when* and phenomenological *what*, there's a fork in the 5W1H road—a metaphor (albeit clichéd) advancing the earlier statement that he either (1) *saw a paranormal ghost* or (2) *experienced an event that—while anomalous—wasn't a ghost but something with a simpler explanation.*

The first exit to examine (let's make it the one to the left) is a continuing smooth and paved thoroughfare I'll label the *first-path*. In terms of *how* his sighting took place, the first-path represents the opinion that "something" was present in the corridor and Tom was able to learn about it through conventional (or those *alternative*) perceptual channels. The very nature of the paranormal is challenging, hence unconventional solutions will need to be sought.

From this point, a *who* and a *why* for Tom can be investigated, and an A–Z encyclopaedia entry on ghosts will assist in this process. These are possibilities limited only by the availability of historical records (and the imagination). Newspaper articles from the early 1920s might tell of a man name Harold committing suicide in room 113 following his financial ruin. He's still not at peace, and as Tom has now confirmed with his own eyes, Harold's visible spirit paces the hotel corridor reliving long forgotten woes. There the road ends, with the *when, where, what, how, why* and *who* all ticked off to the believer's satisfaction.

While superficially satisfying, this conclusion is incomplete because the most interesting thing about Tom's ghost—*what* it actually *is*—requires an additional component to the 5W1H approach. Specifically, the *what* isn't merely a phenomenological description of what happened, but also an explanation of *what is a ghost?* That is, what the ghost is "made from",

[2] Tyrell (1953, p. 8).

and why it can't be experienced all the time like all the other natural objects present in the environment. The discussion so far, including my fictitious A–Z encyclopaedia of the paranormal, has focused on listing topics (including ghosts) and describing their potential origins. Offhanded notions like "floating energy" and "stone tape effect" are very interesting, but require elaboration. To provide a better grasp of *what a ghost might be*, it's necessary to reference the works of those who believe in ghosts and have thought a lot about them, and who provide comprehensive insight into their composition by *extending* and *modifying* existing theories of physics (and other fundamental sciences).[3]

The first-path recognises that there's a greater range of external information in the environment that we typically perceive through our senses, allowing the paranormal to remain beyond normal, but be part of the same continuum as any other experience. The drive is to integrate paranormal things like ghosts into the natural world rather than leave them unexplained. Assuming Tom's ghost story was phenomenologically accurate, and represents a legitimate perceptual experience, the first-path provides plenty of roadway for travel, and that's where we'll head in the next chapter.

References

Bierman, D. (2010). Consciousness-Induced Restoration of Time Symmetry (CIRTS). A Psychophysical Theoretical Perspective. *Journal of Parapsychology, 24,* 273–300.

Sloan, M. C. (2010). Aristotle's Nicomachean Ethics as the Original Locus for the Septem Circumstantiae. *Classical Philology, 105,* 236–251.

Tyrell, G. N. M. (1953/1963). *Apparitions.* New York: Collier Books.

[3]Bierman (2010).

5
Low Energy Ghosts

It's reasonable to assume that a first-path ghost is measurable. Even if they do not consist of substantial matter that can be poked and prodded in a laboratory setting, they must still consist of something closely related—perhaps an organised bundle of information-laden energy containing the history and essence of a person remaining in the "vacuum" of space after physical death. This energy could be something plausible, like *electromagnetic radiation*, which can be seen by Tom through the emission of photons. According to the standard model of physics, these chargeless and massless quanta move around at the speed of light (their original definition is the *substance of light*). And if they possess the appropriate physical parameters, like wavelength, energy and momentum, then they can enter the eye and interact with pigment molecules stored in the retina. The result is a range of conformational (photochemical) changes once the photon wavelength is converted into an equivalent "amount" of chemical energy that is transmitted to higher brain regions for recognition as a visual stimulus.[1] This process represents *what* is happening in Tom's ghost sighting.

[1] Mannu (2014).

T. Jinks, *Psychological Perspectives on Reality, Consciousness and Paranormal Experience*,
https://doi.org/10.1007/978-3-030-28902-7_5

Photons are produced by various atomic and subatomic events, such as the acceleration of atomic particles (e.g., electrons) in material substances like metallic wire, or particle collision (the creation or destruction of particles). Photons can also gain and lose energy through processes such as deflection of other materials, which accounts for one method of altering wavelength that ultimately determines how we experience different colours.

So if a spirit's presence (sentient or non-sentient) has a physical basis and *seen* as photons of light, then why *can't* they be regularly measured like any other energy-emitting process? A retort might be that the spirit is made from a more exquisite type of energy or field (analogous to a magnetic or electrical field) that is not directly seen or felt, that's able to excite electrons in matter, which in turn allow the propagation of electromagnetic waves exciting photons of a type that can stimulate the visual system. It is this conventional energy, not the ghost's true subtle state, that's perceived by the witness through their senses, or measured through the appropriate scientific instruments.

Whatever a ghost is made of, the implicit understanding is that there's not much to actually measure in the first place. One interpretation is that they constitute *very little* energy. If the ghost truly consists of electromagnetic radiation, then these waves possess an extremely low frequency. Indeed, the wavelength of the energy is so vast that it's approaching the infinite—although this is still classified as part of the radio frequency spectrum because there isn't an everyday expression for waves longer than radio waves. This low energy notion of spirits was implicitly understood by nineteenth Century mediums, who often performed their séances in the glow of ruby-coloured light because low-energy red was considered the "spectral wavelength"[2] (it also helped prevent the audience seeing what they were doing!).

What's missing here is an explanation of *how* a once-living person gets to become a low frequency wave when they die, and somehow becomes tangled up in a building's brick, timber or plasterboard according to the stone-tape effect. There's also the problem of how, if the ghost remains

[2] Harvey (2004).

sentient, these waves can store the kernel of the deceased person's characteristics and intellect. However, one could argue there hasn't been adequate technical investigation of ghosts as actual states of energy or field, so any serious investigation that should have been made has *not* been made. And while this claim is tenuous, the low energy theory at least *sounds* intuitive. It accounts for the transient (and mostly powerless) nature of the ghost. Any spirit in a low-energy state won't possess the necessary strength to generate gross physical changes to the surrounding environment, so it's not surprising their presence is short-lived, like a raspy groan or a momentary shadow. In this respect, Tom's feature-filled ghost was unexpectedly robust.

When ghosts do seem to have some influence over matter, it is when they act in the fashion of an *energy sink* and only occasionally harness available energy from the surrounding environment. In this way spirits haunt locations in a continuous but amorphous state that's too subtle to measure with existing technology, or they *are* able to be measured by we just don't know what to look for yet (their presence might be indistinguishable from natural energies and fields). When they sporadically draw on ambient energy introduced to the environment (most likely from the people present) they are able to enter a more energetic, hence perceptible, manifestation.[3] That's how they can present as a more substantial human figure like Tom's ghost, or be seen through their noticeable actions, like the slamming of doors or the levitation, materialisation or throwing of objects.[4]

For this reason, ghost hunters measure temperature alongside electromagnetic field fluctuations in haunted spaces. Lowered temperatures apparently indicate a source of energy (heat) being drained—so the use of electromagnetic field meters, thermometers and barometers is to determine the impact of the ghost on the environment, rather than quantifying the ghost directly.[5]

[3] Jinks (2012).
[4] Harvey (2004).
[5] Belanger (2007, p. 172).

This rationalisation attempts the best of both worlds—to explain the visual or auditory component of the ghost as product a standard perception due to the harnessing of natural energy. However at the same time the source itself is *not* easily measurable and if sentient it can slink away into its low-energy nothingness when it so desires. Or, if it is non-sentient but infused into the built environment as a vision, sound or smell or something able to trigger a non-sensory impression, playback is initiated when the conditions are just right. That depends on the nuances of weather—perhaps the nearness of an electrical storm or a rapid fall in barometric pressure. It could also be the result of a change in the earth's geomagnetic activity,[6] or some other physical variable that might assist in providing or harnessing the necessary energy for paranormal activity.

A non-material, low-energy presence that is intermittently material and measurable matches the historical doctrine of the *aether*. Aether is a cornerstone of the physical world of Plato and Aristotle, where the atmosphere consisted of the invisible ingredients of *aer*, corresponding to atmospheric misty air, *aither* corresponding to the fiery upper air, and *pneuma*, which is the air of breathing.[7]

Indeed;

> From antiquity throughout the Middle Ages, into the Renaissance and the Early Modern periods, different kinds of ethers were invented and invoked to perform a great variety of cosmological and physical functions.[8]

In Newton's time there was widespread acceptance of a *luminiferous* aether that (as the name implies) assisted the transmission of light.[9] This style of aether comprised an invisible, elastic-like substance present in both the "Heavens" and its "Bodies", and the movement generated by its vibrations allowed light (and heat) to spread from one place to another.[10] Moving forward to 1865, and with a deeper scientific toolbox at his disposal, the

[6] Persinger (1979) and Persinger and Cameron (1986).

[7] Pokazanyeva (2016).

[8] Asprem (2011, p. 134).

[9] Merali (2006).

[10] Shaw (2014).

great physicist James Maxwell was able to identify the aether as a concentration of subatomic material located in the space between objects, and within objects which allowed the promulgation of electromagnetic waves and the manifestation of electromagnetic fields.[11] Comparable to Newton's aether, this substantive medium is elastic and in motion, with energy stored and released as a function of the displacement of its components. The transmission of electromagnetic wave therefore corresponds to the motion of aetheric matter.

These concepts interested spiritualists and Theosophist in the late nineteenth century and early twentieth century. Alice Bailey (who was one of the first people to use the expression "New Age" in her writings) saw aether as the source of all the energy in the universe,[12] and human spirits (and things that were *never* human, like nature spirits, elementals, sylphs and fairies) possess aetheric bodies. According to Theosophy's founder Helena Blavatsky, these bodies are smaller, finer and purer than normal physical matter.[13] That's why we can't see them easily. And because these entities are themselves aetheric, they are able to pneumatically and hydraulically compress the aetheric material surrounding them to create force and motion.[14] That's why spirits, for example, can make noises and move things around haunted houses. It also accounts for the mysterious events that take place in the séance parlour. The appearance of spirits during a séance is a consequence of the moulding of aether, although the shape they take is a function of the thought processes of the sitter. The entity's form is fashioned by the content of the witness's "unconscious psychic library".[15]

That said, for this idea to have merit a physical substance like aether should be detectable, like electromagnetic radiation. Unfortunately for aetheric proponents, the famous 1887 experiment by Albert Michelson and Edward Morley found no evidence for this invisible medium, and that's what's been accepted by mainstream science ever since. Briefly, Michelson and Morley shone light at mirrors and measured their return

[11] Ibid.

[12] Asprem (2011).

[13] H. P. Blavatsky (1954/1966); Elementals are either made from these waves, or use these waves to cause physical effects.

[14] H. P. Blavatsky (1877/2012).

[15] Jinks (2012).

speed. Having discovered these beams returned at the same time, they concluded there was no interfering aetheric "headwind" obstructing the passage of the beams.[16] No headwind means no aether, so space must be a vacuum. Two decades later, Einstein's special theory of relativity established that light can travel through this vacuum.[17] That removed the requirement for an absolute frame of reference, in turn eliminating another pillar of aetheric theory. No absolute frame of reference means there's no need for the aether to hold the universe in place.

More recently, the aether has been rejuvenated on the scientific fringe for the purpose of explaining various unresolved problems in physics, such as the composition of *dark matter* and *dark energy*.[18] Similar to Maxwell's formulation, this aether is an ocean of;

> ...tiny non-rigid subatomic cells that have the property of elasticity, the capacity to vibrate, and the ability to attach to and detach from each other.[19]

An alternative modern take on aether is as a field (not a substance) that pervades space-time and affects gravity.[20] This replaces the need to explain mysterious dark matter as a substance.[21]

In yet another contemporary twist, there are various theories which resemble the concept of aether and have been used to account for paranormal phenomena such as ghosts—specifically *what* these entities are made of. One of these theories involves a transparent, infinite fluid of negative energy electrons[22] called the *Dirac Sea* that was advanced in the early 1930s by British physicist Paul Dirac to account for negative-energy quantum states.[23] While not measurable (like the aether), the Dirac sea is mathematically derivable.[24]

---

[16] O'Meara (1996).
[17] Merali (2006).
[18] Shaw (2014).
[19] Ibid., p. 603.
[20] Merali (2006).
[21] Ibid.
[22] Song (2017).
[23] Bian and wang (2016).
[24] Finster (2011).

More to the point, this sea does not consist of conventional negatively-charged electrons, but—as the expression suggests—negative *energy* electrons (or alternatively, anti-matter positrons). This sea pervades the universe, but is not actually made of anything discernible to us because we can only measure positive energy states. However as the 1930s progressed the idea of an indiscernible "sea" made up of an *absence* of particles (negative energy or anti-particles) was replaced with a treatment of positrons as actual particles. As they are no longer any anti-particles, there space returns to a vacuum with nothing actually residing in it.[25]

The Dirac sea is only one potential hidden medium. An associated concept comprises the exotic products of high-energy reactions between electrons and positrons with names like leptons (electrons, muons and taus), hadrons and the massless boson.[26] Unlimited in number, these fundamental particles condense into a single (grounded) state identified as the Bose-Einstein Condensate (BEC). In recent decades the BEC has received a kind of mystical reverence as a *self-organizing-synergetic dissipative structure.*[27] Pervasive but imperceptible, the infusion exists *below* absolute zero (−273 °C), and with the word "zero" being used there's an urge to link the concept to the *zero-point energy field.* That's another vision of an aetheric-like material that;

> ...interweaves all energetic forms of existence [and] perpetuates the dynamic interaction of subatomic particles and energy fields acting as a source by which virtual particles manifest into existence.[28]

Not only is the "super-dense" zero-point energy field the substrate that holds everything in place, according to philosopher Ervin Laszlo it is also "super-rich information field". To Laszlo that's equivalent to the *Akashic Field*. Derived from the Sanskrit word *Akasha*, this field refers to the fundamental composition of the universe from which everything emerge, and into which everything will eventually re-converge.[29]

---

[25] Bian and wang (2016).

[26] Agafonov (2018).

[27] Geesink and Meijer (2016).

[28] Benford (2001, p. 133).

[29] Yong (2008).

That gives Laszlo the confidence to state;

> The A-Field takes its place among the fundamental fields of the universe, joining science's G-field (the gravitational field), EM-field (the electromagnetic field), and the various nuclear and quantum fields.[30]

With such an expansive pedigree, this fundamental physical domain (whether called a sub-Kelvin condensate or an aetheric, zero-point energy or Akashic Field) is a convenient hypothetical space for ghosts and other things to hide. Hence ghostly events are not so much a function of low energy, but *negative* energy. That "solves" the problem of measurability. Laboratory instruments might be able to measure very, very low energy activity, but they can't necessarily measure negative energy that's not *officially* there in the first place.

The appearance of ghosts only become possible when there's enough actual energy provided from the everyday world to allow some of these negative energy particles to be lifted into the positive energy domain. That accounts for how negative energy ghosts can intermittently appear, or be sensed, by normal people. The negative energy field morphs from an immaterial state to a material expression, and this simply represents two sides of the same universal symmetrical coin—we only live in one half of it—the other half is the ghostly anti-matter (negative energy) counterpart.

And it goes deeper than that. Since the positive and negative world are so interlinked, a person expressing physical or psychological stress in the positive world causes a ripple effect in the negative world counterpart with the potential release of detectable energy. This might manifest during times of life crisis as psychic chaos epitomised by PK; the human-centred poltergeist. Or it might involve visions of a ghost.[31] Although ghosts might appear on their own volition, because they are sentient occupiers of the negative energy space they can also be drawn out of this space by the emotional status of the living witness.

And it's not just ghost that are accounted for by negative energy states. Consider a living brain cell (neuron), and the ionised atoms such as sodium

[30] Laszlo (2004).

[31] Borrutta (2009).

and potassium which are in constant flux across the lipid bilayer separating the intracellular from the extracellular environment. These are positive energy particles that make up ordinary, everyday matter.[32]

The more advanced the organism, the more complicated the energy flux and at some stage of complexity the sheer weight of movement (the brain consists of one trillion cells of which approximately 100 billion are neurons[33]) will place strain on the underlying BEC[34] (or Akashic field…or whatever). It's theorised that strain encourages the BEC to generate a squirt of usable energy from the resting negative state. That's just enough to sustain the ionic flow and preserve the energy requirements of life. So even though it's not observable, the BEC is an interconnected "shadow" of this world and everyone in it. From another perspective, the world is the *product* and the BEC is the *blueprint.*[35]

This is the alleged basis of OBEs, NDEs and astral journeys. For various reasons—in the case of an NDE the process of *actually dying*—the traveller will become less a product of their "positive energy self", and more attuned to their negative energy analogue that persists outside of coherent space and time and which is independent of their life status. During this process they become cognisant of the wider negative energy environment around them, which contains the negative energy forms of other (former) people who no longer possess a positive energy body. That's how the traveller meets the departed family and friends, as in this statement from an NDE survivor;

> ….In my NDE there was an overwhelming feeling that I was returning "home" to a much more comfortable and familiar place than the reality I had left.[36]

And when it comes to reincarnation, it's suggested that following death an individual's characteristics in the negative energy realm re-emerge in positive energy world as *something completely different*—perhaps another

---

32 Borutta (2009).

33 Presti (2011, p. 25).

34 Borutta (2009).

35 Ibid.

36 Ibid., p. 197.

living person or any other biological object. That means there's nothing stopping the negative energy person returning as a dog, or even a tree (however disconcerting that outcome might be).

References

Agafonov, A. I. (2018). Massless Composite Bosons Formed by the Coupled Electron-Positron Pairs and Two-Photon Angular Correlations in the Colliding Beam Reaction $e^{-} e^{+} \rightarrow B\gamma\gamma$ with Emission of the Massless Boson. *Advances in High Energy Physics*, 1–18 (Article ID: 6137380). https://doi.org/10.1155/2018/6137380.

Asprem, E. (2011). Pondering Imponderables: Occultism in the Mirror of Late Classical Physics. *Aries, 11*(2), 129–165.

Belanger, J. (2007). *The Ghost Files: Paranormal Encounters, Discussion, and Research*. Franklin Lakes, NJ: Career Press.

Benford, M. S. (2001). Can a Theory Derived from Recent Experimental Data Explain Precognition and Other Mysterious Spiritual Phenomena? *Journal of Religion and Psychical Research, 24*(3), 132–142.

Bian, J., & Wang, J. (2016). A Deduction from Dirac Sea Model. *Physics Essays, 29*(2), 218–221.

Blavatsky, H. P. (1966). *Collected Writings* (Vol. 6). Wheaton, IL: Quest Books (Original work published 1954).

Blavatsky, H. P. (2012). *Isis Unveiled: A Master-Key to the Mysteries of Ancient and Modern Science and Theology. Volume 1: Science*. New York: Cambridge University Press (Original Work published 1877).

Borutta, R. (2009). Paranormal Phenomena, Physics and Dirac Energy. *Journal of Spirituality & Paranormal Studies, 32*(4), 192–200.

Finster, F. (2011). A Formulation of Quantum Field Theory Realizing a Sea of Interacting Dirac Particles. *Letters in Mathematical Physics, 97*, 165–183.

Geesink, H. J. H., & Meijer, D. K. F. (2016). Quantum Wave Information of Life Revealed: An Algorithm for Electromagnetic Frequencies That Create Stability of Biological Order, with Implications for Brain Function and Consciousness. *Neuroquantology, 14*(1), 106–125.

Harvey, J. (2004). The Photographic Medium: Representation, Reconstitution, Consciousness, and Collaboration in Early-Twentieth-Century Spiritualism. *Technoetic Arts: An International Journal of Speculative Research, 2*(2), 109–124.

Jinks, A. (2012). *An Introduction to the Psychology of Paranormal Belief and Experience*. Jefferson, NC: McFarland.
Laszlo, E. (2004). Matter and Mind: The New Holism and the Greater Humanity. In D. Loye (Ed.), *The Great Adventure: Toward a Fully Human Theory of Evolution* (pp. 39–52). New York: State University of New York Press.
Mannu, G. S. (2014). Retinal Phototransduction. *Neurosciences, 19*(4), 275–280.
Merali, Z. (2006). Ether Returns in Bid to Oust Dark Matter. *New Scientist, 191*(2566), 8–9.
O'Meara, S. J. (1996). Space-Time and the 'Ether' Bunny. *Odyssey, 5*(9), 16–21.
Persinger, M. A. (1979). Possible Infrequent Geophysical Sources of Close UFO Encounters: Expected Physical and Behavioral-Biological Effects. In R. F. Haines (Ed.), *UFO Phenomena and the Behavioral Scientist* (pp. 396–433). Metuchen, NJ: Scarecrow Press.
Persinger, M. A., & Cameron, R. A. (1986). Are Earth Faults at Fault in Some Poltergeist-Like Episodes? *Journal of the American Society for Psychical Research, 80*, 49–73.
Pokazanyeva, A. (2016). Mind Within Matter: Science, the Occult, and the (Meta)Physics of Ether and Akasha. *Zygon: Journal of Religion and Science, 51*(2), 318–346.
Presti, D. E. (2011). Neurochemistry and Altered Consciousness. In E. Cardeña & M. Winkelman (Eds.), *Altering Consciousness: Multidisciplinary Perspectives: History, Culture, and the Humanities; Biological and Psychological Perspectives* (Vol. 2, pp. 21–41). Santa Barbara: Praeger.
Shaw, D. W. (2014). Reconsidering Maxwell's Aether. *Physics Essays, 27*(4), 601–607.
Song, D. (2017). Comment on Information and Dualism. *Neuroquantology, 15*(2), 141–144.
Yong, A. (2008). Science and the Akashic Field: An Integral Theory of Everything (Book Review). *Perspectives on Science and Christian Faith, 60*(1), 53.

6

High Energy Ghosts

Another long-discussed physical explanation for ghosts and other paranormal things also provides a "*what is happening*" explanation for the first-path. This idea takes the notions of low energy and low frequency states and flips them; now suggesting invisible entities exist in an incredibly *high* energy state (often referred to as a *high vibrational* state). This is distinct from normal matter, which expressed in Theosophical language;

> ...is the coarsest, slowest-vibrational manifestation of energy[1]

Of course, high vibrational realms (like low energy states) are invisible, unmeasurable and have unexpected properties. However, these vibrational characteristics are *possessed* by non-human entities and are very different to the "normal–vibrational" environment we live in.

A style of high energy entity discussed in the works of John Keel[2] is the paraphysical alien. Recall paraphysical things are beyond the physical, so

[1] Lampis (2004, p. 46).
[2] Keel (1973).

T. Jinks, *Psychological Perspectives on Reality, Consciousness and Paranormal Experience*,
https://doi.org/10.1007/978-3-030-28902-7_6

these entities are no longer extra-terrestrial (outside the normal boundaries of terrestrial earth), but *intra*-terrestrial or *ultra*-terrestrial or some other bizarre term. Like the flying saucers they fly around in, aliens are "transmogrifications of energy" which don't exist quite in the same way normal matter does. Although they ultimately have a material basis, they aren't *permanent* assemblies of matter.[3] Their parts vibrate so rapidly that they emit electromagnetic radiation with an extremely short wavelength—in the "very highest possible point of the frequency spectrum".[4] It's only when these agile and intelligent beings purposely lower their vibrational state that they become visible. That's typical of stories of flying saucers appearing and disappearing in thin air, and aliens who behave in unusual ways, like the flying saucer-ghost from 1960s Saskatchewan described earlier. Because of their incredibly short wavelength, these entities come from the *violet* end of the spectrum[5] not the opposite low energy red range associated with séance spectres. Apparently, that's how they can be differentiated.

Despite any misgivings about humans being able to achieve a higher energy level, it's generally held (by the descendants of Bailey's new age movement) that even living people can attain the status of a paraphysical alien. When the circumstances permit, the parts of a regular physical body can indulge in a vibrational change that allows material transience, and in the case of alien abductees they are able to pass through solid walls. Remember the four fellows and the flying saucer over the Allagash river? In terms of what this *feels* like, psychiatrist and alien abduction researcher John Mack reports that the victims;

> ....feel as if an intense energy is separating every cell, or even every molecule, of their bodies...[they] feel it remarkable they can come back together whole on the other side.....a powerful vibratory sensation in arms, legs and other parts of body[6]

[3]Ibid., p. 182.

[4]Ibid., pp. 191–192.

[5]Keel (1973).

[6]Hancock (2005, p. 396).

Mack argues the abductee actually goes *physically* missing (i.e., disappears) because they're not present where they should be when the abduction occurs (according to friends and relatives).[7] Aliens force the abductee's body to change its vibrational state, and if this is accepted as something that can happen to ordinary people it's easy to suggest the essence of a once-living human can also gain the necessary force to progress to a higher energetic state after they die.

But this is surely going too far? Contemplate how the expressions *high vibration* or *high energy* apply to matter. Very simply, the building blocks of matter are atoms, so on face value the elevated vibration or energy must apply to atomic vibration and an increased atomic energy state. There's quite a lot known about this process, and it doesn't easily conform to paraphysical beings;

> ...Conventional physics tells us that the atoms and molecules comprising matter do vibrate, but this is a very complicated and multi-faceted type of vibration. For example, a solid object is made up of molecules that vibrate relative to one another. Within these molecules are atoms, and the bonds connecting these atoms also vibrate in many different ways. For this reason it's not appropriate to sprout the opinion that some everyday objects vibrate "normally," while others vibrate at abnormally high frequency, and that's the distinguishing feature separating everyday people [from high vibrational beings]. In fact, at normal temperature all molecules in all objects vibrate at an incredibly high frequency [ten trillion hertz that's typical of atoms] simply because they're so small. And this vibration is measured as an average, because not every atom or molecule is doing the same thing at the same time.[8]

This complication is amplified if high energy entities possess a molecular structure;

> ...molecules might be bonded together in a solid object, but when energy is added to the system (such as heat) these bonds are weakened and can break, hence the object melts. If energy continues to be added the fluid

[7] Mack (1993).

[8] Jinks (2016, p. 139).

> material transitions to a gas, and at even greater energies the gas progresses to a plasma or some other non-classical high-energy state. There's no room in this orthodox treatment of matter to accept that increasing (or lowering) the energy in an object creates anything other than this standard phase change.[9]

There also isn't any room for sentience to be retained (let alone generated) in this high energy state, meaning a person subject to these forces will become nothing more than an energetic goo. And they certainly won't be able to resume their everyday human form once the vibrational pressures are removed.

Where does that leave high vibrational beings? To salvage the concept, it could be argued that an entity's atoms aren't vibrating at a frequency of ten trillion hertz but at tremilliatrecendotrigintillion hertz. That's an actual number, although by no means an accurate representation of true atomic vibration.[10] At such an incredibly elevated rate *anything* could happen to matter, including the creation of ghosts, sylphs and ondines.

This "defence" of high vibrational/energy beings is rather weak, so another solution is to go full-circle and match vibrational theory with the aether. Once the concept of molecular bonding and atomic vibration became popular in the nineteenth century,[11] Theosophists used this model of matter to argue that when subjected to special forces, molecules break apart and their vibrational state increases. This is the conventional "high energy" perspective of non-human entities. However, the next Theosophical step doesn't end with a plasma (or some other high energy soup) but rather the less empirical (and more mystical) aether.

Charles Webster Leadbeater was a highly influential Theosophist who guided the magical and ritual implications of Theosophy as a way to practice the occult. In his worldview, "divine power" could be channelled to the everyday world using mystical conduits akin to the way pipes channel energetic steam created in a boiler room[12] (as with the previous reference

[9] Ibid.
[10] Ibid., p. 140.
[11] Durham (2015).
[12] Tillett (2012).

to hydraulics and pneumatics, the Theosophists obviously enjoyed using the technical language of the Victorian era).

Leadbeater was particularly interested in the denizens of the invisible realm who possess a very particular vibrational state. In terms of *how* they can interact with this world, and *how* and *why* they are invisible, the aether plays an important role;

> The phenomenon of disintegration….may be brought about by the actions of extremely rapid vibrations, which overcome the cohesion of the molecules of the object operated upon. A higher rate of vibrations of a somewhat different type will separate these molecules into their constituent atoms. A body reduced by these means to the etheric condition can be moved by an astral current from one place to another with very great rapidity; and the moment that the force which has been exerted to put it into that condition is withdrawn it will be forced by the etheric pressure to resume its original form.[13]

Vibrational disintegration and reintegration of conventional objects allow them to enter a confirmation that facilitates passage through the aether (Leadbeater's astral current) and consequently other matter.

This effect is particularly relevant to séance apports, where a favourite trick was for an object to magically enter or leave a locked box in the middle of a table surrounded by sitters (and the medium). According to spiritualist belief, this was possible because the medium's otherworldy guide could assist in matching the vibrational state of the chosen object to their own state and allow it to become an intangible sludge that could easily transverse any solid barrier.

To try and make sense of this idea, imagine that intense atomic vibrations *induce* aetheric vibrations, meaning the expressions "high-vibration" or "high-energy" don't actually refer to measurable qualities of ordinary matter. They are metaphors for how there can be matter, and how there is something *analogous* but *different* to matter. That's aether, or negative energy or some other corresponding label. You achieve this state by increasing your vibration to a point where you set off an equivalent vibration in the aetheric plane, and your body can travel to (and through) all sorts of

[13] Durham (2015, p. 36).

different places. And for Tom's ghost it's a reverse process. A spirit able to "lower" its aetheric vibration will generate equilibrium with sensible, everyday molecular vibration. Lo and behold, a ghost appears that can temporarily be seen like any other everyday object.

References

Durham, R. B. (2015). *Charles Webster Leadbeater: The Astral Plane, the Devachanic Plane, Clairvoyance, the Inner Life* (Vol. 1 & 2). Morrisville, NC: Lulu Enterprises.

Hancock, G. (2005). *Supernatural: Meetings with the Ancient Teachers of Mankind.* London: Century.

Jinks, A. (2016). *The Disappearing Object Phenomenon: An Investigation.* Jefferson, NC: McFarland.

Keel, J. A. (1973). *Operation Trojan Horse.* London: Sphere Books.

Lampis, R. (2004). The Theory of Reality: Or Why, If We Understand Reality, the Theory of Everything Is an Impossibility. Or How to Establish Basic Principles That Clarify Physical, Mental and Paraphysical Phenomena, and to Dispose of Concepts Such as Non-locality, Place. *International Journal of Humanities and Peace, 20*(1), 43–48.

Mack, J. E. (1993). The UFO Abduction Phenomenon: What Might It Mean for the Evolution of Human Consciousness? *ReVision, 16*(1), 26–32.

Tillett, G. (2012). Modern Western Magic and Theosophy. *Aries: Journal for the Study of Western Esotericism, 12*(1), 17–51.

7

Ghosts and the *How?*

That gives a basic summary of ways the first-path accounts for *what* ghosts and other paranormal things are made from. Broadly speaking, ghosts consist of a pattern of information residing in a low energy, high energy, negative energy (or whatever) conformation. I've suggested that *how* they are perceived is through our conventional senses, but does that necessarily have to be as visible light? Are there other types of energy that the ghost can present as, which can also be observed by the witness?

To better pin down the *how* the energy of a ghost is perceived, let's return to the use of analogies. This style of reasoning examines the similarities between two "things" (such as objects, processes or systems) and argues that further (hidden) similarities also exist between them. For instance, the resemblance of telepathy to telegraphy is a useful analogy, although it lacks the robustness of direct evidence hence isn't the strongest method of explanation. Nevertheless ghosts aren't exactly robust themselves, so analogous reasoning might be the best (and only) method to find out more about how they're detected.

What's a good analogy to account for how people see ghosts? If people only occasionally see (or hear, or smell) normally imperceptible ghosts, this

T. Jinks, *Psychological Perspectives on Reality, Consciousness and Paranormal Experience*,
https://doi.org/10.1007/978-3-030-28902-7_7

might simulate the way instruments can only measure concealed aspects of the environment when they are switched to the right setting. This seems an appropriate comparison, since I've mentioned how "ghost hunting" instruments are used to measure the electromagnetic properties of an alleged ghost. If instruments can do the job, why can't people? And what might an instrument possess that's also utilised by a witness?

Before attempting an answer, it's relevant to point out that the human senses receive a *lot* of information from the outside world, to the extent that the amount of information approaches the infinite.[1] Focusing on vision (since Tom saw his ghost), the human retina has the reliable ability to detect the energy of just a single photon.[2] Therefore, the visual system is proficient at *receiving* subtle energies and transmitting this information contained within to higher cortical regions, although there is a minimum threshold for the amount of photons required to actually be dimly *aware* of the stimulus—roughly six.[3] Indeed a just-detectable flash requires approximately 100 photons to be perceived.[4]

However, this is only referring to photons of *visible* light. There's a lot of electromagnetic radiation outside this narrow range. The human retina contains three varieties of colour sensitive cones which respond the visible spectrum—either in the blue range with an absorbance peak at approximately 420 nanometres (nm), in the green range with an absorbance peak at approximately 530 nm, and in the red range with an absorbance peak at around 565 nm.[5] *Officially* humans don't have the ability to *see* non-visible photons below 420 nm or above 565 nm wavelengths, even though there's a lot of them travelling around the environment—more than half of the sun's radiation impacting earth is in the longer infrared (IR) range beyond 565 nm.[6]

There's also evidence of alternatives to the normal absorption of light by the pigments of retinal cells, and at wavelengths greater than 800 nm the photosensitive *rods* become the chief detecting cell allowing IR to be "seen"

---

[1] Howard (2011).
[2] Pizzi et al. (2016).
[3] Irwin (2006).
[4] Field and Sampath (2016).
[5] Gentili et al. (2016).
[6] Palczewska et al. (2014).

by some people as white light in the periphery of the visual field. Above 1000 nm investigators have reportedly perceived IR as a "pale greenish light". In fact, this alternative absorption process (involving very different molecular processes) might allow visual experience at wavelengths as high as 1300 nm.[7] That creates a lot of visual possibilities.

At the other, lower end of the spectrum there is ultraviolet (UV) light in the range of 10 to 400 nm.[8] Numerous terrestrial invertebrates and vertebrates are sensitive to UV, and some varieties of mammals (rodent such as mice and rats, and also gerbils and gophers) possess cones that contain UV-sensitive pigments[9] allowing radiation in the 350 nm range to be detected.[10] Nonetheless it's generally accepted that humans cannot detect UV. This is not because human retinal cells aren't able to be stimulated by shorter UV wavelengths—they can. Rather, UV radiation never reaches the rear of the eye as it is first absorbed by the cornea and lens.[11] Interestingly, when cataracts are treated using UV transparent lens transplants, patients sometimes report a blue or purple glow corresponding to their newly-acquired ability to perceive in the ultraviolet range.[12]

This leaves *some* scope for people to experience normally invisible radiation as a visual stimulus—particularly IR. There are also plenty of other forms of energy *not* in the form of electromagnetic radiation which can elicit a visual experience. For instance, cosmic rays are high energy charged particles like atomic nuclei and protons that emanate from deep space—perhaps the supernova product of exploding stars.[13] They are associated with an elusive phenomena reported by astronauts (and cosmonauts) in orbit called *light flashes*. These particles are "seen" with eyes opened or closed following a period of dark adaptation, and are most commonly described as spots, blobs and streaks of white, blue and occasionally yellow, red and pale green "light".[14]

---

[7]Ibid.

[8]Reinhuber (2017).

[9]Jacobs (2010).

[10]Gentili et al. (2016).

[11]Ibid.

[12]Williams (2014).

[13]Limoli (2017).

[14]Bidoli et al. (2000).

The origins of LF are a mixed bag. The stimuli might be electromagnetic after all; produced by fluorescence of the lens of the eye or even Čerenkov radiation.[15] The latter is produced when charged particles like cosmic rays move more rapidly than photons inside a medium other than a vacuum.[16] The vitreous humour of the eyeball is fluid filled, so Čerenkov radiation might be produced *within* the eyeball to be perceived as an subtle visible stimulus.

Alternately, charged heavy nuclei energy particles of the cosmic rays might pass straight through closed eyelids and initiate a visual signal directly in the retina. Or in the case of protons, they produce "knock on" particles following nuclear interactions in the eye which are perceived as a visual catalyst.[17]

And some visual sensations might not even depend on photons or other energies impinging on the retina. *Entoptic* is a word of Greek origin meaning "within vision", and entopic imagery originates from the visual pathways within the brain itself, rather than retinal stimulation by outside energies.[18] These sensations might derive from scraps of biological material like blood cells or blobs of protein floating in front of the retina, or when retinal blood vessels are "seen" overlaid on the field of vision.[19] There's also the mechanical induction of comparable light forms known as *phosphenes* that occur when pressure is applied to the eyeballs beneath closed eyelids, or during a migraine.[20] Mostly we experience this pressure as a brief flash of what appears to be white light, but phosphenes can take on a wider range of forms, such as;

> … spirals; exploding stars; glimmering and shining dots or spots; wispy clouds; wheels; tunnels; funnels; parallel lines; wavy lines; wave patterns, pulsating waves; dotted lines; zigzags, checkerboards; honeycombs; spider webs; dot patterns; circles within circles; crosses; thin meandering lines, like lightning; geometric shapes, like triangles, squares, pentagons, etc.

15 Budinger et al. (1972).
16 Altschul (2017).
17 Casolino et al. (2003).
18 Labib (2017).
19 Bradley et al. (1998).
20 Arnulf (2013).

> They may swirl, pulse, superimpose, fragment, shine, move in different dimensions, or morph into other images.[21]

Entopic imagery is also possible when the retina is circumvented and higher visual pathways become directly activated. For instance, a blow to the head can cause phosphene "stars" and other patterns to be perceived, and similar experiences are reported during electrical and magnetic stimulation of the scalp adjacent to the occipital lobes, which hold the visual cortex (the processing destination for visual stimuli).[22]

Then there are more controversial possibilities for experiencing unconventional visual phenomena. Is it possible for people to perceive *gravitational waves*, or the dozens of elementary particles *that aren't* photons? While some of these waves and particles have been recently measured and appear to be authentic (for instance gravitational waves),[23] many are still elusive enough to be better viewed as hypothetical and mathematical concepts. Things like gravitons or quarks and antiquarks are often said to be *built* from mathematics. Therefore, the notion of perceptually interacting with these *parasensory* energies isn't even a field of discussion other than in *extremely* fringe pseudoscientific literature.

Presumably these theoretical fields and waves and massless particles permeate space but never excite human receptive organs. They travel straight through people as though they aren't there, which just reinforces the difference between a universe of mathematical components, and a macroscopic one that's physical and measureable and where people reside. The same can be said about the aetheric plane and its constituents which are imperceptible and have no impact on any level of the visual system. That is, unless there's a conversion of aetheric waves to sensible energy where the normally *unseen* can now be *seen* as a ghost or any other extraordinary inhabitant of this fantastical realm.

---

[21] Sayin (2014, p. 429).

[22] Ibid.

[23] Cossins (2018).

References

Altschul, B. (2017). Why Cerenkov Radiation May Not Occur, Even When It Is Allowed By Lorentz-Violating Kinematics. *Symmetry, 9* (11), 250–260.

Arnulf, I. (2013). Dream Imagery, Rapid Eye Movement Sleep Behavior Disorder, and Hallucinations. *Sleep and Biological Rhythms, 11*(Suppl. 1), 15–20.

Bidoli, V., Casolino, M., De Pascale, M. P., Furano, G., Morselli, A., Narici, L., et al. (2000). Study of Cosmic Rays and Light Flashes on Board Space Station Mir: The Sileye Experiment. *Advances in Space Research, 25*(10), 2075–2079.

Bradley, A., Zhang, H., Applegate, R. A., Thibos, L. N., & Elsner, A. E. (1998). Entoptic Image Quality of the Retinal Vasculature. *Vision Research, 38*(17), 2685–2696.

Budinger, T. F., Lyman, J. T., & Tobias, C. A. (1972). Visual Perception of Accelerated Nitrogen Nuclei Interacting with the Human Retina. *Nature, 239*(5369), 209–211.

Casolino, M., Bidoli, V., Morselli, A., Narici, L., De Pascale, M. P., Picozza, P., et al. (2003). Space Travel: Dual Origins of Light Flashes Seen in Space. *Nature, 422*(6933), 680.

Cossins, D. (2018). Total Wipeout: The First Gravitational Waves Validated Einstein's Theory Of Gravity—The Ones to Come Could Destroy It. *New Scientist, 238*(3183), 32–35.

Field, G. D., & Sampath, A. (2016). Behavioural and Physiological Limits to Vision in Mammals. *Philosophical Transactions of the Royal Society B, 372*, 20160072. https://doi.org/10.1098/rstb.2016.0072.

Gentili, P. L., Rightler, A. L., Heron, B. M., & Gabbutt, C. D. (2016). Extending Human Perception of Electromagnetic Radiation to the UV Region Through Biologically Inspired Photochromic Fuzzy Logic (BIPFUL) Systems. *Chemical Communications, 52*(7), 1474–1477.

Howard, R. G. (2011). If Paranormal Phenomena Are Information Received by the Conscious Psyche, Then What Is the Conscious Psyche? *Journal of Spirituality and Paranormal Studies, 34*(2), 62–72.

Irwin, K. D. (2006). Seeing with Superconductors. *Scientific American, 295*(5), 86–92.

Jacobs, G. H. (2010). Recent Progress in Understanding Mammalian Color Vision. *Ophthalmic and Physiological Optics, 30*(5), 422–434.

Labib, B. (2017). Reflecting on Entoptic Phenomena: Vivid Images of the Retinal Vasculature Often Appear to Our Patients During Exam. Are They a Cause for Concern? *Review of Optometry, 154*(3), 24–25.

Limoli, C. L. (2017). Deep-Space Deal Breaker. *Scientific American, 316*(2), 54–59.

Palczewska, G., Vinberg, F., Stremplewski, P., Bircher, M. P., Salom, D., Komar, K., et al. (2014). Human Infrared Vision Is Triggered by Two-Photon Chromophore Isomerization. *Proceedings of the National Academy of Sciences of the United States of America, 111*(50), E5445–E5454.

Pizzi, R., Wang, R., & Rossetti, D. (2016). Human Visual System as a Double-Slit Single Photon Interference Sensor: A Comparison between Modellistic and Biophysical Tests. *PLoS ONE, 11*(1), 1–24.

Reinhuber, E. (2017). Phasmagraphy: A Potential Future for Artistic Imaging. *Technoetic Arts: An International Journal of Speculative Research, 15*(3), 261–273.

Sayin, H. U. (2014). Does the Nervous System Have an Intrinsic Archaic Language? Entoptic Images and Phosphenes. *Neuroquantology, 12*(3), 427–445.

Williams, C. (2014). All-Seeing Eyes. *New Scientist, 221*(2961), 40–41.

8
Filters and Tuners

Having established the basic parameters of human visual perception, it's time to return to the analogy of how ghostly energy can be comprehended, and the notion of the human visual system as an instrument. You might remember from earlier discussion, orthodox perception of ordinary things in the world is a function of receiving photons from all parts of the visible spectrum. Evidence also suggests that visual experience may be possible outside this range—particularly in the IR. And there's the further proposal that stuff that's *not* electromagnetic in nature might also be perceivable, like cosmic rays and other peculiar things (albeit in special environments, such as deep space). However there's no evidence that mathematical (or mystical) parasensory stimuli can be "seen" by anyone, although it's tempting to imagine this possibility!

This implies there are many forms of environmental energy that *could* be detected by people, but the reason why they're *not* detected is because the brain possesses a *sensory filter*. This filter controls the entry of information, meaning people are unaware of the bulk of energetic data shimmering around them because it is "sieved out" at the very earliest stages of perception. Only those specific nanometres of visible light are given free passage

T. Jinks, *Psychological Perspectives on Reality, Consciousness and Paranormal Experience*,
https://doi.org/10.1007/978-3-030-28902-7_8

through the filter to form the basis of ordinary experience. This idea is historically popular, perhaps the most famous proponent being Aldous Huxley. In *The Doors of Perception* (1952) Huxley pondered his personal observations following ingestion of various psychotropic drugs[1] (meaning they affected his psychological state), most prominently the alkaloid[2] *mescaline* derived from the peyote cactus (and other cacti varieties), and the ergot alkaloid *lysergic acid diethylamide* (LSD). For Huxley;

> The essential nature of the mescaline experience....was an experience 'just of light, of everything flooded with light... a kind of luminous living geometry'...... With mescaline and LSD, light, colour, and the significance of ordinary physical objects were all dramatically altered and intensified.[3]

Huxley frequently corresponded with psychiatrist Humphry Osmond about his novel experiences with mescaline and LSD. It was Osmond who fashioned the word *psychedelic* in 1956 by combining the Greek *psyche* (mind—i.e., cognition, memory and imagination[4]) and *delos* (to make visible or clear)[5] to refer to natural chemical agents ingested through the nostrils, eaten, drunk or smoked which possess particular psychological properties strongly associated with anomalous experiences.[6] Other labels for the effect of mescaline and LSD (and similar drugs) include *hallucinogens, psychodysleptics, entheogenics* (meaning bringing forth the inner God or divine[7]) and *psychotomimetics* (since these drugs mimic "real" dysfunctional psychosis.[8])

In particular, Huxley (and Osmond) attempted to understand;

[1] Psychotropic refers to *any* drug effect on psychological state, therefore includes caffeine, alcohol, and nicotine (Presti 2011, p. 24).

[2] Chemical substances derived from plants, possessing a ring structure and with nitrogen as a component.

[3] Hansen (2014, p. 681).

[4] Williams (2005).

[5] Carhart-Harris (2013).

[6] Rioux (1996).

[7] St John (2016).

[8] Presti (2011, p. 33).

> ...how psychedelics opened doors to other worlds, or revealed layers in perception that remained invisible to the conscious mind[9]

Huxley came to the conclusion that the brain possesses, or acts like, a filter.[10] This filter is not there to *produce* experience, but gate-keep and protect the individual from being overwhelmed *by* experience. In an oft-quoted paragraph he echoes the statement of Cambridge philosopher, Dr C. D. Broad;

> .... the function of the brain and nervous system and sense organs is in the main eliminative and not productive. Each person is at each moment capable of remembering all that has ever happened to him and of perceiving everything that is happening everywhere in the universe. The function of the brain and nervous system is to protect us from being overwhelmed and confused by this mass of largely useless and irrelevant knowledge, by shutting out most of what we should otherwise perceive or remember at any moment, and leaving only that very small and special selection which is likely to be practically useful.[11]

Broad in turn derived these ideas from French philosopher Henri Bergson's much earlier 1896 work *Matter and Memory*[12] where the brain possesses a barrier mechanism that restricts entry of material from the outside world, but also controls the awareness of the accumulation of memories from an individual's internal world. The ultimate purpose of the barrier is to prevent the individual from being overpowered by swathes of distracting information that is of no constructive value. In Huxley's opinion, the material that gets through the filter is only a;

> ...very small and special selection which is likely to be practically useful.....[to] stay alive on surface of this particular planet....[and] what comes through the reducing valve is consecrated as genuinely real by local language.[13]

---

[9] Dyck and Farrel (2018, p. 247).

[10] Brabant (2016).

[11] Huxley (1952/2011, p. 11).

[12] Luke et al. (2012).

[13] Huxley (1952/2011, p. 11).

That explains Huxley's closing remark—people are animals and their purpose is survival, so it makes sense for unlimited information;

> ...to be funnelled through the reducing valve of the brain and nervous system. What comes out at the other end is a measly trickle .... which will help us to stay alive on the surface of this particular planet.[14]

But there are times where all this additional information can manifest in unusual ways. Bergson addressed the SPR in 1913 and told the audience that while the past isn't forgotten, the purpose of the brain is to concentrate on the present, which means a necessary restriction of the "field of consciousness". However at the moment of death the filter is diminished and all those "forgotten" past memories are experienced in the life review.[15]

This early conceptualisation of the sensory filter identifies inconsequential material as ordinary, everyday stimuli. That could be scenes or sounds or anything else that might activate the senses. Hence the filter will block out an uninteresting conversation from a group of people standing nearby; or prevents you calculating the exact number of branches on the tree you're walking past; or allows you to be oblivious to how many clouds are in the sky above your head—and so on.

Should this filter be bypassed or malfunction, then people would be forced to consume a great deal more of this additional information than is normally allowed. That could be to the point where *everything* environmental is processed leading them to remember *everything* about *everything*. According to Bergson this might be happening at some deeper level (hence the life review), but in terms of how an ordinary mind works that's not desirable. One would assume there needs to be the cognitive capacity to deal with such an amplified experience, or there would be some kind of mental breakdown! The filter is there for a very good purpose, and without it working properly the outcome will be catastrophic for the psyche. To push the analogy a little further, it might mean poisonous contaminants will be introduced into the mind (in the style of a faulty tap filter) causing

---

[14] Ibid., p. 11.

[15] Poller (2015).

serious psychological dysfunction, or the *container* of the mind will burst at the seams (in the style of a pressure or volume regulator).

Huxley reasoned that drugs like mescaline acutely disturbed the filter for a short period of time resulting in extra-informational experience that are *beyond normal* (Huxley didn't use the word "paranormal", but the expression fits the circumstances). Although this doesn't necessarily mean the filter is allowing *everything* in. It might be slightly relaxed, whereby only a small trickle of additional material is permitted. Nor does this extra-information refer solely to routine material that *could* be perceptible if granted awareness (like the group conversation, or the number of branches on a tree). There is a possibility it resembles all that normally imperceptible energy that people don't usually sense—stuff homologous to IR or UV, cosmic rays or perhaps more exotic energies like gravity waves. Such energies are not normally observable as they possess little if any survival value. Regardless, the phenomenological outcome of "seeing" this new energy would be strange enough (on top of what's normally seen) to be interpreted by the witness as a paranormal experience.

Regarding Tom's ghost sighting, it's not necessary that the ghost changes its conformation for him to passively see its human form, the way he reflexively "sees" a chair or a coffee mug. A change in Tom's sensory filter might allow him to *actively* perceive the ghost. However the filter analogy is perhaps too coarse in this factual context, as mechanical filters are either *in place*, or they're *not in place*. That is, they either let everything in (so Tom will be swamped with information), or refuse him access to anything outside of what he is allowed to see.

It's better to think of a *loosening* of perceptual control, which requires an additional (analogous) component to be attached to the filter. That component is the *tuner*.

In a resting state, a tuner fixed to a filter is normally neither fully open, nor fully closed. It sits somewhere in-between. In this respect the theoretical filter/tuner system is similar to other biological homeostatic apparatus where it's most effective for a variable to be "somewhere in the middle"—neither too high or too low. If the state or value of what is being monitored creeps into the high range, mechanisms can be enacted to bring the variable back to level. For example, if body temperature increases cutaneous blood vessels will dilate and the skin will flush, bringing blood to the surface and allowing the passage of heat into the atmosphere. A similar mechanism is enacted if the variable falls too low. Should blood pressure drop, blood vessels will contract and the heart rate will increase to counteract the fall.[16]

Tom will see the ghost of the man haunting the hotel corridor only if his filter/tuner complex is dialled to a more "open" position, easing the filter's restrictive activity. This would allow *some* additional information from the external world to access dynamic brain processes. The source of the ghost could be one of many possibilities grounded in the imperceptible realm of energy. Some places have active spirits of various spectral wavelengths or vibrational states, other places are stone-taped, leaking inert energy that's stored in in the structure the way that an asphalt pavement gives off heat after sunset. In Tom's case, the playback option seems most viable as there was nothing else in the context of his sighting that suggests purposefulness or intelligence to the entity. There was no "eye to eye" contact, or attempts to by the phantom to communicate with him. With his tuner temporarily dialling down his filter, he became privy to the energetic replay of the man's passage, which is available to be seen by anyone, all day, every day, until the display eventually runs out of steam.

The idea that people "see" normally unseeable energy is not unreasonable to paranormal investigators (as opposed to mainstream scientists). Images of ghosts are allegedly captured using a camera with full spectrum capability, also known as VNIR (short for *visible and near infrared*). These record information just beyond the visible spectrum—usually in the 350 to 1000 nm range which encompasses some UV and some IR.[17] Hence the fuzzy blue shape discovered on a full spectrum photograph is precisely what a witness reports seeing when circumstances dictate that their internal filter/tuner complex shifts to a different "station" on the dial.

By the same argument, the filter/tuner analogy applies to hearing any of the normally inaudible frequencies that exist outside the conventional human auditory range, best represented by awareness of *infrasound*. Sound

[16]Wickens (2009).

[17]Reinhuber (2017).

waves are oscillations in (and of) atmospheric pressure,[18] and the set-standard for the lowest frequency of sound audible to humans is between 16 and 20 Hertz (Hz) while the upper limit is taken to be in the vicinity of 16–20 Kilohertz (kHz).[19] However unlike the top end of the audible range, which falls away sharply, the bottom-end of the spectrum declines more gradually. Therefore, depending on *how* the sound is being heard the minimum detectable value can be well below 16 Hz. That might be as low 1.5 Hz when listening to noise through earphones.[20] For this reason, a great deal of natural (and human-generated) sounds "typically" considered imperceptible can actually be heard after all, in the right conditions. Examples include weather events such as wind and storms, the actions of ocean waves, and the low hum of vehicles and machinery[21] (*truly* inaudible infrasound occurs at the much lower 0.01–2 Hz range,[22] and they cannot be heard by anyone).

When barely audible infrasound is detected by listeners, the outcome can be psychologically disturbing;

> ...a family complained of disturbance at night, and consequent effects on health, allegedly caused by noise from a boiler house. Measured levels were higher at times of disturbance, but well below the median threshold. For example, at 10 Hz where the threshold is 97 dB, a measured level of 35 dB caused disturbance, whilst a level of 15 dB did not. Separate laboratory measurements of the complainants' low-frequency thresholds showed these to be close to the median, leaving the precise mode of detection still unexplained, although it was noted that the higher levels, and consequent annoyance, were associated with windy conditions. The complainants' descriptions of the noise and its effects on them included "very low frequency, hum, drone, intermittent pulsating, pain in the legs and in the area of the stomach.[23]

Some environmental infrasounds possess a mysterious origin. A well-known example is reports of "The Hum", a persistent low frequency

[18] Parsons (2012).

[19] Møller and Pedersen (2004); although detecting noise is also a function of its decibel level.

[20] Leventhall (2007).

[21] Parsons (2012).

[22] Leventhall (2007).

[23] Ibid., p. 135.

droning which is "heard" by a select group of people in various localities around the world, but which is not detectable by auditory measuring equipment. It can result in stress, anxiety and consequently poor life quality.[24] Since the stimulus can't be detected by instruments it might not actually be infrasound at all. The source might be something entirely different, like electromagnetic radiation that's mistakenly perceived by the sufferer as sound, although there's very little evidence of how this could manifest.[25]

In terms of ghosts and infrasound, there's a common conception that spirit noises are a by-product of "their" use of electromagnetic energy during the materialisation process. That's described using another clever analogy—ghosts are not just seen, but heard in just the same way that atmospheric energy simultaneously produces the *light* of lightning and the *sound* of thunder.[26] Maybe that's what is recorded as the *electronic voice phenomena* (EVP),[27] where meaningful words and statements are allegedly captured by voice-processing devices (like the standard digital voice recorder, but also answering machines, cellular phones, televisions and computers) and discovered on playback—either accidentally or intentionally if the purpose is spirit communication. The message is most often in response to a person's interactions, so they are taken as coming from a sentient entity—for instance a deceased family member.[28]

EVPs are referred to as *infrasonic* by some investigators, hence detecting a sub-threshold ghost's voice is possible when a person's biological tuner attached to an auditory filter is "dialled down". That's analogous to what digital voice recorders do—although the counterargument is that these types of recorders are not incapable of recording or measuring infrasound, so the comparison is unsuitable.[29] Criticisms aside, the notion of people hearing things in the fashion of a voice recorder set at the correct tuning

[24] Leventhall 2007

[25] Ibid.

[26] Parsons (2012).

[27] A more contemporary label for EVP is *instrumental transcommunication* (ITC)

[28] Elliot (2018) and Segal and Talbott (2016).

[29] Parsons (2012, p. 152), whose criticism is that "There seems to be a generally poor understanding of the original work by Tandy, and of the technical constraints in making infrasound measurements, and this has led to misunderstanding of any actual relationship between infrasound and paranormal experiences and accounts".

is very popular. In Tom's case, he mentioned that he had an urge to "look up" before he saw his ghost. That might be an indication of infrasound at work, attracting his attention and causing a reflexive movement to the source of the subtle noise—*seen* as a visual ghost. He only happens to remember the apparition component of the ghost experience, while the putative sound component never entered awareness.

And as to where the voices (or noises) come from, rather than comprehensible infrasound more radical investigators including the pioneers of EVP Friedrich Jürgenson and Kostantin Raudive point to the origin of communication as something even stranger; paranormal transmissions from a spectral "radio" station[30] residing what Raudive called "the opposed world".[31]

Regardless of whether a ghost is perceived visually or audibly, the hypothesis that a biological filter/tuner system can be manipulated to allow the reception of parasensory information requires two assumptive leaps. The first is that Tom possesses this contraption in the first place, and it's set at "baseline" to allow perception of conventional environmental external information. There's no strong evidence for such a system, although circumstantially the example of *hypermnesia*—extreme clarity of memory—might functionally indicate that people possess something resembling a filter that has been "tuned down". Mnemonists have the ability to remember long lists of information (like numbers, names or dates), and one such person, Solomon Shereshevsky possessed;

> ...memories so comprehensive and so intrusive that they often find them more of a hindrance than a blessing. It is rather as though the sufferers lack some filtering mechanism that others possess.[32]

The second assumption for a filter/tuner complex, and applied to Tom's story, is that on that particular afternoon in the hotel corridor he experienced a brief change in the mechanism's gating, permitting entry of an

---

[30]Wright (2012).

[31]Cardoso (2012) (in other words, an aetheric-style realm).

[32]Gauld (2012, p. 224).

unexpected burst of external energy. In the case of a spontaneous paranormal experience like Tom's ghost, it might be assumed the action is accidental rather than controlled by an external source such as the "ghost" itself. This is now going beyond the *how* of the 5W1H model and touching the *why* the event happens. Was this an accidental fluctuation due to circumstance, allowing him to see ghosts that just happen to be everywhere?

Evan accepting these two assumption, there's still the difficulty of accounting the appearance of Tom's ghost. Tom didn't merely see a hovering purple smudge of UV radiation cross the corridor, or a glow of white light from the IR end of the spectrum. He saw a detailed, man-like figure. It's one thing to apply the filter analogy to a ghost sighting, but something else entirely to take what's known about how this energy might be perceived, and explain how it then becomes the fully formed image or a man walking with fluid movement and possessing hue, contrast, and depth. I'll provide some possibilities for the transformation of *energy into object* later in the book. However for the moment let's spend some time examining the operation of this conceptual filter-system, and how it might apply to the witnessing paranormal "things".

References

Brabant, O. (2016). More Than Meets the Eye: Towards a Post-materialist Model of Consciousness. *Explore, 12*(5), 347–354.

Cardoso, A. (2012). A Two-Year Investigation of the Allegedly Anomalous Voices or EVP. *Neuroquantology, 10*(3), 492–514.

Carhart-Harris, R. L. (2013). Psychedelic Drugs, Magical Thinking and Psychosis. *Journal of Neurology, Neurosurgery, and Psychiatry, 84*(9, e1).

Dyck, E., & Farrel, P. (2018). Psychedelics and Psychotherapy in Canada: Humphry Osmond and Aldous Huxley. *History of Psychology, 21*(3), 240–253.

Elliott, C. (2018, April). Meaning in Randomness. *Psychologist*, 22–25.

Gauld, A. (2012). Living Consciousness: The Metaphysical Vision of Henri Bergson (Book Review). *Journal of the Society for Psychical Research, 76*(909), 222–227.

Hansen, D. (2014). Heavenly and Hellish—Writers on Hallucinogens. *The Psychologist, 27*(9), 680–682.
Huxley, A. (2011). *Doors of Perception and Heaven and Hell.* London: Thinking Ink (Original work published 1952/1954).
Leventhall, G. (2007). What Is Infrasound? *Progress in Biophysics and Molecular Biology, 93*(1–3), 130–137.
Luke, D., Zychowicz, K., Richterova, O., Tjurina, I., & Polonnikova, J. (2012). A Sideways Look at the Neurobiology of Psi: Precognition and Circadian Rhythms. *Neuroquantology, 10*(3), 580–590.
Møller, H., & Pedersen, C. S. (2004). Hearing at Low and Infrasonic Frequencies. *Noise and Health, 6*(23), 37–57.
Parsons, S. T. (2012). Infrasound and the Paranormal. *Journal of the Society for Psychical Research, 76*(908), 150–174.
Poller, J. (2015). Beyond the Subliminal Mind: Psychical Research in the Work of Aldous Huxley. *Aries, 15*(2), 247–266.
Presti, D. E. (2011). Neurochemistry and Altered Consciousness. In E. Cardeña & M. Winkelman (Eds.), *Altering Consciousness: Multidisciplinary Perspectives: History, Culture, and the Humanities; Biological and Psychological Perspectives* (Vol. 2, pp. 21–41). Santa Barbara: Praeger.
Reinhuber, E. (2017). Phasmagraphy: A Potential Future for Artistic Imaging. *Technoetic Arts: An International Journal of Speculative Research, 15*(3), 261–273.
Rioux, D. (1996). Shamanic Healing Techniques: Toward Holistic Addiction Counseling. *Alcoholism Treatment Quarterly, 14*(1), 59–69.
Segal, T. L., & Talbott, V. (2016). Close Encounters of the Familiar Kind. *Journal for Spiritual and Consciousness Studies, 39*(2), 120–129.
St John, G. (2016). The DMT Gland: The Pineal, the Spirit Molecule, and Popular Culture. *International Journal for the Study of New Religions, 7*(2), 153–174.
Wickens, A. (2009). *Introduction to Biopsychology* (3rd ed.). Harlow, UK: Pearson Education Ltd.
Williams, R. J. (2005). *The Symbol of the Hybrid Human/Alien Child in the Abduction Phenomenon: Rebirthing Within the Psyche and the Psychosomatic Imagination.* Proceedings of the Eighth Australian and International Religion, Literature and the Arts Conference 2004. http://hdl.handle.net/2123/1263.
Wright, S. H. (2012). An Interview with Anabela Cardoso: Anomalous Electronic Voices Received via Instrumental Transcommunication. *Neuroquantology, 10*(3), 515–518.

9
Multiple Filters?

Real-life tuners dial in both directions, so assuming that a loosening of the tuner corresponds to an "open" setting and the entry of subtle information, what might happen if the filter/tuner complex was tuned *down* from the halfway mode, into a "closed" position? Envisaging the actions of an actual tuner system, one would expect much less to get through, so presumably at its most extreme an individual would perceive nothing but "blackness" due to an absence of any information entering the brain.

That's where a *single* filter analogy has its limitations. Embellishing Bergson's idea of external information versus memory information, it is more effective to posit a system with two filters and a single tuner; filter/tuner A and filter/tuner B. Keeping it simple, when the tuner is dialled in one direction from the half-way point, Filter A is opened and a little more material from the external world is let into the brain. When the tuner is dialled in the opposite direction, Filter B is now opened and a little more material from the *internal* world enters the brain.

Furthermore, relaxation of Filter B is accompanied by a simultaneous *tightening* of Filter A, resulting in the prevention of worldly information from entering awareness. That leaves greater capacity for endogenous

T. Jinks, *Psychological Perspectives on Reality, Consciousness and Paranormal Experience*,
https://doi.org/10.1007/978-3-030-28902-7_9

content to be processed. In its simplest form this material is physiological; subtle signals presumably taking the form of normally undetectable body sensations (rather than detectable ones like pain or the feeling of heat or cold that can potentially be ignored). Comparable to the detection of UV and IR light, not even directed attention is powerful enough to allow perception of these biological signals. Indeed, in mainstream neuroscience, attention is also viewed as a filter which acts to limit the brain being overwhelmed with information. Stimuli compete for the attention *of* attention so as to be recognised.[1]

Examples of these imperceptible stimuli are the processes that occur at the neural level, like ligand binding at protein receptors resulting action potential, or the activity of enzymes and hormone release from the endocrine glands[2] Apparently these processes are sensed in the fashion of emotions or gross motor movements! (although this idea is certainly not a widely held belief).

The notion that people can experience indiscernible biological activity is loosely connected to the claims of psychiatrist Wilhelm Reich and his *Bions.* This energy source is apparently found in all living and non-living things, and the energy released by Bions is *Orgone*. Indeed in the middle years of the twentieth century Reich spent a great deal of time trying to harvest Orgone with the help of his patent Orgone accumulator (albeit unsuccessfully).[3]

In Reich's opinion, the heathy dispersion of Orgone energy in the body is necessary to avoid a variety of physical and mental illnesses, and this formed the basis of his therapies.[4] Indeed, a thorough training in Reichian psychotherapy allows appreciation of the internal milieu, and roughly corresponds to the "loosening" of the hypothetical Filter B. When a person detects normally silent somatic states there is no context for them to be rationalised, hence these feelings are projected onto the "outside world"

[1]Ligeza et al. (2017).

[2]Howard (2011).

[3]Self (1997); The orgone accumulator consisted of a box large enough for people to sit in, made of alternating wood and steel which was said to capture environmental orgone and assist therapeutic outcomes for the patient.

[4]Briehl (1995).

and potentially interpreted as paranormal.[5] For example, standing on a low bridge over a stream, neo-Reichian psychotherapist Nick Totton give the example of how he was able to detect the current of water beneath him as a "fizzing" sensation in his feet and legs. Totton argues his comprehension of "organ function" (or in the current context, a release of filter B), allows variety of unusual perceptions channelled through the peripheral nervous system.

Then there are the other types of information feasibly garnered through the relaxation of Filter B and the concurrent tightening of Filter A. However these are mental rather than physical—Bergson's "memories" and perhaps other, stranger things.

Stranger things? It's been speculated that undiscovered information transmitting mechanisms draw their material from external sources[6]—but not through conventional sensory systems like vision or hearing. This alternative sensory organ (or organs) directly transduces a range of energies (perhaps even those just outside the visual spectrum) into a multi-sensory impression. That's implicit in many definitions of the paranormal;

> An exceptional, anomalous, or 'paranormal' experience is a rare experience, provoked or appearing spontaneously, considered by the person who reports it as the result of a *non-ordinary interaction with the environment* [emphasis added].[7]

This is the basis of the proposition that psi ability is legitimate and a function of the transmission of information through a non-standard sensory channel when the tuning is correctly set. This is known as the signalling or information-theory approach; one of a number of perspectives on psi.[8] Previously I suggested that psi (the sixth sense) loosely conveys the idea that what is *seen* is not necessarily "see-able" through the eyes (in the case of vision)—rather the information is "seen" using some other method. This extrasensory information apparently falls below the threshold of detection

---

[5] Totton (2003).

[6] Howard (2011).

[7] Rabeyron and Caussie (2016, p. 55).

[8] Carr (2008).

for electric and magnetic radiation—even gravity, so an alternative detection channel has been proposed that is sensitive to these very weak signals and forces. One suggestion centres on *cryptochromes*, blue photoreceptor proteins responding to signals in the 400–500 nm range of the visual spectrum. It's suggested that these can detect weak magnetic radiation *as well as* photons of light, and their activation can occur in even when light is *not* present.[9]

Therefore, if non-standard information reaches the brain (and body) as a magnetic signal it will chemically activate the cryptochromes. If there is adequate summation of this signal by cryptochromes over a large area, the product can be distilled into something perceivable. That takes place in the two regions where the protein is most concentrated; the retina (presumably assisting with unusual visual experience) and also the *pineal gland*.

The function and purpose of this pinecone-shaped gland has been of great interest for hundreds of years.[10] The organ is the basis of what René Descartes called "the seat of the soul",[11] and (unavoidably returning to Theosophy), Blavatsky stated that while to physiologists and biologists it was no more than "a horny gland with a little sand in it", it was in fact;

> ...the very key to the highest and divinest consciousness in man – his omniscient, spiritual and all-embracing mind. This seemingly useless appendage is the pendulum which, once the clock-work of the inner man is wound up, carries the spiritual vision of the ego to the highest planes of perception, where the horizon open before it becomes almost infinite.[12]

This remains a popular New Age concept, but from a more grounded parapsychological perspective the pineal gland is simply a receiver of alternative

[9] Shiah (2012).
[10] St John (2016).
[11] Ibid.
[12] Blavatsky (1889).

information. It is either the filter/tuner system, or has a filter/tuner system attached, allowing it to transduce unconventional signals into sensed experience. This characterises the sixth sense.[13]

So if the *how* is a filter/tuner mechanism, *why* would the tuner component ever change its position? In the heyday of psi investigation, an understanding of "psi-conducive states" centred on *relaxation.* Successful psi performance was positively correlated with electromyogram recordings and autonomic nervous system measurements. According to parapsychologist William Braud, these measurements correspond to "the subjective state of relaxed awareness and of the physiological state of reduced arousal".[14]

Braud continues;

> Slowed [brain] rhythms characterized two out-of-the-body experiences (which are sometimes accompanied by psi) .... Meditative states, which have traditionally been associated with heightened psychic sensitivity, are characterized by low arousal.....and sensory [external] "noise level" (including body tension) is minimized. Physically reducing external stimuli is one way of eliminating distraction, of allowing persons to attend to weak psi signals In a series of experiments involving different state-altering procedures (hypnosis, biofeedback training, sensory isolation)....subjects whose attention exhibited the greatest shifts from external (sensory) stimuli toward internally generated stimuli scored significantly better on ESP tasks than did those subjects whose attention remained externally oriented.[15]

In Tom's case, perhaps he was in an especially relaxed state at the moment of opening his room door? Under normal circumstances, any "psi information" that's out there is well beneath awareness—swamped by overwhelming sensory activity. And assuming causation rather than mere correlation, any relaxation-induced sluggishness of the higher cognitions—encouraged

[13]Meijer and Korf (2013); Alternately, rather than surmise that the pineal gland (or some other receptive device) is responsible for the experience, the signal might be induced directly in the CNS. This is equivalent to the way electricity (the visual experience) is induced in a coil through the actions of a moving magnet (the subtle signal). The inducting field might be magnetic or electric, or as has been suggested something as fundamental as "immersion in external quantum fields that penetrate our organism" and which conceivably act directly on the human brain to initiate a visual (or other sensory) experience.

[14]Braud (1975).

[15]Ibid., p. 144.

by mediation or laboratory techniques such as Ganzfeld[16]—will allow the tuner-dial to be released from its rigidly fixed position. Filter A can now focus on energy outside the officially visible spectrum, or let energies sneak in through alternative sensory pathways linked to the pineal gland. Tom has a relaxed temperament, and might easily and regularly "tune out" of the everyday to "tune in" to the parasensory, particularly after a tiring day of having to think hard about work and interact with difficult colleagues. Returning to his hotel room, Tom witnessed something quite unexpected because there just so happened to be plenty of imperceptible historical energy swirling around the corridor, compared with the dearth of activity his new townhouse complex, or his gym or favourite busy coffee shop (where he also "tunes out", but never reports any unusual sightings). The circumstances of energy presence, relaxation and filter/tuner status collide in an unexpected way, and hey presto, a ghost sighting!

Unfortunately it's impossible to determine which style of ghost Tom is seeing, as there is no way of identifying the settings of his putative brain filter/tuner. Consider this fictional scenario; Tom's workmates Chris and Sam just happen to accompany Tom to his room before they all go to dinner, and as Tom unlocks his door all three look up the moment the ghost walked across the corridor. Only Tom and Sam see the apparition. The selective nature of this ghost-sighting example is not uncommon, and the outcome is frequently attributed to an ESP event rather than consensus perception. Here's an informative story from 1930s England;

> Fourteen people had come to tea one 1932 afternoon in a London suburb, with Mrs Blanche Hornsby; and four saw a man who looked to them as if he were something to do with the law. The first their hostess knew about it was when they asked afterwards who he was; and she had to tell them that no such man had been there.[17]

But why *can't* the ghost be a product of veridical perception, if filters are involved? A mix of individual differences (like relaxation, biological filter

---

[16]The technique where participants are placed in a sensory deprivation environment, with white noise played and ping-pong ball halves placed over their eyes in order to reduce external input, hopefully allowing subtle psi signals to be detected.

[17]Inglis (1985/1986, p. 205).

disposition, weather or geomagnetic activity in sensitive people) leads to a transient relaxation of *either* Filter A *or* Filter B for Tom and Sam but not for Mike, hence the stimulus could be internally sourced or externally derived. Ultimately all that can be proposed is that the ghost sighting is triggered by a legitimate parasensory experience of *uncertain* nature, and moderated by the *unknown* operational status of a *hypothetical* filter/tuner. That's not helpful when attempting to formulate a 5W1H solution for Tom's experience, but with the assistance of psychedelic drugs anything could happen (as we'll explore in the next chapter).

References

Blavatsky, H. P. (1889). *Dialogue on the Mysteries of the After Life: On the Constitution of the Inner Man and Its Division.* Theosophical Articles, Vol. II, Retrieved September 5, 2018, from https://theosophy.org/Blavatsky/Articles/DialogueOnTheMysteriesOfTheAfterLife.htm.

Braud, W. G. (1975). Psi-Conducive States. *Journal of Communication, 25*(1), 142–152.

Briehl, W. (1995). Wilhelm Reich. *Psychoanalytic Pioneers, 1*(1), 430–439.

Carr, B. (2008). Can Psychical Research Bridge the Gulf Between Matter and Mind? *Proceedings of the Society for Psychical Research, 59*(221), 1–96.

Howard, R. G. (2011). If Paranormal Phenomena Are Information Received by the Conscious Psyche, Then What Is the Conscious Psyche? *Journal of Spirituality and Paranormal Studies, 34*(2), 62–72.

Inglis, B. (1985/1986). *The Paranormal: An Encyclopedia of Psychic Phenomena.* London: Paladin.

Ligeza, T. S., Tymorek, A. D., & Wyczesany, M. (2017). Top-Down and Bottom-Up Competition in Visual Stimuli Processing. *Acta Neurobiologiae Experimentalis, 77*(4), 305–316.

Meijer, D. K. F., & Korf, J. (2013). Quantum Modeling of the Mental State: The Concept of a Cyclic Mental Workspace. *Syntropy, 1*, 1–41.

Rabeyron, T., & Caussie, S. (2016). Clinical Aspects of Out-of-Body Experiences: Trauma, Reflexivity and Symbolisation. *L'Évolution Psychiatrique, 81*(4), e53–e71.

Self, W. (1997). The Edison of the Toolshed Orgasm. *Spy Magazine, 11*(5), 34.

Shiah, Y.-J. (2012). A Possible Mechanism for ESP at the Initial Perceptual Stage. *Journal of Parapsychology, 76*(1), 147–159.
St John, G. (2016). The DMT Gland: The Pineal, the Spirit Molecule, and Popular Culture. *International Journal for the Study of New Religions, 7*(2), 153–174.
Totton, N. (2003). Introduction. In N. Totton (Ed.), *Psychoanalysis and the Paranormal: Lands of Darkness* (pp. 1–14). London: Karnac.

10
The Receiver

Despite the limitations of pinpointing how a filter/tuner complex leads to a predictable ghost sighting, if you're comfortable with the analogy to account for anomalous perception, then you can explore different applications for the filter and tuner components, or even invent new ones. That might involve conceiving of the tuner not as attached and subservient to a sensory filter, but having free rein over the type of information entering the senses. And this isn't restricted to "here and now" energy of photons bouncing off the materials of the hotel corridor, or infrasound, or playback of a phantom recording stored in the hotels walls, or even the IR given off by a spirit presence. It could be ventured that something resembling an aether, or a thick invisible soup, surrounds us and holds a record of information from all times in the past (or all times in the future), or *anything, anywhere*. In terms of us accessing this content, it is commensurate to tuning into a different television station to the one you are watching now, but rather than re-using the word "tuner" I'll use the word *receiver*. This analogy is particularly popular when describing the outcome of psychedelic drug use.

T. Jinks, *Psychological Perspectives on Reality, Consciousness and Paranormal Experience*,
https://doi.org/10.1007/978-3-030-28902-7_10

Albert Hoffman, a pioneer in the original synthesis of LSD in 1938, describes his personal discovery of its psychedelic properties while riding his bicycle home from his laboratory after accidently infusing it through the skin;

> In the afternoon of April 16, 1943, when I was working on this problem, I was seized by a peculiar sensation of vertigo and restlessness. Objects, as well as the shape of my associates in the laboratory, appeared to undergo optical changes. I was unable to concentrate on my work. In a dream-like state I left for home, where an irresistible urge to lie down overcame me. I drew the curtains and immediately fell into a peculiar state similar to drunkenness, characterized by an exaggerated imagination. With my eyes closed, fantastic pictures of extraordinary plasticity and intensive color seemed to surge toward me. After two hours this state gradually wore off.[1]

Hoffman also applied the receiver analogy to the effects of LSD, writing that the experiences;

> …may be explained by the fact that the brain, the seat of the receiver, becomes biochemically altered. The receiver is then tuned into another wavelength than that corresponding to normal, everyday reality[2]

Rick Strassman is a medical doctor and academic well-known for his book *DMT: The Spirit Molecule* (2001). He describes the stories of his research volunteers who, in the early-to-mid 1990s and in controlled laboratory conditions, were intravenously administered with the alkaloid *N,N*-dimethyltryptamine (DMT)—a metabolic derivative of the essential amino acid tryptophan (which is the precursor to the important brain neurotransmitter serotonin and a closely related product, the hormone melatonin).[3]

We gave a solution of DMT fumarate by means of a 30-second infusion, followed by a 15-second flush of the line with sterile saline. Psychological responses to DMT peaked at about 2 minutes after completing the saline

---

[1] Smith (2007, p. 1).

[2] Hoffman (1979/1983, pp. 196–197).

[3] Szara et al. (1966).

flush, began resolving thereafter, and were barely noticeable by 30 minutes. Quite marked was the "rush" of IV DMT. Nearly all subjects reported this within 5–10 seconds from the beginning of the infusion. This was an overwhelming torrent of experience that completely replaced previously held mental contents. A sense of mounting inner tension and vibration, accompanied by a high-pitched sound, was associated with intense visual hallucinatory effects. This rush climaxed in the feeling that consciousness had separated from the body. Once the rush began resolving, DMT effects were rather similar to those reported in the literature and described by our initial cohort of DMT respondents. Volunteers reported what they believed were "freestanding" realities, "more real than real," an apperception of planes of existence coexisting with this one. Most unsettling, for both us and our volunteers, were descriptions of contact with "beings"—conscious, intelligent, interactive entities that, no matter how hard we tried to interpret as "something else," remained fixed in the subjects' minds as undeniably existent as such. In fact, some of the "alien abduction" literature is extraordinarily similar to these types of DMT experiences.[4]

Strassman opined that alcohol and other non-psychotropic drugs might "adjust the brightness", "contrast" or "colour" of human perception, giving what's experienced a slightly different character to normal. However, other types of alkaloids appear to change stations completely.[5] These include the tryptamine psilocin, from the psilocybin mushroom, and the chemically related DMT-expressing saliva of the *Bufo* toad genus (in its 5-hydroxy-DMT form called *bufotenine*).[6] There is some evidence that both these chemicals are endogenous to human biochemistry, for instance found in small traces the blood, cerebrospinal fluid and urine.[7]

Ingestion of DMT allows the consumer to encounter the much clichéd "other realm" teeming with fantastical entities. Indeed, Strassman's suggested that DMT's ability to alter an individual's receiver allows them to see something akin to the aether (or zero-point energy field, or whatever term

[4]Strassman (2005, p. 67).

[5]Ibid.

[6]Blainey (2015).

[7]Barker (2012).

is preferred). To Strassman that might be dark matter—the unaccounted 95% of universe's mass;

> Some of the most speculative and far-reaching conclusions we arrived at regarding DMT's properties revolve around explicating the frequency, intensity, and "reality" of these types of sessions. For example, if these "encounters" are perceptions of freestanding, independent phenomena, rather than being "only" brain generated, this suggests a "receiving" rather than solely "data-generating" role of the brain in its relationship to consciousness. In addition, we considered possible "locations" of the content of these experiences and refer to contemporary cosmological theories regarding parallel universes and dark matter.[8]

Unsurprisingly, with its controversial role as an additional sensory channel the pineal gland is heavily implicated in the receiver analogy. Alongside the chryptochromes, another pineal chemical is *serotonin*, which is converted to melatonin during night-time (melatonin regulates the circadian rhythm of the sleep cycle).[9] Through the actions of pinolene—another molecule speculated to have a pineal origin—and with the assistance of supporting enzymes, melatonin can be converted to endogenous DMT.[10]

As to the experiential difference between the filter/tuner analogy and the receiver analogy, it may be best to view the tuner as a device allowing the filter to "relax" along a continuum. I've previously suggested the loosening Filter A allows greater entry of material from the "outside world", but that also means increased amounts of parasensory information *in addition* to anything normally discerned.[11] The implication is that extraordinary content is overlayed on the "everyday" scene. Hence UV light is visualised as a floating silver orb hovering above actual carpet in the room you are in, or infrasounds are made out to be snippets of voices blending with background noise you're straining to hear.

---

[8]Strassman (2005, p. 68).

[9]Gonzalez et al. (2012).

[10]Roney-Dougal et al. (2013); DMT production in the pineal gland is speculation, although it's been found in pineal gland microdialysate of rodents.

[11]The analogies become complicated when discussing the respective states of Filter A and B, so I'll just concentrate on the single-filter configuration.

The receiver analogy is different, in that this theoretical apparatus switches across discrete stations, resulting in total content *replacement.* To give a television analogy, if everyday normality is set on TV Channel 8, when the receiver shifts to Channel 9 there's a completely new show to watch, and the same goes for Channel 10, 11 and so on. In this way, the pure DMT experience might differ to a chemically variant psychedelic drug as they prefer entirely different channels.[12]

So in the hotel corridor, and just for a moment, Tom "changed channels" and perhaps encountered some other time in history (a kind of "time slip") thanks to a complete shift of his internal receiver. That's described by Paul Deveraux when;

> …for a few passing moments, someone in this time for unknown reasons obtains a glimpse of a person, object or scene of some other time….especially the ghost houses, spectral landscapes, and phantoms in old fashion dress, particularly when they seem to be walking on an earlier ground level.[13]

Deveraux calls this a *coming-and-going*, epitomised in this famous tale from early twentieth century England;

> A celebrated coming-and-going was a fairground seen by the writer Edith Oliver in the huge megalithic complex of Avebury in Wiltshire. On a rainy late afternoon in 1916 she saw coconut shies, swingboats, crowds of people and torches and flares amidst fairground booths. Because of failing light and the rain, she did not bother with it any further and left the area. It was some years later that she learned that though the Avebury fair had been an annual event, the last one had been held over half a century prior to the time she had seen the fairground. It seems that she also saw standing stones around which the fairground was situated that are no longer extant.[14]

A receiver switched from the normal channel will result in a fully engrossing time-slip phenomenon, where you can see, hear, feel, smell and taste

[12] Strassman (2001, p. 327).
[13] Deveraux (2001, p. 187).
[14] Ibid.

a living scene indistinguishable from the everyday. Let's assume this experience comprises some kind of complex array of energy or fields that are simultaneous with (or embedded in) the energy that represents the "here and now" consisting of what once was, or will be. There are countless—or as Hoffman puts it, infinite—numbers of channels to surf, each telling a different story.[15]

But not all stations are historical recordings of what comes before, and what might be still to come. Some will be completely different to anything remotely conceived as normal. What is experienced is still information, although it has lost all meaningful environmental associations. The resulting perception is more bizarre than a scene of the past, and well represented by the "stars all around" reported by Nadia the astral traveller. Outwardly the astral journey does seem to resemble a discrete dial-in process. Although astral journeying (and NDEs and OBEs) implies the "projector" leaves and hovers over their physical body before traveling to other terrestrial or non-terrestrial spaces, based on their phenomenological *what happened* there's no requirement for them to have actually left their physical body at all. Nadia's "journey" *seemed* like she travelled to the stars—as do similar stories where there's an initial sense of vibration during the separation process, and frequently the physical body is seen by the traveller as separated from them at the start of their journey. However the impression of bodily separation might be conceived as a side-effect of a "channel switch". In fact, it is simpler to suggest the sources of experience are merely perceived by the traveller when their inbuilt receiver is adjusted.

Of course there are problems treating Tom's sighting as an alteration in his biological receiver. A *coming-and-going* is fitting for tales where the spectral scene contains non-sentient objects like Edith Oliver's coconut shies and swingboats. The experience also fills the *entire* visual field. If Tom had experienced was a similar style of "time-slip" from some period in the past, then why are the inanimate objects included in the scene—walls and carpet and other fixtures—the same *then* as *now*. In defence, perhaps the modern corridor was refurbished in the same style as in the past? Or did Tom tune into a vision of the corridor from only last week? That might mean the ghost is just a n ordinary hotel worker, alive and

[15] Hoffman (1979/1983).

well and still working downstairs at the reception counter. That said, the fact the ghost was walking in the stub of a corridor implies there was once additional structure to the building, but that's mere speculation requiring historical research to back it up.

Filter/tuner analogies and receiver analogies (or a combination of both) are first-path methods to at least provide a model for paranormal experiences—the *how* they happen. They are based on the assumption that witnesses are mostly passive observers and collectors of information, although they may have the ability to initiate the experience through self-induced states such as relaxation (and the ingestion of drugs). That's slightly more empowering than the basic idea that ghosts (and other unusual things) come and go as they wish, and people who witness them make no contribution to the process. The analogies have more difficulty pinning down the *why* and *who* of a sighting, but do support their rarity; in order for the *who* to manifest (e.g., an active spirit or some other entity) there needs to be the right state of flux, energy level or field conformation, while the witnesses' filter, tuner or receiver needs to simultaneously occupy a matching setting. The *why* might actually be pure chance, or strengthened by particular psychological states and a very energetic space (a location being the setting of a past, intense human emotional activity).

That's a lot to accept. The majority of mainstream physicists don't actually believe in an *actual* aether—it's a mathematical curiosity—and it takes a leap of faith to identify ghosts as consisting of genuine energy. That's even accepting the word "energy" is no more than a metaphor for something unmeasurable. What type of energy *doesn't* have a measurable or fundamentally physical source? The answers I've provided so far are tenuous, and require the stretching of non-classical theories of physics to snapping point.

That's where a new path—*the second-path*—stakes its claim as an alternative route in the 5W1H journey, with a much simpler *what, how, why* and *who.*

References

Barker, S. A., McIlhenny, E. H., & Strassman, R. (2012). A Critical Review of Reports of Endogenous Psychedelic N, N-Dimethyltryptamines in Humans: 1955–2010. *Drug Test Analysis, 4,* 617–635.

Blainey, M. (2015). Forbidden Therapies: Santo Daime, Ayahuasca, and the Prohibition of Entheogens in Western Society. *Journal of Religion and Health, 54*(1), 287–302.

Deveraux, P. (2001). *Haunted Land: Investigations into Ancient Mysteries and Modern Day Phenomena.* London: Piatkus Ltd.

Gonzalez, S., Moreno-Delgado, D., Moreno, E., Perez-Capote, K., Franco, R., et al. (2012). Circadian-Related Heteromerization of Adrenergic and Dopamine D4 Receptors Modulates Melatonin Synthesis and Release in the Pineal Gland. *PLoS Biol, 10*(6), e1001347. https://doi.org/10.1371/journal.pbio.1001347.

Hoffman, A. (1979/1983). *LSD: My Problem Child: Reflections on Sacred Drugs, Mysticism, and Science.* Saline, MI: McNaughton and Gunn.

Roney-Dougal, S. M., Ryan, A., & Luke, D. (2013). The Relationship Between Local Geomagnetic Activity, Meditation and Psi Part 1: Literature Review and Theoretical Model. *Journal of the Society for Psychical Research, 77*(911), 72–88.

Smith, D. E. (2007). Lysergic Acid Diethylamide: An Historical Perspective. *Journal of Psychoactive Drugs, 1*(1), 1–5.

Strassman, R. (2001). *DMT: The Spirit Molecule: A Doctor's Revolutionary Research into the Biology of Near Death and Mystical Experiences.* Rochester: Park Street Press.

Strassman, R. (2005). Hallucinogens. In M. Earleywine (Ed.), *Mind-Altering Drugs: The Science of Subjective Experience* (pp. 49–85). Oxford: Oxford University Press.

Szara, S., Rockland, L., Rosenthal, D., & Handlon, J. H. (1966). Psychological Effects and Metabolism of N, N-Diethyltryptamine. *Archives of General Psychiatry, 15*(3), 320–329.

Part III

The Second-Path

11
Defining Reality

The 5W1H road's next exit, the right-hand fork, starts with and acceptance of Tom's phenomenological experience (the *what*), but quickly terminates in a dead-end where the *what, how* and *why* are deeply interconnected, and there is no necessary requirement for a responsible *who*, other than it being the witness themselves.

I apologise for stepping through this so pedantically but the next point is important; what I'm really saying is that there are two paths so far when examining paranormal experiences.

The first-path implies Tom's ghost was authentic; a negative, low or high energy form, a telepathic signal, a past life memory, a time-traveller, or something we haven't even discussed yet. Most importantly the event *actually* occurred, in space and time, and independent of Tom. He is merely a passive observer. The agent is an energy, field or environmental state that's able to be perceived directly or indirectly through conventional senses—or perhaps via another perceptual channel—but at the same time is *elusive*, therefore difficult to measure with current technology (if measurable at all). The upshot is that not only does a phenomenological *what* transpire but also a physical *what*; what physical change occurred in the brain, and

T. Jinks, *Psychological Perspectives on Reality, Consciousness and Paranormal Experience*,
https://doi.org/10.1007/978-3-030-28902-7_11

what it is made of? Although much more difficult to pin down, there is scope for responsible *whos* and proposed *whys* that explain the purpose of his sighting, and they change depending on circumstance (and personal preference).

On the other hand, the second-path[1] follows an orthogonal direction, whereby Tom's ghost sighting may be phenomenologically valid, but it was not genuine in the sense of being independent of his self. This implies there is only a phenomenological *what* in common. And as to *how* Tom saw his apparition, it cannot be a function of conventional (or unconventional) perception. This requires alternative speculation about the way something can be *give the* impression of perception when nothing was actually perceived. As to *why* he saw a ghost, that question can be answered with adequate investigative effort. Finally there are no responsible *whos* for the second-path, other than identifying Tom as the "who"....*who it happened to*.

Harking back to the definition of experience as an event that has been lived through, on reflection this second-path effectively negates the legitimacy of a paranormal experience. According to the earlier statement, experiences are only valid if they involve perceiving events that happen in the world, and through the senses. If what you see could not have taken place in the sensed environment, then then it can't have been an event, hence it cannot have been experienced. From this perspective, a *paranormal experience* should be re- labelled a paranormal *pseudo-experience.*

So if paranormal experiences aren't actually experienced, what are they? Kenneth Drinkwater and colleagues use the alternative label of general *subjective* paranormal experiences (GSPE), since these now refer to an exceptional or special experience beyond scientific comprehension that people only *believe*, *label* or *consider* paranormal.[2] And from a formal definition given by the parapsychological society (1989), paranormal experiences are;

[1] For convenience I will use "second-path" (and any other path) anthropomorphically, as though it were an individual rather than a theoretical perspective.

[2] Drinkwater, et al. (2013, 2017).

> …*apparent* anomalies of behaviour and experience that exist apart from currently known explanatory mechanisms that account for organism-environment and organism-organism information and influence flow.[3]

For both of these descriptions I've used italics to emphasise revealing words. In Drinkwater's definitions they are *believe, label* and *consider.* Tom only believes, labels or considers something he's experienced as paranormal. That doesn't imply the experience ever happened. Likewise, the parapsychological society's *apparent* anomalies don't automatically qualify as authentic anomalies.

Until now I've tried to avoid using the words "real", "not real" and "reality" as far as possible—substituting it with terms such as "authentic" or "legitimate". This can't be sustained indefinitely as the *reality* is the foundation to any discussion of the paranormal, regardless of whether the first- or second-path is taken.

So what does it mean for something to be labelled "real" or "not real"? Reality is *so assumed* that there's no strong drive to define it outside academic journals with a focus on philosophy (and perhaps for a wider audience, articles and books on cosmology). In the manner of "paranormal" there's an assumption that everyone knows what reality is, and what it means for something to be real. Indeed, the concept of reality is so encompassing its definition will be vastly more complex than that of a tree or a car, or even a ghost or a UFO. A single sentence definition of reality can only resemble a statement such as this;

> What is reality to a rational mind? The answer is that whatever is observable by the senses and/or measurable in some way belongs to the domain of reality.[4]

So if you like the idea of Tom's ghost being real, then you know implicitly what that entails—an entity hanging out somewhere in *reality*, and Tom sees it just like any other real thing that's seen or heard or smelt, felt or tasted.

---

[3]Irwin (2007, pp. 8–9).

[4]Lampis (2004, p. 43).

Contemplating reality is a metaphysical pursuit, which put simply is consideration of what the universe is—assuming there *is* something called the universe which describes everything that is out there, beyond "us". The next step is to contemplate how universes can operate in an infinite variety of possible ways, but we might judge that one type of universe (our universe) is represented by a particular conglomerate of rules. Examples include the law of gravity and the law of thermodynamics.

You could end your metaphysical consideration there, or go further on the basis of observation and reason to conclude that these rules apply to, or operate on, *something*. That's the subject matter of *ontology*. Ontology is contemplation of the elements of the universe; those things that are *acted on* by the concerns of metaphysics. The stuff of the universe might be what we call matter; visible objects and other physical things that are solid or substantive. These ontological components of the universe, or reality, conform to metaphysical laws.

Metaphysics and ontology are philosophical pursuits, so the outcome of this musing doesn't necessarily have to result in you accepting a law of gravity, or that something called "matter" is what reality is made from. However these examples are the agreed ones applicable to the *real* universe that's around us, not some theoretical universe where everything's either topsy-turvy or merely a mathematic curiosity (and this includes notions of negative energy realm and the aetheric plane). Therefore the conclusion is singular—reality conforms to fixed laws and consists of matter that behaves in accordance with these laws.

This is the basis of the dominant assumption in the Western world since the seventeenth century[5] which takes both the rules and the components of reality and turns them into a productive story, *materialism*. Despite all the metaphysical and ontological posturing that can be indulged, it's very difficult *not* to accept that objective reality is made of hard and tangible matter. And when you dig down as far as you can into this matter—the very bottom—there'll still be discrete particles that constantly exist, and which behave in a predictable fashion.[6] That is surely less taxing (and far more common in contemporary discourse) than imagining reality

[5]Brabant (2016) and Frank (1957).

[6]Barušs (2010).

from the viewpoint of metaphysical *infinitism*,[7] which posits there are *no* fundamental particles and no constituent parts. How can such esoteric philosophising be of any *practical* use?

A materialistic model of reality is best be conceived of as *Classical* or *Newtonian*, and is not only grounded in the idea of individual and unique objects existing independently of one another and surrounded by empty space[8] (certainly not aether), but also progressing along a single unidirectional time scale, from past to present only. This conforms to a 3S-1t model of dimensionality, with the three spatial dimensions (length, breadth, and height) maintained in this present moment.[9] These four dimensions are the four corners of the standard models of physics, and what most scientists (and most people) would conceive of all of reality, or all of what exists. As summarised long ago by philosopher Norman Campbell, materialism;

> …is necessary not so much in order that science should be true as in order that there should be any science.[10]

While such a treatment of reality sounds reasonable, it's widely understood that this *isn't* the way the world works. There's more to matter than simply *matter*, and that's implicit in the quantum treatment of reality, which is an integral component of the first-path. In terms of a background;

> Quantum theory was developed to describe the behavior of the microscopic world of the atom and its constituents. At this scale the masses of the particles and the forces between them make the precise positions of these particles difficult to measure and the energies involved in their interactions are broken into tiny discrete packages instead of the continuous flow we normally experience in the larger dimensions we inhabit. In one observation an entity can be a particle and in another observation it appears as a wave. All the entity's possibilities are described by its "wave function" and

---

[7]Morganti (2015).
[8]Borutta (2009).
[9]Neppe and Close (2015).
[10]Campbell (1921, p. 399).

> the interaction between wave functions determine what outcome will be observed.[11]

Hence subatomic particles possess a dual nature; their behaviour can also be modelled by an equation analogous to the description of the propagation of a light wave.[12] These particles also have the quantum property of *indeterminacy*.[13] They are in an indeterminate state (or an infinite number of states) until they are observed and are "collapsed" to occupy a single realised state.[14] On face value this puts pressure on the materialistic assumptions of reality, because the particles that make up matter are not only characterised by the state they are in at the moment of observation, but also every other vacant and *un*realised state they are not occupying. Are all these states real? The occupied (realised) state might be considered real, in Chemistry Professor Lothar Schäfer's words;

> ….because it possesses observable, empirical, or measurable properties, specifically an observable probability of presence. Similarly, an unoccupied state is said not to be real in the same way, because its probability of presence is not observable, and nothing else is measurable there.[15]

That seems straightforward, but he then adds a fascinating qualifier;

> Not to be real in a measurable way is not meant to say that such a state is not part of reality.

This identifies a distinction between what Schäfer calls virtual states that are *quantum-ontological potentialities*, and actual states which are factually real. However since virtual states can actualise in space and time to become real, they can be considered a "nonmaterial part of reality".

---

[11]Borutta (2009, pp. 192–193).
[12]Schäfer (2006).
[13]Darby (2010).
[14]Schäfer (2006).
[15]Ibid., pp. 574–575.

That's one anti-materialist position. Another is the theory of relativity (both special and general), which retains the notion of independent objects residing in empty space, but;

> ...modifies the interactions and geometries when speeds begin to approach the speed of light since the speed of light is an invariant parameter regardless of the frame of reference.[16]

The uneasy relationship between relativity and materialism has been recognised for 100 years. In 1921 Philosopher Herbert Carr discussed whether the principal of relativity overturns the doctrine of matter as being real.[17] His argument was that materialism is a causal theory of reality, and when something is perceived to be real it's assumed to have an independent cause. However this might not be applicable if the universe (the "Minkowski-Einstein universe" to be precise) is comprised of observers in different frames of reference. Reality then becomes the coordination of observation—and in this framework the observer is just as important as the observed so there's no primacy for physical, causal reality. In more extreme terms, it might be argued that the universe isn't objectively real other than as an "order of arrangement" of objects perceived in it, with time lacking independent existence apart from the order of events by which it is measured.[18] How can the observing scientist objectively measure the physical world, and determine its materialistic foundations, when they are themselves a part of that world?[19]

What follows is arguably the three most important problems relativity theory bequeaths to materialism.[20] The first is that the universe isn't Euclidean and three dimensional (with time as an added extra) but a four-dimensional non-Euclidean (curved) space time continuum. Secondly, there's the formulaic demonstration that matter can be converted into radiant energy, which is nonmaterial (that is, the $E = MC^2$ equation), so matter isn't necessarily fixed in perpetuity. Thirdly, the configuration of

[16]Borutta (2009).

[17]Carr (1921).

[18]Frank (1957).

[19]Ibid.

[20]Ibid.

spacetime can affect matter and vice versa, an example being matter distorting the curvature of space (through gravity), and gravity influencing the motions of heavenly bodies.

However, it's generally taken that quantum theory and relativity don't negate materialism. The materialist approach is highly effective in making sense of everyday reality, and it's implicitly held by most people for the very reason that it works so well. Returning to Campbell's straightforward argument from 1921, reality also includes the *reality of common sense*, which claims a universality of experience, regardless of whether you're a materialist, idealist or physicist (of a quantum or relativistic persuasion). That's the reality of everyday science—a "space-time energy *sensible* reality".[21] There might be a deeper reality of metaphysics, which indulges in speculation over the nature of reality, but in no way supplants what reality *is*, which is self-evident.

Everyday experiences, and by extension Tom's pseudo-experience of a ghost, take place in a sensible world. Of course there's a quantum level of this same world that's mathematically mappable and empirically justifiable, and a relativistic aspect that involves measurements from different locations. However, all the uncertainty these perspectives entail is "evened out" by the time it reaches Tom as a visual experience. There might be quantum or negative energy ghosts *on paper*, but they're not seen as a dark-suited man in a hotel corridor.

Here's a thought-experiment highlighting why this "evening out" is considered the only reasonable option, and incorporates both quantum theory and relativity.[22] General relativity maps the relationship between gravity and time, and the product is not static. Time runs faster as gravitational fields weaken. With this as the background, envisage a one-atom clock in a *superposition*. That's not the style of superposition where there is a cancelling out of combined wave-forms (like two "out of sync" ocean waves colliding with one another, resulting in calm water). Rather it's the quantum definition referred to previously which involves a sub-atomic particle holding potential in any one of an infinite number of states. The single atom is radioactive, in an energetic state, and as it decays a particle

[21] Schäfer (2006).

[22] Brooks (2015).

is emitted in one of two directions. Following observation the particle will assume a realised state, but in orthodox quantum theory it's in a superposition of these two states or wave functions until this observation takes place.[23]

The particle therefore has the potential of being in two places at the same time. Returning to the one-atom clock example, this timepiece could be recording time simultaneously in two positions in the universe. Those positions could be experiencing different effects of gravity, hence the time they record will be necessarily different (the time lines diverge). This situation places the system in a quandary as the positions of both expressions of the wave function (the particle's *superpositional* state) could potentially be determined. Consequently, the system will unavoidably decohere to avoid discovery. The conclusion is that in a universe of gravity, the effect on time will ensure that quantum superpositions will not manifest in the everyday world of ordinary observation.[24] The reality that's observed is *sensible*. It's not quantum reality, and works in a way a materialistic universe would be expected to operate.

Likewise, none of the concerns of relativity will undermine seeing things that exist in the universe with our eyes (discounting Tom's "ghost", which is assumed to be a non-real pseudo-experience[25]). In a similar way to quantum mechanics, all the strange things that the theories of relativity demonstrate through mathematical equations (and to some extent empirical evidence) occur in very different circumstances to how people go about their everyday lives. For example, relativity has a lot to say about the effects on the observer's mass when they are approaching the speed of light relative to a particular frame of reference. As light speed is approached there will be an unrestricted increase in the observer's inertial mass.[26] However that's not the speed that people are travelling at in *this* frame of reference. Our working experience of reality is something detected through the five

[23]Brown (1985).

[24]Brooks (2015); "...superpositions are impossible for objects composed of more than a certain number of particles because of a phenomenon called *spontaneous localisation*, which suggests that the distribution of mass – its density – is what matters".

[25]That is, the ghost cannot be the energetic expression of matter.

[26]Riggs (2013).

sense that's made up of exactly the things we assume it to be made up of—materials that can be reduced to basic, predictable elements. Our working experience of reality is something detected through the five sense that's made up of exactly the things we assume it to be made up of—materials that can be reduced to basic, predictable elements. This is reductive materialism, and the next step is to examine how such an approach can explain Tom's ghost more effectively than anything attempted by the first-path.

References

Barušs, I. (2010). Beyond Scientific Materialism: Toward a Transcendent Theory of Consciousness. *Journal of Consciousness Studies, 17*(7–8), 213–231.
Borutta, R. (2009). Paranormal Phenomena, Physics and Dirac Energy. *Journal of Spirituality & Paranormal Studies, 32*(4), 192–200.
Brabant, O. (2016). More Than Meets the Eye: Towards a Post-materialist Model of Consciousness. *Explore, 12*(5), 347–354.
Brooks, M. (2015). The Secret Life of Reality. *New Scientist, 225*(3002), 26–29.
Brown, J. R. (1985). von Neumann and the Anti-realists. *Erkenntnis, 23*(2), 149–159.
Campbell, N. R. (1921). Metaphysics and Materialism. *Nature, 108*(2717), 399–400.
Carr, H. W. (1921). The Metaphysical Aspects of Relativity. *Nature, 106*(2677), 809–811.
Darby, G. (2010). Quantum Mechanics and Metaphysical Indeterminacy. *Australasian Journal of Philosophy, 88*(2), 227–245.
Drinkwater, K., Dagnall, N., & Bate, L. (2013). Into the Unknown: Using Interpretative Phenomenological Analysis to Explore Personal Accounts of Paranormal Experiences. *Journal of Parapsychology, 77*, 281–294.
Drinkwater, K., Dagnall, N., Grogan, S., & Riley, V. (2017). Understanding the Unknown: A Thematic Analysis of Subjective Paranormal Experiences. *Australian Journal of Parapsychology, 17*(1), 23–46.
Frank, P. (1957). Philosophical Uses of Science. *Bulletin of the Atomic Scientists, 13*(4), 125–130.
Irwin, H. J. (2007). Science, Nonscience and Rejected Knowledge: The Case of Parapsychology. *Australian Journal of Parapsychology, 7*(1), 8–32.

Lampis, R. (2004). The Theory of Reality: Or Why, If We Understand Reality, the Theory of Everything Is an Impossibility. Or How to Establish Basic Principles That Clarify Physical, Mental and Paraphysical Phenomena, and to Dispose of Concepts Such as Non-locality, Place. *International Journal of Humanities and Peace, 20*(1), 43–48.

Morganti, M. (2015). Dependence, Justification and Explanation: Must Reality Be Well-Founded? *Erkenntnis, 80*(3), 555–572.

Neppe, V. M., & Close, E. R. (2015). The Concept of Relative Non-locality: Theoretical Implications in Consciousness Research. *Explore: The Journal of Science and Healing, 11*(2), 102–108.

Riggs, P. J. (2013). Why a Spaceship Cannot Reach the Speed of Light from the Perspective of the Spaceship's Rest Frame. *Latin-American Journal of Physics Education, 7*(4), 604–608.

Schäfer, L. (2006). A Response to Ervin Laszlo: Quantum and Consciousness. *Zygon: Journal of Religion and Science, 41*(3), 573–582.

12
The Observer and Consciousness

Reductionist materialism represents sensible everyday reality without the need to refer to non-classical theories. That's nicely summarised by this single sentence;

> The incompatibility of the Classical, Relativistic, and Quantum physical models of physical reality are reconciled by limiting their application to the appropriate scale they have been derived from.[1]

However one valuable perspective non-classical theories provide (albeit reluctantly[2]) involves the importance of the observer, either as a tool to "collapse" reality into one of many potential states, or as an entity whose perception of reality is not in a fixed frame of reference. Even if these concepts don't directly impose on how you see or smell the flowers in your

[1] Borutta (2009).

[2] Ibid.; Classical, Relativistic, and Quantum physical models of physical reality "strive to eliminate or minimize the role of the observer in the expression of the physical laws they propose" (p. 193).

T. Jinks, *Psychological Perspectives on Reality, Consciousness and Paranormal Experience*,
https://doi.org/10.1007/978-3-030-28902-7_12

garden or read the book in front of you, they still suggest that what is observed (materialistic reality) is done so by the observer (you).[3]

The observer also plays a crucial role in the first-path and accounts for the elusiveness of a ghost sighting. The probability of seeing an energetic ghost must coincide with the probability of the observer's filter or tuner being in the correct configuration. Nevertheless, the word "observer" still needs to be handled carefully, since there's no superficial reason why the observer of any system has to be human, or sentient—especially when referring to non-classical systems. Recall that quantum particles have wave properties but do not express both characteristics simultaneously. This can be expressed in terms of not knowing their position and velocity simultaneously (the Heisenberg uncertainty principle in quantum mechanics). Therefore;

> Any information gained about the quantum particle's position irrevocably alters its momentum (and vice versa) in a way that is fundamentally uncertain.[4]

Expressed in terms of Thomas Young's well-known interference experiment with light waves, when a particle is projected onto a double-slit screen it cannot be said to traverse one slit or the other (signifying its position) if the particle's interference effects are measured (signifying its momentum). The trajectory of the particle, and wavelike interference it creates, are not independently measurable characteristics but complementary features of a quantum system.[5] If another piece of technology in the form of a particle detector is set up to observe the slits hence the particle's position, the interference pattern will be destroyed.[6]

Made clearer using another thought experiment, imagine a Geiger counter positioned in a laboratory for the purpose of measuring atom

[3]Although as Borutta (2009) states; "This produces the paradox of the 'measurement problem' and may lead one to the objection expressed by Einstein about having to believe that the moon may not exist when we are not looking at it. Another problem is the non-local nature ('spooky action at a distance') of the phenomena at the atomic scale, which has been confirmed by experiment over and over at the most sensitive levels of detection" (p. 193).

[4]Kocsis et al. (2011, p. 1170).

[5]Ibid.

[6]Brooks (2015).

decay. This particular piece of equipment uses flashes of light, rather than audible clicks, to record whether atomic decay has or has not occurred. These flashes are then recorded by a camera, which relays the image on a monitor to an experimenter in another room. In this arrangement, the camera is the first-in-line as the observer and represents the collapse of the superposition of the atom's decay as either a light flash, or no light flash. However, the next-in-line experimenter watches this outcome on the monitor, so are they *both* observers, and where does the superposition take place? Where is the cut-off between what's being measured, and what is doing the measuring?[7] The entity ultimately responsible for collapsing the wave function is a philosophical conundrum known as the *von Neumann chain*.

Neils Bohr's view was positivistic in nature[8]; an epistemology where knowledge is only achievable through that which is "immediately observable".[9] The cut is between the atom and the macro-measuring device hence the Geiger counter is responsible, being the first component in the chain. The concern here is that this is incompatible with the "general unity of science"—an idea of *one* universal theory for *one* world. For this reason, John von Neumann's argument is that *everything* along the chain of observation is in superposition—from the atoms themselves, through the mechanical measuring device to the eyes of the human observer. But there is something beyond the eyes positioned at the end-of-the line, which is sentient and responsible for collapsing the wave function hence forcing *decoherence* onto the system—and that is the mind and *consciousness.*[10]

Materialism is a position implicitly held by most people (whether they realise it or not), and yet these same people would acknowledge that as well as the brain (the central nervous system or CNS) there is also "something" called consciousness. The etymological root of the word is "to know with",[11] and a rudimentary definition is that it involves;

[7] Brown (1985).

[8] Ibid.

[9] Leahey (1987, p. 144).

[10] Leahey (1987).

[11] Presti (2011).

> …the capacity to be aware—aware of mental processes such as perceptual experiences (visual, auditory, olfactory, gustatory, tactile, temperature, pain), mental images, thoughts, emotions, and one's sense of self[12]

These mental states perceive, or interact with, the material world but seem to be fundamentally different to material things. That's a very *Cartesian* position;

> Descartes took it upon himself to doubt the very existence of minds and mental states, but proceeded famously to conclude that such doubt involved the doubter in logical incoherency. In the aftermath of the Cartesian philosophy, until very recently, no philosopher doubted the existence of minds and mental states; the philosophical questions were about the nature of these entities and states, the existence of which was taken to be obvious.[13]

This position is *dualism*, which identifies the physical brain as objectively real, but the mind as purely subjective and not open to empirical observation—hence the difficulty in defining what it is. There are several varieties of dualism,[14] however the one that best suits this mind/brain conceptualisation is *substance dualism* where the material and mental are two distinct domains or substances;

> Any substance with mental properties lacks material properties and any substance with material properties lacks mental properties.[15]

Descartes is identified as substance dualist because of the content of his writings;

> …on the one hand I have a clear and distinct idea of myself, in so far as I am simply a thinking, non-extended thing; and on the other hand I have a distinct idea of body, in so far as this is simply an extended, non-thinking

[12] Ibid., p. 21.
[13] Hannan (1993, p. 165).
[14] *Explanatory dualism* and *property dualism* are two further examples.
[15] Rodríguez Pereyra (2007).

> thing. And accordingly, it is certain that I am really distinct from my body, and can exist without it[16]

To reiterate, the body is made of substance and extended in physical space, but the mind is *non*-physical and *not* extended in space.[17] Furthermore, while the physical body operates *deterministically* through rational universal laws, it can be manipulated by the mind to generate behaviour. Not surprisingly, that's performed through the physical/mental interface of the pineal gland (his "seat of the soul").

Similarly, von Neumann's colleague at Princeton University, Eugene Wigner, had no problem with substance dualism, since the link in the chain where reality collapses occurs between the (physical) brain and the (non-physical) mind. This Wigner cited as evidence of dualism;

> To make material reality go into the definite states which are observed, something non-material which does not obey the quantum mechanical laws of superposition must exist.[18]

The dualistic notion of matter and mind as separate but parallel processes is a foundation assumption of parapsychology.[19] Dualism is implicit in the very idea that the human essence survives physical death and perseveres in the environment in the absence of the body.[20] This is a straightforward, first-path way of pondering reality—in waking life it feels as though consciousness is separate but parallel to the material world. Frederick Myers summarised this distinction as;

> …all cognizable Mind is as continuous as all cognizable matter[21]

16 Ibid., p. 71.
17 Bensley (2003).
18 Brown (1985, p. 151).
19 Bensley (2003).
20 Arnette (2012).
21 Myers (1903/2007, p. 701).

The importance of consciousness in the study of the paranormal is also found in an explanation for psi that's more unconventional than suggestions of information transmission mediated by subtle energies. We take for granted the passage of time as *asymmetric*, in the sense that time passes in one direction, from past, to present, to future. But there's nothing in any physics textbook to say this has to be the case, since "all the equations of physics are time-symmetric".[22] And time asymmetry can apparently be made time *symmetry* through the actions of consciousness. So in a laboratory a stimulus like an emotional picture that's shown to participant might evoke a slightly-delayed measure of skin conductance measured by electrodes, and seen as a rising bump on the screen of the recording device (the "retarded" wave). And an equal time *before* the stimulus, a small bump known as the "Advanced" wave may also be observed, hence the principle of "what happens after, happens before". More specifically;

> ...at any moment a signal has determinants that are form past as well as from future boundary conditions[23]

That simply means for any event, and the outcome flows two ways (in a two-dimensional time structure). And this also means the practical value of such psi-effects are not as impressive as we might hope or expect. Psychologist Dick Bierman continues with an enjoyable paradox;

> The advanced wave cannot be used to derive some behaviour to prevent the retarded wave from occurring. So if we have a precognitive dream where candles are foreseen to set fire in the house the next day...[we]..remove all candles form the house in order to prevent the fire from happening.[24]

And if the candles are now removed, there won't be a fire, and no retarded wave (the consequence) and also no advance wave (the dream)! Bierman suggests it might be of some delicate advantage is if you suffer a surprise physical attack, since the advanced wave in the form of arousal will occur shortly before the event, and *might* just offer some preparatory advantage.

[22] Carr (2008, p. 25).
[23] Bierman (2010, p. 283).
[24] Ibid., p. 283.

While this and other possibilities are fascinating, dualism has been unfashionable in psychology and philosophy for well over a century, and in contemporary writings it is regarded as only of historical interest.[25] The primary reason dualism is dismissed amounts to this; substances as radically different as matter and non-matter can have no purchase on each other, so how can the supposedly immaterial soul and the material body work harmoniously to compose a smoothly functioning human being? In particular (and despite Descartes's reference to the pineal gland), how can consciousness, the immaterial soul's unique property, *cause anything to happen in the material brain or be influenced in turn by that brain?* There is no good argument for their Interaction, or one causing change in the other.

> Causation in the material world works by energy transfer of some sort: transfer of motion, electrical energy, gravitational force. But pure consciousness could not give off energies of these kinds, so how does it cause changes in anything? How does it reach out and touch something? What mediates its alleged causal powers?[26]

The mind is therefore nothing more than the "ghost in the machine".[27] In reverse, there's the question of the *explanatory gap*—the need to explain how a real, material brain can create non-material mental states. That is;

> ....how the brain—something physical—can generate experiences, which are qualified as mental or psychological phenomena and are considered by some as non-physical phenomena.[28]

The opinion that dualism is unworkable leads to various methods of equating of consciousness with brain activity, and this is the direction taken by the second-path. At its most extreme, *eliminative materialism* proposes that functioning material brains are all that exist, and that mental states such

---

[25] Stafford Betty (2004).
[26] McGinn (1999, p. 92; quoted in Stafford Betty 2004).
[27] Ryle (1949, pp. 15–16; quoted in Stafford Betty 2004).
[28] Musacchio (2005, p. 404).

as beliefs, desires and attitudes not only *can't* cause human behaviour,[29] they are not even legitimate talking points of ontology the way material things are. Instead;

> ...they are relics of an outmoded 'folk theory' of human psychology, a theory that is rapidly proving false in the face of developments in the empirical sciences of brain and cognition.[30]

Folk psychology is the first-path, common-sense psychology that ordinary people practice when they rationalise their subjective mental states by evoking hollow concepts like consciousness and the mind.[31] To the eliminative materialist, these states do not refer to anything real, hence should be *eliminated* from discourse. In the manner of Tom's ghost, consciousness cannot be effectively described (and poorly at that), let alone be explained. That means there's no need for it to be considered.

A less hard-line view is *reductive materialism*, which doesn't eliminate the concept of consciousness, but links it directly with brain activity. This is not dualism, but *monism*. Brain activity originates from the energetic activity of atoms and molecules in neurons, which in biological language includes processes such as ionic flux across neural membranes resulting in synaptic transmission. This is the foundation of consciousness, as suggested by a Professor H. W. Chase as far back as 1912;

> Consciousness is not present everywhere in the animal series, but first appears when a certain degree of nervous complexity is reached, surely in the mammal, perhaps before.[32]

Therefore, when brain activity is sophisticated enough, consciousness is inevitable. But more importantly, they are one and the same thing, or *fully*

[29] Taylor (1994).
[30] Hannan (1993, p. 166).
[31] Meijer and Korf (2013).
[32] Chase (1912, p. 20).

identical.[33] The metaphor of mind as fire is helpful here; the flame emanating from a match-head might appear to possess a unique quality—something intangible compared to the physicality of the match-stick. However it is entirely derived from the chemical reactions of the tip's phosphorus ingredients.[34] In the same way, consciousness might *appear* to be separate to neural activity, but is merely a different way of perceiving that very same activity. To paraphrase the philosopher John Searle, this style of monism explains the experience of pain as being nothing more than having one's C-fibres stimulated (a type of neuron that specifically relays pain signals to the brain).[35] Consciousness might be something able to be deliberated, but it has no role other than to *be experienced.* All that's required to understand the mind is a thorough measurement of the dynamic changes in brain activity.

A useful outcome of this view of consciousness is that *all-brain* activity—the product of everything that's happening at all levels of the CNS—is incomprehensibly complicated. That means consciousness as a whole doesn't have to be explained, modelled or even predicted until there is adequate means to do so. Limited understanding of certain states of consciousness is still possible, for instance the mapping of C-fibre stimulation to account for the experience of pain. However *whole-consciousness*, as a product of *all-brain* activity, will have to wait for the moment. Neurobiologist David Presti calls this stance the *Standard Model in the Neuroscience of Mind*;

> ... physical processes in the brain (and body)...provide a sufficient explanatory framework for mentality and that someday, when enough is known about the structure and function of the brain, it will become clear just how this is so. Importing a term used to describe certain widely accepted explanatory scenarios in cosmology and in elementary-particle physics, I will call this notion the Standard Model in the neuroscience of mind—that a more complete understanding of physical processes in the brain and body will be sufficient to determine a solution to the mind–body problem. Standard Model arguments include (but are not limited to) stances such

---

33 Meijer and Korf (2013).

34 Stafford Betty (2004).

35 Stafford Betty (2004, p. 374).

> as: Mental experiences are somehow identical to neurobiological processes…mental experiences somehow emerge from underlying neurobiological processes…and mental experiences are higher-order interpretations of neurobiological processes.[36]

These last two propositions; that mental experiences "emerge" from, or are "higher order interpretations" of brain activity, are preferred over the strict monist treatment of consciousness as *identical* to this activity. Eliminative materialism is simply not intuitive. You might believe mind is identical to brain activity, but that's not how you tend to treat your own thoughts and feelings—they seem more significant than being identical to the outcome of masses of ions passing back-and-forth across the lipid bilayers of countless neurons.

While the arguments against eliminative materialism can be complex, a straightforward reason is provided by another philosopher, Crispin Wright. The approach has;

> …. given us not even the most rudimentary explanation of how it might be practical to dispense with the concepts of CP [commonsense psychology], nor the slightest inkling of a way of thinking about the province of (what we now conceive of as) human rational agency which might supplant them.[37]

References

Arnette, J. K. (2012). Recollections of Death: A Medical Investigation. *Journal of Scientific Exploration, 26*(1), 207–211.

Bensley, D. A. (2003). Can Minds Leave Bodies? *Skeptical Inquirer, 27*(4), 34–40.

Bierman, D. (2010). Consciousness-Induced Restoration of Time Symmetry (CIRTS): A Psychophysical Theoretical Perspective. *Journal of Parapsychology, 24*, 273–300.

---

[36] Presti (2011, p. 37).

[37] Wright (1993, p. 316).

Borutta, R. (2009). Paranormal Phenomena, Physics and Dirac Energy. *Journal of Spirituality & Paranormal Studies, 32*(4), 192–200.
Brooks, M. (2015). The Secret Life of Reality. *New Scientist, 225*(3002), 26–29.
Brown, J. R. (1985). von Neumann and the Anti-Realists. *Erkenntnis, 23*(2), 149–159.
Carr, B. (2008). Can Psychical Research Bridge the Gulf Between Matter and Mind? *Proceedings of the Society for Psychical Research, 59*(221), 1–96.
Chase, H. W. (1912). Consciousness and the Unconscious. *Psychological Bulletin, 9*(1), 20–25.
Hannan, B. (1993). Don't Stop Believing: The Case Against Eliminative Materialism. *Mind and Language, 8*(2), 165–179.
Kocsis, S., Braverman, B., Ravets, S., Stevens, M. J., Mirin, R. P., Shalm, L. K., et al. (2011). Observing the Average Trajectories of Single Photons in a Two-Slit Interferometer. *Science, 332*(6034), 1170–1173.
Leahey, T. H. (1987). *A History of Psychology* (2nd ed.). Englewood Cliffs, NJ: Prentice-Hall.
Meijer, D. K. F., & Korf, J. (2013). Quantum Modeling of the Mental State: The Concept of a Cyclic Mental Workspace. *Syntropy, 1,* 1–41.
Musacchio, J. M. (2005). The Ineffability of Qualia and the Word-Anchoring Problem. *Language Sciences, 27*(4), 403–435.
Myers, F. (1903). Phantasms of the Dead. In M. Ebon (Ed.) (1978), *The Signet Handbook of Parapsychology* (pp. 273–285). New York: Lombard Associates.
Presti, D. E. (2011). Neurochemistry and Altered Consciousness. In E. Cardeña & M. Winkelman (Eds.), *Altering Consciousness: Multidisciplinary Perspectives: History, Culture, and the Humanities; Biological and Psychological Perspectives* (Vol. 2, pp. 21–41). Santa Barbara: Praeger.
Rodríguez Pereyra, G. (2007). Descartes's Substance Dualism and His Independence Conception of Substance. *Journal of the History of Philosophy, 46*(1), 69–89.
Stafford Betty, L. (2004). Mind, Paranormal Experience, and the Inadequacy of Materialism. *International Philosophical Quarterly, 44*(3), 373–392.
Taylor, K. A. (1994). How Not to Refute Eliminative Materialism. *Philosophical Psychology, 7*(1), 101–125.
Wright, C. (1993). Eliminative Materialism: Going—Concern or Passing Fancy? *Mind and Language, 8*(2), 316–326.

13

The Emergent Mind

So what other options for consciousness are available that might allow at least some discussion points?

The emergent or *epiphenomenal* mind is a monist compromise that avoids eliminating the elements of consciousness entirely, or identifying them solely with brain function.[1] Putting aside for the moment the seriousness of the explanatory gap, it's reasonable to conclude that we may be aware of our immediate conscious experience, but this is a side-effect of the true underlying cause—brain activity.[2] Returning to the fire metaphor, consciousness is the visual impression of flame. While it isn't directly equivalent to the underlying chemical reactions, it is still a product of those reactions. And in the same way chemical reactions—not the visual flame—create the consequences of fire (e.g., heat and the influence over other materials), consciousness as a by-product of brain activity has no causal effect on behaviour.

[1]Pockett (2004).

[2]Brabant (2016).

T. Jinks, *Psychological Perspectives on Reality, Consciousness and Paranormal Experience*,
https://doi.org/10.1007/978-3-030-28902-7_13

The advantage is that epiphenomenal consciousness can *at least* be deliberated and discussed independently of what the brain is doing. These elements are of course not material (i.e., made of actual physical *things*), but are the thoughts, affects, sensations, motivations and interests[3] that interact with one another in consistent and predicable ways and provide the substance of experience. Going further, these conscious states are "attached" to separate individuals, and consist of elements similar to those discussed by early psychologists including Edward Titchener;

> The ideas of the primitive mind are, as it were, photographic copies, life-likenesses, of the perceptions which go before them. Thus the idea of a landscape would be in part a picture-idea, the look of stream and hills and trees; in part a sound-idea, the idea of splashing water and rustling boughs; in part a tactual idea, the 'feel' of springing grass and moving wind; in part a smell-idea, a remembered freshness and fragrance of air and flowers. The life-likeness is, of course, never perfect: the idea is weaker, passes by more quickly, and is more sketchy, than the perception that corresponds to it: but the qualities of the perception are found again in the idea. Produced in perception, they are reproduced in idea.[4]

Titchener's statement gives the impression of dualism, but it's really a type of clever monism because it's not implying the mental states are separate from the physical ones—they're just able to be discussed as *though* they were. Such a conclusion is found elsewhere in Titchener's writings;

> "Just as we translate words and sentences from one language into another, so may the nervous system translate a perception into more" familiar terms, — into an idea which has none of the qualities that were contained in the perception, but replaces them by other and more familiar qualities that mean the same thing.[5]

Furthermore, the actual experience of these elements is *elementary.* In other words, it is erroneous to speak of conscious as the primary state,

[3]Blum (2013).
[4]Titchener (1899, p. 122).
[5]Titchener (1899, p. 123).

and awareness of the elements of consciousness as the secondary state—that's what gives the false impression of dualism. Rather, to be self-aware *is what it means* to *be* conscious. Consciousness is something that's directly experienced, a singular event—to *be* conscious is *to experience.* There's no separate conscious state that *then* processes the mental, physiological and physical elements within.[6] People are not aware of being aware—they are simply aware, full stop.

The next step is to assume that people can only be attentive to a limited range of mental elements at any one time. Self-awareness is equivalent to a spotlight arcing across the darkened space of consciousness briefly capturing accessible mental states for consideration in the form of an experiential stream.[7] Some of these states are the individual's autobiographical memories of life events—which coupled with their emotional valence allows the impression of self-continuity across the lifespan.[8] Overall, attention to the elements of consciousness generates the *sense-of-self.*

Another useful term for the sense-of-self is the *phenomenological ego*, which represents the internal image of a person-as-a-whole, the "I" as it manifests as conscious experience. The phenomenological ego allows the individual to interact with their internal world of mental elements, as well as the external world of material things, leading to the first-person perspective with;

> ...feelings of agency, ownership ("mineness") and immediacy ("nowness"), spatial perspective....as well as the feeling of being embedded in our bodily sensations.[9]

And finally, these ingredients are distilled into *qualia*, or the "essence of experience".[10] This is;

> ...the first person perspective seemings associated with specific sensations, such as those that we experience as colors, smells, musical tones, etc. Qualia

[6] Kokoszka (1999–2000).
[7] Brabant (2016).
[8] Hennig-Fast et al. (2008).
[9] Preller and Vollenweider (2018, p. 229).
[10] Kramer (2002, p. 658).

> are sometimes defined as what something feels like, so emotions and desires such as fear, love, hunger or thirst are also qualitative experiences. In general, non-sensory experiences are not called qualia, but there is no reason why they should not be. Like sensations, all emotions, desires and most conscious feelings have properties similar to those of … qualia.[11]

Qualia is a unique combination of the mental elements of consciousness, physiological sensations (from the body) and external perceptions. These include the birds and trees and buses and every other thing that possess texture and temperature and firmness, but which also move in certain ways, display shades of light and are attached to memories that frequently generate affective states.[12]

In answer to *why* consciousness and qualia emerge in the first place, "mind as an epiphenomenon" doesn't need to provide a purpose for non-causal mental states. They simply happen. For instance, the *feeling* of pain is an offshoot of what pain *really* is—the activation of various pain receptors transmitting their signals through the spinal cord to the appropriate processing regions of the brain (mainly via those C-fibres John Searle mentioned earlier).

However the idea that the mind has no purpose or function is somewhat unsatisfying. That's why epiphenomenalism tends to be a creeping position, with liberties taken as to the nature of consciousness; it is more than the accidental sum of the products of complex neural operations. These biological operations are dynamic both within the system and in response to external information. They are also in constant flux, and this volatile mix results in an "unpredictable" product—mind consisting of meaning, intentionality, qualia and consciousness that is greater than the sum of its parts.[13] Therefore, while still elementary, mind is irreducible to underlying neurobiological activity.[14] Expressed philosophically, there is no necessary *supervenience* of mind and brain, since they don't possess the same properties.[15]

---

[11] Mussacchio (2005, p. 404).

[12] Tschigguel and Hunter (2008).

[13] Kramer (2002).

[14] Meijer and Korf (2013).

[15] Yoshimi (2012).

A formulisation of the emergent mind is *cerebral mental field* (CMF) theory, where mind is a product of billions of nerve cells working simultaneously, and the field that is generated by this activity is entirely unpredictable. It is also non-physical, but how this comes about is not known—it just happens! Specifically, mind;

> ... must simply be regarded as a new fundamental "given" phenomenon in nature, which is different from other fundamental "givens," like gravity or electromagnetism.[16]

Perhaps the definitive treatment of consciousness as an emergent property is *neurophenomenology*, which tries to fill the explanatory gap separating subjective experience from the electrophysiological and metabolic processes of the brain.[17] That's achieved by "personalising" the objective measures of these biological processes (e.g., brain imaging techniques such as fMRI and PET, alongside physiological variables such as heart rate, respiration and galvanic skin response) through a detailed assessment of first-person descriptions of subjective experience.[18] Instead of starting from the measurement of brain activity and treating mind as being epiphenomenal to, or even emergent from (but irreducible to) these processes, the approach is the other way around. One begins with the irrefutable fact of "first-person experience" and works backwards to determine;

> ...how nonphysical conscious experience could arise in this physical world; why it would arise only from some particular physical processes and not from others that from the outside appear to be entirely equivalent; and *how it could act back on – or play a functional role in – the physical world.*[19]

This exposes the popular inclination, even from strict materialists, to jettison the meaning of "epiphenomenal" and endow the mind not only autonomy, but the ability to influence the physical brain. This perspective is found in health psychology as far back as the 1950s, with Hans Seyle's

---

[16]Libet (2006, p. 324)

[17]Solomonova and Wei (2016).

[18]Ibid.

[19]Hut and Shepard (1996).

General Adaptation Syndrome proposing that psychological stressors can generate biological change in the adrenal cortex, and thymus (affecting immunity), and cause gastric and duodenal ulcers.[20] A later therapeutic field derived from GAS is *psychoneuroimmunology*, which holds that mental stress affects the immune system through specific actions of neuropeptides circulating throughout the body.[21]

Functionally, this relationship is expressed through *biofeedback*;

> ….a process that enables an individual to learn how to change physiological activity for the purposes of improving health and performance. Precise instruments measure physiological activity such as brain waves, heart function, breathing, muscle activity, and skin temperature. These instruments rapidly and accurately provide "feedback" information to the user. The presentation of this information— often in conjunction with changes in thinking, emotions, and behavior— supports *desired physiological changes a process that enables an individual to learn how to change physiological activity* for the purposes of improving health and performance.[22]

One of the reasons for holding this position is the apparent absurdity of discounting the potency of mental states in changing physiological and behavioural states. Regardless of whether the product is directly reducible or irreducible to brain activity, if mind is merely an accident and only microscopic *physical* mechanisms can cause macroscopic changes in behaviour, then it becomes untenable to assume that anyone *else* has a mind (or consciousness)—besides yourself due to some quirky accident. If consciousness has a more purposeful role and has the ability to *determine* behaviour—something beyond a coincidence—then it makes more sense to assume consciousness is a state possessed by everyone. That's the position most people, and medical science, takes for granted.

There's tacit acceptance that these mind-body approaches *seem* to resemble Cartesian dualism.[23] And to maintain the façade of monism, some nimble backpedalling can return the mind to its brain origins. Using clever

---

[20] Dacher (2014).

[21] Ibid., p. 149.

[22] Austad (2018, p. 265).

[23] Ibid.

semantics, you could argue that when mind is said to influence the body, as with psychoneuroimmunology, it's actually *brain* affecting *body*. We're just using the word "consciousness" as a convenient way to represent a brain state that's different to a body state. Once your audience is reassured of your monist credentials, you can return to discussing consciousness as though it is something mental, and above all causative!

References

Austad, C. S., & Gendron, M. S. (2018). Biofeedback: Using the Power of the Mind–Body Connection, Technology, and Business in Psychotherapies of the Future. *Professional Psychology: Research and Practice, 49*(4), 264–273.

Blum, H. (2013). Dissociation and Its Disorders. *Psychoanalytic Inquiry, 33*, 427–438.

Brabant, O. (2016). More Than Meets the Eye: Towards a Post-materialist Model of Consciousness. *Explore, 12*(5), 347–354.

Dacher, E. S. (2014). A Brief History of Mind-Body Medicine. *International Journal of Transpersonal Studies, 33*(1), 148–157.

Hennig-Fast, K., Meister, F., Frodl, T., Beraldi, A., Padberg, F., Engel, R. R., et al. (2008). A Case of Persistent Retrograde Amnesia Following a Dissociative Fugue: Neuropsychological and Neurofunctional Underpinnings of Loss of Autobiographical Memory and Self-Awareness. *Neuropsychologia, 46*(12), 2993–3005.

Hut, P., & Shepard, R. N. (1996). Turning 'the Hard Problem' Upside Down and Sideways. *Journal of Consciousness Studies, 3*(4), 313–329.

Kokoszka, A. (1999–2000). Altered States of Consciousness: A Comparison of Profoundly and Superficially Altered States. *Imagination, Cognition and Personality, 19*(2), 165–184.

Kramer, M. (2002). The Biology of Dream Formation: A Review and Critique. *Journal of the American Academy of Psychoanalysis, 30*(4), 657–671.

Libet, B. (2006). Reflections on the Interaction of the Mind and Brain. *Progress in Neurobiology, 78*(3–5), 322–326.

Meijer, D. K. F., & Korf, J. (2013). Quantum Modeling of the Mental State: The Concept of a Cyclic Mental Workspace. *Syntropy, 1*, 1–41.

Mussacchio, J. M. (2005). The Ineffability of Qualia and the Word-Anchoring Problem. *Language Sciences, 27*(4), 403–435.

Pockett, S. (2004). Does Consciousness Cause Behaviour? *Journal of Consciousness Studies, 11*(2), 23–40.
Preller, K. H., & Vollenweider, F. X. (2018). Phenomenology, Structure, and Dynamic of Psychedelic States. *Current Topics in Behavioral Neurosciences, 36,* 221–256.
Solomonova, E., & Wei, S. X. (2016). Exploring the Depth of Dream Experience. *Constructivist Foundations, 11*(2), 407–416.
Titchener, E. B. (1899). Idea and the Association of Ideas. In E. B. Titchener (Ed.), *A Primer of Psychology.* London: Macmillan.
Tschigguel, W., & Hunter, M. (2008). Awakening in Hypnosis. *Contemporary Hypnosis, 25*(1), 39–45.
Yoshimi, J. (2012). Supervenience, Dynamical Systems Theory, and Non-reductive Physicalism. *British Journal for the Philosophy of Science, 63*(2), 373–398.

14

Scientific Reality and Errors in Consciousness

The monist treatment of the mind/brain dilemma is the basis of the second-path, and I'll provide it with the label *scientific reality*.[1] While some liberties can be taken discussing mental states as independent and causative "things", ultimately consciousness is only meaningful as a complete product of the brain (even though the product might have certain special qualities not directly reducible to brain activity, and hold potential influence over behaviour). Matter is still dominant, as it's the only sensible product of reality that can be said to actually *exist*. The material world as what is defined as *real*, and consciousness is "in the head"[2] and subservient to the dynamic workings of this world (specifically, the CNS). Nevertheless, thoughts and feelings and other mental elements can be addressed separately to their physical origins, and for this reason consciousness can be said to "exist" in a parallel but diminished sense.[3]

[1]Not to be confused with *scientific realism*, which involves a more complicated discussion of the achievements of science therefore doesn't have much in common with Tom's story.

[2]Tart (1998).

[3]Compare that with an alternative treatment of consciousness, idealism, where there is no objective reality outside the mind so physical reality is simply mental reality. This perspective serves no

T. Jinks, *Psychological Perspectives on Reality, Consciousness and Paranormal Experience*,
https://doi.org/10.1007/978-3-030-28902-7_14

However scientific reality offers no special treatment to mental states. Like physical processes, they need to conform to the set of rules called *science* that plot how everything that exists interacts in a stable and predictable manner. That's the "scientific" part of scientific reality. There are many scientific rules, developed and refined from the works of earlier *positivists* like Ernst Mach. Although Mach was quite radical in the sense that he didn't believe in atoms because they've never been directly observed,[4] his ideas (and those of other positivists) have been highly influential in consolidating the scientific process during the twentieth century. These processes include, but are not limited to;

> ...empirical adequacy, falsifiability (the capacity of making risky predictions, which are passive of being tested), predictive accuracy, broadness of scope (capacity to solve the greatest possible number of problems), simplicity, theoretical integration, theoretical ordering, and heuristic power (capacity to predict new kinds of phenomena, ability to lead to new discoveries).[5]

Events are observed in the world and these lead to hypotheses—general statements tested by deduction to determine their accuracy.[6] In this way positivist-style rules combine with materialism and subservient consciousness to create a powerful and universal position that, in Charles Tart's, words "works quite well most of the time".[7] Scientific reality neatly wraps up the metaphysics, ontology, mereology (speculating on the *parts* of reality[8]) and all the other "-ologies" underlying what reality is—or should be.

There might be the vagaries of space, time and matter at the quantum and relativistic corners of this reality, or hurdles to discussing mental states as autonomous entities. However *practically speaking* reality is made up of

productive purpose if it denies reality, as it is incompatible with the straightforward comprehension of how the world seems to be (real), and to operate. Likewise the notion of "pure consciousness" in the Platonic sense is merely of anachronous interest—so this has no place in our quest for understanding what reality is.

4Leahey (1987).

5Moreira-Almeida and Lotufo-Neto (2017, p. 285).

6Poynton (2012).

7Tart (1998).

8Morganti (2015).

real objects that behave in foreseeable ways, and consciousness is irrefutably aware of the nature of these objects.

Taking the second-path is no impediment to accepting Tom believed he saw a man walk across a hotel corridor, but the experience wasn't a ghost, or anything paranormal. Paranormal things apparently contravene the laws of nature, their legitimacy would imply all scientific findings to date are precarious, and they possess no sound theoretical basis.[9] Consequently, *what* Tom saw is determined instead by the three alternative ingredients of scientific reality; (1) materialism, (2) consciousness derived (emergent or otherwise) from material interaction, and (3) the scientific process.

From this frame of reference, it's patently clear that Tom's alleged ghost sighting doesn't conform to the scientific process. It lacks empirical adequacy and falsifiability. There's no predictive accuracy, it isn't simple (the concept of life-after-death certainly isn't simple), it cannot be theoretically integrated with existing scientific principles and perhaps most tellingly, it isn't experimentally measurable.

At least, that's the view the mainstream second-path adopts, despite all the quantum (and other) possibilities suggested by the first-path that argue for a theoretical basis to paranormal events, accompanied by quite extensive empirical support—particularly for psi. In fact, when examining a good portion of the history of laboratory-based psi-findings, over many years, Lund University Psychology Professor Etzel Cardeña comes to the following conclusion;

> … meta-analyses, conducted on studies using different protocols and by different researchers, provide cumulative vertical and horizontal support of psi. Vertical in the sense that across time different protocols have continued to produce positive results beyond what would be expected by chance, and with increasing methodological rigor; horizontal in the sense that there is support for psi across research areas.[10]

The results for psi apparently matches meta-analyses of social science experiments in general, which is encouraging considering psi itself is more a product of "real-life circumstances" than a laboratory setting where the

[9] Cardeña (2018).

[10] Ibid., p. 672.

stimuli are "impersonal …. of little or no consequence",[11] and where the talent isn't necessarily expressed in all people equally—unlike the assumptions of experimental psychology. Outside the differentiating power of the independent variable, participants in an everyday psychology experiment are treated as homogenous organisms. Indeed psychology expects they are, or the issue of confounding variables interfering with the validity of the experiment might undermine too many research findings for comfort!

And returning to ghosts, aren't these as elusive as psi? I've suggested the first-path recognises that ghosts might be physically real, but multiple variables have to simultaneously hold the correct values for the spectre to create a visual impression on the observer, beginning with attracting their attention. Had Tom been overly distracted when figuring out how to open his door he might have missed the urge to look up, leaving the ghost unnoticed. That's despite the timing being perfect, the right balance of physical variables required for the entity's manifestation being present, and the ghost's *desire* to be seen. None of these make much difference if Tom is looking in the wrong direction all along!

A spontaneous ghost might not need to conform to the demands of the scientific process in the way common objects and events can be subject to rigorous examination. However the second-path considers spontaneity a feature of everyday objects and events as well. Imagine an altercation between two people on a quiet street and a sole witness watching from a distance. Fists might fly and the perpetrators run off in different directions, leaving the witness waiting for the police to arrive so they can provide a statement. As a method of characterising lived experience, second-path would not discount the witness' personal testimony in the absence of hard evidence that the fight took place. And yet there is the implicit acceptance that had a camera been placed strategically on the street, recording 24 hours a day 7 days a week, then all facets of the dispute would be captured. The same would be true if a camera (perhaps a highly advanced model able to record full spectrum energy) had been positioned above Tom's hotel corridor and netted the few seconds of the ghost as it crossed Tom's path.

Then again, there really *wouldn't* be any footage showing a dark figure, because it's not the spontaneity of ghosts and their inability to conform the

[11] Cardeña (2018).

scientific process that's the problem, but the other expectations of scientific reality that also aren't met. That's why all the contemporary musings about the role of non-classical physics underpinning paranormal phenomena, and any supportive evidence, can be side-stepped. The concept of a "ghost" is simply not allowed by materialism. When the expensive camera positioned for weeks on end in the hotel corridor doesn't detect anything out of the ordinary, it's not because of a ghost's elusiveness of spontaneity, but because it's not there. It isn't real. It doesn't exist. Not even reference to quantum theories salvage Tom's ghosts since these claims don't affect reality on *this* scale, where people are big and ghosts and other paranormal things like flying saucers or Bigfoot are (apparently) just as big as people. Some muted concessions might be made to *very* minor laboratory-based psi-events, but there's no absolutely no room for consideration of hotel corridor ghosts that are the size of real people.

So if Tom's ghost doesn't conform to the scientific process, and also doesn't correspond to anything real, then what is it? There's one point so far untouched; *consciousness derived from material interaction* (that is, neurobiology). Granted Tom had a phenomenological experience, but this encounter with the paranormal must be an *error* in consciousness. Having familiarity with the various viewpoints on what consciousness is, it is now possible to evaluate how such errors could occur—but that's a question for the neuroscientist and psychologist to answer, not the physical scientist.

Going easy on Tom, let's assume that *what* he saw was indeed real, but certainly not ghostly. Starting from the external source, it's quite possible he did perceive the presence of an energetic stimulus in the corridor emanating photons of light, and these photons interacted with his retina(s). This draws on a *bottom-up* model of visual processing styled on James Gibson's work.[12] Perception is hierarchical, with links in the same neural chain successively registering the visual information that's being received from lower centres and passing this information to higher CNS regions—ultimately leading to global scene recognition.[13]

The information is also subject to strong neural influence as it passes up the chain. An example would be the automatic highlighting of *contrast*

[12] Gibson (2002).

[13] Greaney and MacRae (1992).

between and object and its background.[14] The scene is represented as an array of neural activity in the retina consisting of the object's edges which are bright in comparison to the adjacent surroundings which can be much darker. Receptors detecting an edge will be highly active because they are strongly illuminated, whereas receptors adjacent to the edge—in the darker zone—will be less active. These receptors are interconnected through mutual lateral inhibition, which refers to "sideways" suppressive influence over the surrounding network. The illuminated neurons on the "bright" edge of the scene will only receive moderate suppression from the adjacent neurons in the darker zone, while they dispense considerable suppression themselves. This one-sided activity results in the perception of the object as standing-out in sharp contrast to the background, and is an illustration of how a visual scene is altered (or "touched-up") by neural mechanisms very early in the chain of perception.[15]

It's still important to recognise that from the perspective of the second-path, the progression of a visual signal, its distribution and analysis, occurs *as a single process* (monism). Initially a higher brain region called the lateral geniculate nucleus of the thalamus receives input from the optic nerve, sorting this information as a function of scene features and relaying these features higher again to the primary visual cortex (the striate cortex), and beyond that to visual associate areas of the extrastriate cortex. Two streams of information assist with scene creation; a ventral stream projects to the inferior temporal cortex allowing colour and the identity (form) of the object to be determined, coupled with its three dimensional conformation and its relationship to background. Simultaneously the dorsal stream projects to the posterior parietal cortex to determine where objects in the scene are located (spatial perception) and their speed and direction if they are moving.[16]

Each stage the burgeoning narrative is subject to additional neural analysis until a final product emerges that is experienced not so much "in consciousness" or "by consciousness", but *as* consciousness. Although words like "distribution", "analyse", "determine", "process" and "responsible" are

[14]Pinel and Barnes (2017, p. 171).

[15]Biederman (1987).

[16]Carlson and Birkett (2017).

used to convey scene-generating operations from the retina to the striate and extrastriate cortices, they don't imply the activity of a separate consciousness (or many consciousness's) reflecting on the information received at each stage of the hierarchy. This is evidenced by the way damage to the different level of the visual system result in specific deficits to the perceived scene, and these are experienced as *absences from consciousness.* For instance, *agnosia* is the inability to recognise sensory impressions (not just vision), and *prosopagnosia* is a specific type of agnosia following extrastriate lesions where the sufferer is unable to identify familiar faces despite other visual abilities remaining intact, like colour vision, visual-motor control, binocular vision and spatial attention.[17]

Scientific reality also provides flexibility to the visual system, allowing what we see to be more than the automatic product of a bottom-up scene generation mechanism. There are additional cognitive influences over striatal and extrastriatal regions that assist visual recognition. These "volitional" *top-down* systems originate in the frontal cortex.[18] With access to existing stored information, they are able to assess the raw perceptions entering via the eyes and make judgements about what they are.[19] Here we shift from describing the bottom-up biological mechanisms, to considering the way the basic visual ingredients are manipulated and matched with "stored" representations for the purpose of identification.[20]

This assimilation might be quite simple to conceive without resorting to dualism. For example, the ability to recognise a simple object like a nail from a range of similar-sized and shaped objects (match-sticks, bobby pins and screws) might be conceived as analogous to rolling differently-sized balls across a sloping table covered in holes of various diameters. As the balls travel they are automatically sorted—buckets under the table will collect the different sizes that fall through the matching holes. Equivalently, the image of a nail will into a conceptual "nail bucket", but not the match-stick, bobby pin or screw bucket. That's how we know it's a nail. However genuine object recognition is not this simple. It is based on

[17]Rosenthal and Behrmann (2006).

[18]Ligeza et al. (2017).

[19]Gregory (1970).

[20]Clement (2007/2008).

probabilistic matching rather than direct, bottom-up matching, and uses more sophisticated procedures. This is better explained using another line of analogous reasoning which comfortably jumps between biological and mind vernacular;

> ....when an incoming input does not strongly implicate a particular object, a stored model of the best-matching shape is activated... [therefore] conscious recognition is not immaculate but the result of a coincidence between bottom-up processes originating from retinal inputs and top-down processes located in higher systems of visual memory. When the processes match, the subject experiences seeing the object as a unified percept.[21]

Accordingly, anything a person witnesses will have been "created" by the bottom-up hierarchy, and expressed as consciousness *phenomenal* experience. And yet his outcome is not foolproof. There can be initial confusion on what the scene represents, requiring top-down processes to compare the immediate perception to stored representations of similar objects. In an abstract *central assembly* an image of the scene is generates as the fusion—or "compromise"—between the bottom-up sensory and top-down memory inputs. An expression of what is termed *the binding hypothesis*, this new image is a *bound* product and becomes the "conscious report" available to the percipient, albeit delayed due to the short time it takes for binding to operate.[22] Divorced from any biological grounding, its' easy to imagine how an individual can subsequently manipulate this newly recognised image, deconstructing it and concentrating on features such as colour, texture or size.

So *how* do the errors in perception emerge? Realistically there's no reason to suspect Tom suffers from damage to his striate or extrastriate cortices, or any subordinate visual system pathway. He does not report chronic deficits or changes in the features of perceptual experience, let alone see ghosts or other weird things regularly, which might be expected of chronic injury to his bottom-up (or top-down) visual system. That leaves two further possibilities. The first involves *transient* physiological dysfunction of

---

[21] Ibid., p. 247.

[22] Tomkins (1992; quoted in 2007/2008).

bottom-up processes which generated a one-off ghostly figure that's accidently processed and integrated into conscious awareness.[23] The second is that Tom's ghost was a cameo top-down psychological error—his visual pathways were functioning correctly but has been "tricked" into recognising a human figure that he subsequently labelled as a ghost for want of a rational explanation.

The second possibility has the most potential to explain his sighting. Basically, cognitive systems support the abstract central assembly (or any comparable psychological construct) to create the conscious visual experience of a scene, and these appraise and make reflective judgements on this experience. If the scene is ambiguous, appears only briefly in foveal (central) vision, or lingers in the periphery of vision, then these systems might struggle to find a recognisable solution. The might be required to force an interpretation on the scene, resulting in an *illusion*.[24] Well known (contrived) examples of visual illusions emanating from this confusion include the Necker cube, the rabbit-duck illusion, the Kanizsa triangle and Escher's drawings.[25]

However illusions can also be generated by "real world" stimuli, which better represents Tom's ghost and other paranormal pseudo-experiences. Psychologist James Sully wrote about their cause in 1897;

> The illusion arises in each case from our ignoring the variation, and treating a given sensation under all circumstances as answering to one objective cause.[26]

Hence;

> The well-known experiment of first plunging one hand in cold water, the other in hot, and then dipping them both in tepid, is a startling example of this organized tendency. For here we are strongly disposed to accept the palpable contradiction that the same water is at once warm and cool.

---

23 A concept I'll return to later in the book.

24 Bottom-up theories of visual processing are founded on *direct perception*, and while they may account for rapid judgments of environmental conditions and the creation of realistic visual scenes they don't handle illusions well.

25 E.g., Culicover and Jackendoff (2010).

26 Sully (1897 [1927], p. 546).

The difference between a real-world misperception, and a real-world illusion, is that the former is "idiosyncratic" and involves only a "transitory mistake" in perceiving the environment, whereas the latter is a more *persistent* error of an ambiguous environmental object.[27] Real-world Illusions are also generated by ambiguous and inexplicable stimuli subject to pressure from multiple top-down factors which may lead to a paranormal attribution.[28] Returning to Sully's early consideration on the topic;

> We may not be distinctly anticipating any one kind of object, but are nevertheless in a condition of subexpectation with, reference to a large number of objects. Accordingly, when an impression occurs which answers only very roughly to one of the associated images, there is a tendency to superimpose the image on the impression. In this way illusion arises.[29]

This is illustrated nicely in the following story of Diane's ghost sighting;

> I was outside in the garden and looked up at the second-storey window at the rear of what we knew as *Nan's haunted house* and saw the face of a woman staring at me through the bottom pane. I moved closer and looked up again but the face was gone. I went inside the house and there was no-one there. It was definitely a ghost.[30]

Assuming this ghost *was* an illusion, for Diane to make this particular paranormal attribution there needs to be prior beliefs *in* and expectations *about* the reality of ghosts (beliefs being a function of higher order, top-down processes). These are coupled with the human tendency to spontaneously organise unrelated concepts into meaningful relationships, a predisposition known as *apophenia*.[31] The transitory mix of ambient lighting, structural components of the window (such as inconsistent glass thickness)

---

[27] Collerton et al. (2016).

[28] Irwin et al. (2013).

[29] Sully (1897/1927, p. 548).

[30] Diane, Personal communication 8 October 2014.

[31] Fyfe et al. (2008) and Alcock (2004); this related to pareidolia, "an illusion involving misperception of an external stimulus; an obscure stimulus is viewed as something clear and distinct....This perception, or misperception, does not involve conscious effort or any particular mental set, and the illusion does not vanish even when one pays closer attention to the stimulus because it is so ambiguous that it has no objective meaning at all" (Alcock 2004).

and features the room backdrop (like shadows cast by a ceiling lamp) are coerced into a brief but stable human face. Additional personal traits help encourage the illusion, such as *suggestibility*, where a witness's cognitive characteristics make them particularly susceptible to incorporating unrelated information into object or scene interpretation.[32] In the case just mentioned, Diane identified the house as "haunted", so she will allowed her existing *perceptual set* to drive false recognition of indistinct shapes in the upper-story window.

Additional cognitive factors may sustain the illusion, including fear of the paranormal and *intolerance* of ambiguity; if Diane or Tom were in the presence of an obscure stimulus they would be more inclined than other people to turn it into something understandable, rather than leave it unrecognised. *Demand characteristics* present in a location can also make an impact. These are external cues ranging from creepy darkness to vague sounds and fragile smells. Together these variables are automatically corralled by the percipient's "central assembly" and encourage a paranormal translation of the scene's meaning—in the case of ghost sightings potentially fulfilling the percipient's need for evidence of the afterlife (that's said to particularly occur in depressed, elderly and widowed individuals, who seem to see ghosts a great deal).[33]

Real-world illusions may also be auditory, explaining other varieties of paranormal experience;

> A common example of apophenia occurs when people are in the shower, and mistakenly think that they hear their door bell or telephone ringing. The white noise produced by the shower contains a broad spectrum of sounds, including those that make up ringing bells. The ear picks out certain sounds from the spectrum, and we "detect" a pattern corresponding roughly to a bell.[34]

According to the second-path, EVPs aren't otherworldly tune-ins from Jürgenson and Raudive's spectral radio station, but simply the product of top-down alterations of the ill-defined noise played back through a digital

---

[32] Mazzoni and Vannucci (2007) (similarly termed the *misinformation effect*).

[33] Parra (2018).

[34] Alcock (2004, p. 1).

recorder (or real-time "spirit box"), identified as an *auditory* pareidolia.[35] A similar rationalisation can be applied to the infrasounds that allegedly hold the voices of ghosts. Infrasounds aren't just heard through the ears. They;

> … may also be detected through skeletal bones, bones within the ear, resonance within organs and body cavities, and tactile senses …. The inability of most people to 'hear' infrasound means that its effects upon a person are largely unexpected and therefore more likely to be attributed to other causes, and in some instances where the percipient is in a haunted location or involved in the pursuit of ghost-hunting such effects are frequently blamed upon a paranormal agent or cause.[36]

After seeing his ghost, Tom didn't shrug his shoulders and convince himself that what he'd seen was merely his own shadow briefly projected onto the end of the corridor (however unlikely that would be). Instead his pre-existing belief in hauntings, and personal susceptibility to demand characteristics, meant he allocated a faulty attribution to the visual experience, one that might have been subtly generated by the eeriness of the second floor's tired Edwardian décor and musty smell, the low lighting and the knowledge that he was completely alone on the second floor.[37] These top-down influences occurred simultaneously with a visual percept formed by bottom-up sensory data impacting on striatal and extrastriatal pathways. The corridor scene is now "presented" to Tom's consciousness as something externally *seen*, in combination with a (non-conscious) verdict about what the moving black shape *was*.

Once such a paranormal attribution has been made, they can become habitual;

> ….once attached, labels can sustain paranormal experiences by initiating a reactive process, which effects subsequent perception/interpretation of additional ambiguous phenomena. In the case of paranormal attributions,

---

35 Elliott (2018).

36 Parsons (2012, p. 155).

37 Lange and Houran (1998).

this manifests as a self-reinforcing attentional bias, which facilitates the perception of flurries of further paranormal observations.[38]

References

Alcock, J. (2004, December 21). Electronic Voice Phenomena: Voices of the Dead? *Skeptical Inquirer*. Retrieved August 16, 2018, from https://www.csicop.org/specialarticles/show/electronic_voice_phenomena_voices_of_the_dead?/specialarticles/evp.html.

Biederman, I. (1987). Recognition-By-Components: A Theory of Human Image Understanding. *Psychological Review, 94*(2), 115–147.

Cardeña, E. (2018). The Experimental Evidence for Parapsychological Phenomena: A Review. *American Psychologist, 73*(5), 663–677.

Carlson, N. R., & Birkett, M. A. (2017). *Physiology of Behaviour* (12th ed.). Harlow: Pearson.

Clement, P. (2007/2008). Dream Imagery as a Result of Emotions Matching with Images. *Imagination, Cognition and Personality, 27*(3), 245–257.

Collerton, D., Taylor, J.-P., Tsuda, I., Fujii, H., Nara, S., Aihara, K., et al. (2016). How Can We See Things That Are Not There? Current Insights into Complex Visual Hallucinations. *Journal of Consciousness Studies, 23*(7–8), 195–227.

Culicover, P. W., & Jackendoff, R. (2010). Quantitative Methods Alone Are Not Enough: Response to Gibson and Fedorenko. *Trends in Cognitive Sciences, 14*(6), 234–235.

Drinkwater, K., Dagnall, N., Grogan, S., & Riley, V. (2017). Understanding the Unknown: A Thematic Analysis of Subjective Paranormal Experiences. *Australian Journal of Parapsychology, 17*(1), 23–46.

Elliott, C. (2018, April). Meaning in Randomness. *Psychologist*, 22–25.

Fyfe, S., Williams, C., Mason, O. J., & Pickup, G. J. (2008). Apophenia, Theory of Mind and Schizotypy: Perceiving Meaning and Intentionality in Randomness. *Cortex, 44*(10), 1316–1325.

Gibson, J. J. (2002). A Theory of Direct Visual Perception. In N. Alva & E. Thompson (Eds.), *Vision and Mind: Selected Readings in the Philosophy of Perception* (pp. 77–91). Cambridge: MIT Press (Original work published 1972).

[38]Drinkwater et al. (2017, p. 26).

Greaney, J., & MacRae, A. W. (1992, May). The Order of Visual Processing: Top-Down, bottom-Up, middle-Out, or None of These? *Bulletin of the Psychonomic Society, 30*(3), 255–257.

Gregory, R. (1970). *The Intelligent Eye.* London: Weidenfeld and Nicolson.

Irwin, H. J., Dagnall, N., & Drinkwater, K. (2013). Parapsychological Experience as Anomalous Experience Plus Paranormal Attribution: A Questionnaire Based on a New Approach to Measurement. *Journal of Parapsychology, 77*(1), 39–53.

Lange, R., & Houran, J. (1998). Delusions of the Paranormal: A Haunting Question of Perception. *Journal of Nervous and Mental Disease, 186*(10), 637–645.

Leahey, T. H (1987). *A History of Psychology* (2nd ed.). Englewood Cliffs, NJ: Prentice-Hall.

Ligeza, T. S., Tymorek, A. D., & Wyczesany, M. (2017). Top-Down and Bottom-Up Competition in Visual Stimuli Processing. *Acta Neurobiologiae Experimentalis, 77*(4), 305–316.

Mazzoni, G., & Vannucci, M. (2007). Hindsight Bias, the Misinformation Effect, and False Autobiographical Memories. *Social Cognition, 25*(1), 203–220.

Moreira-Almeida, A., & Lotufo-Neto, F. (2017). Methodological Guidelines to Investigate Altered States of Consciousness and Anomalous Experiences. *International Review of Psychiatry, 29*(3), 283–292.

Morganti, M. (2015). Dependence, Justification and Explanation: Must Reality Be Well-Founded? *Erkenntnis, 80*(3), 555–572.

Parra, A. (2018). Perceptual-Personality Variables Associated with Entity Encounter Experiences. *Australian Journal of Parapsychology, 18*(1), 23–48.

Parsons, S. T. (2012). Infrasound and the Paranormal. *Journal of the Society for Psychical Research, 76*(908), 150–174.

Pinel, J. P. J., & Barnes, S. (2017). *Biopsychology* (10th ed.). Boston: Pearson.

Poynton, J. (2012). The Out-of-Body Experience: The History and Science of Astral Travel (Book Review). *Journal of the Society for Psychical Research, 76*(907), 116–120.

Rosenthal, O., & Behrmann, M. (2006). Acquiring Long-Term Representations of Visual Classes Following Extensive Extrastriate Damage. *Neuropsychologia, 44*(5), 799–815.

Sully, J. (1927). Illusions and Hallucinations. In W. S. Taylor (Ed.), *Readings in Abnormal Psychology and Mental Hygiene* (pp. 546–550). New York: D. Appleton & Company (Original work published 1897).

Tart, C. T. (1998). Six Studies of Out-of-Body Experiences. *Journal of Near-Death Studies, 17*(2), 73–99.

15
No (Ghostly) Illusions

The notion that Tom's ghost was no more than a misattributed illusion might be supported by clever scientific language, but it faces three challenges. Firstly, demand characteristics are a popular *theoretical* influence affecting the recognition of mundane visual stimuli. There's some empirical evidence that individuals' perceptions of their surrounds may be influenced by very unique and atmospheric locations,[1] but the claim isn't necessarily applicable to every place and situation. And demand characteristics are difficult to find in Tom's narrative. He *didn't* report musty smells or eerie décor when narrating the story. Apparently the corridor was very clean and with a very modern refurbishment. Tom's been to plenty of hotels, many a lot older, darker and spookier than this one, and hasn't seen a ghost. More pointedly, these putative demand characteristics can't be incorporated into Tom's interpretation based on his prior beliefs. He doesn't particularly *believe* in ghosts, even now. At least, not to a point where he's confident to say "ghosts exist" and are the residue spirits of dead people who haunt locations including lonely hotel corridors.

---

[1]E.g., Houran et al. (2002).

T. Jinks, *Psychological Perspectives on Reality, Consciousness and Paranormal Experience*,
https://doi.org/10.1007/978-3-030-28902-7_15

As a retort, not only is the top-down construction of a scene automatic and beneath awareness (like those bottom-up basic contrast, motion or figure-ground mechanisms), so are the cognitive decision-making processes (like demand characteristics) affected by pre-existing belief. Tom might *say* he doesn't believe in ghosts, but deep-down he knows what they are and thinks they are scary. He can't prevent these ingrained ideas from influencing the hidden analytics conducted by his top-down brain when interpreting an ambiguous corridor setting. He unconsciously absorbed information through the two lenses of *belief* and *personality* which took this material and fashioned it into the *conscious* perception of a walking man—even down to the details of a pale shaven face. While this seems a sensible (psychological) conclusion, it does imply some form of "crypto-perception" which is not entirely compatible with the scientific process. It's difficult to support notions of such veiled cognitive processes creating paranormal illusions, while simultaneously dismissing the role of unconscious "forces" in other contexts, such as Freudian psychoanalysis (where universals and primal ideas are active but operate beneath awareness, occasionally leaking into consciousness and producing psychological disorder).

Secondly, it's all very well to suggest that subtle cognitive processes act to mould perception of a scene into an illusion—an inaccurate interpretation of reality. However, there's very little information about how these responsible brain region(s) manage to operate so quickly. After all, what Tom believed—or *knew*—he saw was to him an immediate and salient presentation in consciousness. It was an *actual man* that resembled past experiences of *real* men walking across his visual field. The ghost and a real man were essentially *indistinguishable conscious states*. How do these clever top-down processes rapidly feed into the bottom-up visual pathways to shape a convincing human figure in motion? It's difficult enough to account for *basic* vision (such as the way movement perception as a single feature is managed by the extrastriate cortex[2]), let alone real-world illusions. The standard explanation is mental rather than biological, and it centres on process rather than timing. Indeed there's no pressing requirement to match the exact neural substrates with the appropriate cognitive

[2]E.g., Carlson and Birkett (2017, p. 201).

resources, other that perfunctory reference to the prefrontal cortex and other association areas that regulate cognitive capabilities.

A third reason why Tom's sighting is unlikely to be an illusion is the most straightforward. To experience an illusion, there must be perception of an actual stimulus—one that is able to be distorted by all those complicated top-down mental mechanisms that dabble in scene recognition. Something *had* to have been present at the corridor junction for Tom to receive a projection moulded onto this visual field. The nearest stimulus that might have formed the basis of the ghost sighting is a dark object of a similar height and bearing, crossing the corridor. And yet Tom's is adamant there was nothing on further inspection that could account for a person (or animal, or anything else "normal" or "natural") performing that movement without *still* being trapped in the corridor's dead-end.

It's routine to see things "out of the corner of the eye" that are misinterpreted, like a shrub in twilight that you momentarily mistake for a crouching dog, or a sunbaking snake that turns out to be a gnarly but benign twig. The error is quickly recognised when you turn your head and the stimulus enters foveal visual. That's commonplace evidence for the authenticity, and pervasiveness, of "real world" illusions. Nevertheless it's rare for the source of the illusion *not* to be discovered, the notable exception being something fast-moving that rapidly leaves the location. That's typical of objects seen traversing the sky in daytime and construed as "flying saucers". It seems to be silver-coloured, but it is flying too fast (and in the edge of vision) to be able to discern any other feature. On reflection it *looked* like an ellipsoid flying saucer—with a cupola on top, even though the object *must* have been conventional aircraft or drone viewed from an ambiguous angle, reflecting the sun and giving the impression of an unidentified flying object (which by *definition* it was). This really can't be said to apply to Tom's ghost, since he was able to make close inspection of the scene immediately after the sighting.

What about less obvious sources for an ambiguous visual stimulus? I earlier mentioned a stationary shadow as a potential explanation for the source of Tom's ghost, since there's no requirement for an illusionary stimulus to be moving—perhaps top-down processes generated the effect of *apparent* motion across the gap in the corridor?[3] Tom wouldn't be convinced by this argument, since he says he couldn't reproduce the same effect in the corridor with his own shadow. And if it wasn't his shadow then it must have been thrown by another object, which brings the problem back to square one—what was *this* object, if not Tom himself? Of course, this assumes Tom actually conducted a thorough investigation in the first place—which is another opening for the sceptic. We have to take Tom at his word that he even bothered experimenting with explanations for the culprit, and found nothing suitable. He might be wrong in assuming no person (or cat, or bird) could have entered through the fire escape door, or out through the window.

Alongside Tom's poor chronicling his experience, he might also have *reconstructed* memory of his "shadow". After all, there are plenty of ways for memories to lose accuracy as time passes.[4] This includes the *transience* of autobiographical memories. They simply deteriorate over time, and Tom told me the story many months after it happened. On the other hand, the sighting was highly unusual, and novelty has been shown in *some* circumstances improve the memory of events,[5] although only vividness of the event and confidence in recall is enhanced, rather than precision.[6] And that's assuming what Tom witnessed was something substantial in the first place. He was only attentive of *whatever it was* for a short period of time, so it's feasible the event was never effectively coded. If the only valid information retained was darkness and movement (e.g., from a momentary shadow cast by a bird or cloud through the cross-corridor window), so subsequent recollection will be vulnerable to all sorts of post-event distortion.

And that's another way beliefs and attitudes influence experience, but on this occasion they don't have to act immediately on the sensory impression. There's plenty of time for Tom's (putative) belief in ghosts to act on the memory the bird or cloud's shadow as he regales his friends with the story over dinner. He now says he saw a solid, walking man. This is the dominant

[3] E.g., Heller and Davidenko (2018).

[4] Schacter et al. (2003).

[5] Lanciano and Curci (2011).

[6] Talarico and Rubin (2007).

psychological position for a ghost sighting, however cognitive psychology can be a two-edged sword. Various studies of memory creation show that pre-existing beliefs actually *assist* event recall if they are consistent with that event (as opposed to not holding a belief, or having a contrary belief).[7] In defence of Tom, perhaps he did see man's figure, and because he believes in ghosts the accuracy of recall is better than if he didn't believe in ghosts!

But the psychological rationalising doesn't stop there. Maybe Tom wasn't solely responsible for creating his ghost sighting? While he might be overly suggestible, it isn't demand characteristics or prior beliefs that warp his memory but the influence of his colleagues at dinner arguing that a ghost is the best and only explanation for what took place upstairs. This inexorably turns Tom's memory of the event to something paranormal. Following this friendly instruction, his freshly-distorted recollection is strengthened through *confirmation bias*—another ad hoc cognitive process that modifies past memories so they conform to *current* beliefs. Apparently that's important for preserving the witness' "self-stability."[8] Specifically, aspects of a memory that match present beliefs are honed and embellished, but those parts that contradict these beliefs are attenuated.[9] As the story is told and retold, these processes are reinforced, to the point where Tom's memory of an ambiguous shape now becomes a story of a dark-suited man.

In summary;

> There may be any number of cognitive errors occurring that cause the individual to misinterpret the event [an anomalous experience], be they caused by individual knowledge, beliefs, misjudgements of probability, misperception, misremembering, confabulation, or the context of the situation.[10]

Perhaps Tom *did* see something fleeting and ambiguous at the corridor end, the source of which can only be speculated. The shadow of a bird or a cloud seems a reasonable *what (the ghost) was.* Following rapid interplay between bottom-up and top-down neural systems, this sensory input was

[7] Read and Rosson (1982).
[8] Ibid.
[9] Taylor (2013).
[10] Luke (2011, p. 356).

turned into conscious perception of the most logical possibility—a walking man. This *how and why (the ghost) happens* is initially fragile, but the experience is reinforced over time as a "real" memory through social input and subsequent cognitive distortions.

But there's a catch. Give Tom this explanation and he'll maintain a strenuously objection to downplaying his story. He'll say the man *wasn't* in his periphery, and he stayed long enough in his foveal vision to make out some quite detailed features, down to the shaven cheeks and sallow skin. There was *nothing* ambiguous about the ghost. And if this was all made up "on the spot" by his unconscious brain playing with a mere shadow, then how can *anything* he sees be trusted?

Tom is adamant; he knows what he saw and it was a *real* ghost. There was no accidental re-working of his tale at a later time. He has no patience for disrespectful "explanations" founded on psychobabble. If that's Tom's reaction, then the second-path can trade blows. *Sure* Tom saw exactly what he reported. His memory of the event was outstanding. There was no illusion, so no ambiguous stimulus need be identified. That's because what he saw wasn't even there. Instead, Tom *hallucinated* his ghost.

Any definition of hallucination has the same basic ingredient. Hallucinations involve an object or event[11] that gives the impression of being sensed (seen, heard, smelled or touched….or even tasted) which has no material basis since there has been no sensory input or external stimulation.[12] This means hallucinations are different from true perceptions. They are "false perception" or "quasi perceptions"[13] the same way paranormal experiences are pseudo-experiences. Nevertheless they seem "compellingly" real.[14] Those that are indistinguishable from reality are classified as *true hallucinations*, and *pseudohallucinations* occur when the witness understands their vision isn't real, but it still possesses the "compelling" characteristics of authentic perception.

And hallucinations aren't just a special kind of illusion. James Sully as more to say about this distinction;

---

[11]A dynamic interaction of multiple objects.

[12]E.g., The Concise Oxford Dictionary (p. 485).

[13]Nanay (2016).

[14]Ibid.

> Imagination, instead of waiting on sensation, usurps its place and imitates its appearance. Such a "subjective" sensation produced by a powerful expectation might, perhaps, by a stretch of language, be regarded as an illusion, in the narrow sense, in so far as it depends on the suggestive force of a complete set of external circumstances; on the other hand, it is clearly an hallucination in so far as it is the production of the semblance of an external impression without any external agency corresponding to this.[15]

This becomes the crux of the second-path; the ghost (or any other witnessed paranormal event) as the product of pure hallucination.

References

Carlson, N. R., & Birkett, M. A. (2017). *Physiology of Behaviour* (12th ed.). Harlow: Pearson.

Heller, N. H., & Davidenko, N. (2018). Dissociating Higher and Lower Order Visual Motion Systems by Priming Illusory Apparent Motion. *Perception, 47*(1), 30–43.

Houran, J., Kumar, V. K., Thalbourne, M. A., & Lavertue, N. E. (2002). Haunted by Somatic Tendencies: Spirit Infestation as Psychogenic Illness. *Mental Health, Religion and Culture, 5*(2), 119–133.

Lanciano, T., & Curci, A. (2011). Memory for Emotional Events: The Accuracy of Central and Peripheral Details. *Europe's Journal of Psychology, 7*(2), 323–336.

Luke, D. (2011). Anomalous Phenomena, Psi, and Altered Consciousness. In E. Cardeña & M. Winkelman (Eds.), *Altering Consciousness: Multidisciplinary Perspectives: History, Culture, and the Humanities; Biological and Psychological Perspectives* (Vol. 2, pp. 355–374). Santa Barbara: Praeger.

Nanay, B. (2016). Hallucination as Mental Imagery. *Journal of Consciousness Studies, 23*(7–8), 65–81.

Read, S. J., & Rosson, M. B. (1982). Rewriting History: The Biasing Effects of Attitudes on Memory. *Social Cognition, 1*(3), 240–255.

Schacter, D. L., Chiao Chiao, J. Y., & Mitchell, J. P. (2003). The Seven Sins of Memory: Implications for Self. *Annals of the New York Academy of Sciences, 1001*, 226–239.

[15] Sully (1897/1927, p. 549).

Sully, J. (1927). Illusions and Hallucinations. In W. S. Taylor (Ed.), *Readings in Abnormal Psychology and Mental Hygiene* (pp. 546–550). New York: D. Appleton & Company (Original work published 1897).

Talarico, J. M., & Rubin, D. C. (2007). Flashbulb Memories Are Special After All; in Phenomenology, Not Accuracy. *Applied Cognitive Psychology, 21*(5), 557–578.

Taylor, A. K. (2013). Leveling and Sharpening. In A. K. Taylor (Ed.), *Encyclopedia of Human Memory* (Vol. 1, p. 618). Santa Barbara, CA: ABC-CLIO LLC.

16

Hallucinations

The second-path has a fine pedigree, since hallucinations have long been attributed to ghost sightings. In the early nineteenth century Sir Walter Scott discounted apparitions as illusions, but took the view that under certain circumstances the senses could become "deranged" and this provided a natural solution to why people reported seeing their "empty and false forms" and hearing their "idle sounds".[1] He certainly wasn't the first to use this rationalisation.

Before tackling the idea that Tom hallucinated his ghost, it's worth familiarising ourselves with some basic hallucinations. These affect *all* the senses. For instance a simple somatosensory (touch and pressure) hallucination is *formication*, which refers to the mistaken impression that there are tiny insects crawling *under* the skin (formica is Latin for ant).[2] These tactile hallucinations are mostly associated with schizophrenia, and can involve more than bugs, in shown in this story from a patient named Faith;

[1] Inglis (1985/1986, p. 182).

[2] Dowben et al. (2017).

T. Jinks, *Psychological Perspectives on Reality, Consciousness and Paranormal Experience*,
https://doi.org/10.1007/978-3-030-28902-7_16

> Faith, when queried about the lacerations on her arms and wrists, did elaborate that beginning about 7–10 days before she cut herself that she had initially felt insects crawling on her skin that later transformed into mice, which then morphed into rats that burrowed into her body.[3]

A similar tactile hallucination is *delusional parasitosis*, where the sufferer feels they are covered with (non-existent) parasites on the surface of the skin. They hold onto this belief despite seeing no evidence for the infestation and in order to purge themselves of the pests they remorselessly pick at their skin "to the point of self-mutilation".[4] Yet another style of tactile hallucination involves the sensation of pressure from imaginary hands, as in this example from a stroke patient who believed that;

> …someone had pushed him by his neck, trying to make him fall down, later pushing him on the stairwell in his own house. As the patient said: "That morning, it was as if someone put his hands on the back of my neck. I grabbed with my hand, as if I was holding his finger. I asked my wife 'is there anyone?'" His family reported that the patient had opened the door of his house and was searching for someone on the staircase, angrily shouting "Someone pushed me from the bed to the ground".[5]

The focus of the stroke was around the junction of the temporal, parietal and occipital lobes, which might be taken as close enough to the somatosensory ("touch") cortex to account for the tactile hallucinations.[6]

Simple auditory hallucinations are also common in schizophrenia (75% of patients with schizophrenia experience auditory hallucinations, as do 20–50% of people with bipolar disorder[7]). They're also frequently found in patients with brain tumours and seizures of the auditory cortex in the temporal lobes. While the former might generate crude sounds, seizure activity often leads to the perception of spoken words and phrases which provides a more sophisticated and ongoing hallucinatory storyline;

---

[3] Ibid., p. 220.

[4] Ibid.

[5] Akinci et al. (2016, p. 80).

[6] There's no consideration that the patient was actually attacked by a legitimate poltergeist!

[7] Choong et al. (2007).

> I met Annie* while running a workshop that she was attending in a professional capacity. She confided she trusted me because I dealt with psychological matters and told me she heard voices, cheerful ruminative voices from people of her age and background whom she did not know. She called the voices the 'gossipers'. The voices were never threatening or accusatory, they never told her to do things and in many ways they seemed good company. The voices did not impede the support that she gave to patients. I asked where the voices came from and she explained that they were in her chest, but sounded up through her head. They weren't there all the time, but did appear if she was tired or fed up.[8]

Then there are simple visual hallucinations; impressions like the phosphenes I've previously mentioned that persevere regardless of whether the eyes are open or closed.[9]

> … forms and patterns are more typical than [an extended] drama. Elementary hallucinations— generated by a host of clinical and laboratory-induced states—range from simple spots to the flashing serrated arcs [and] kaleidoscopic patterns.[10]

Apart from ocular pressure, phosphenes can be elicited through electrical stimulation of the retina or electric or magnetic stimulation of the cortex. In the latter case, the greater the stimulation the more complex the experience; simple phosphenes can morph into wedge-shapes and butterfly patterns, as well as grid-like and hexagonal arrangements.[11]

For the second-path, seeing the *human aura* is a practical expression of this more complicated style of visual hallucination, frequently and mistakenly labelled as paranormal. The following story conveys parapsychologist Michaeleen Maher's own aura experience in 1969;

> I caught sight of the stranger who had just stepped out of the car. He was about ten or twelve feet away and was turning towards me. Suddenly, in a spontaneous flash of light, I witnessed a gorgeous, nearly indescribable

[8] Price (2016, p. 51).
[9] Parra (2015).
[10] Billock and Tsou (2012, p. 744).
[11] Ibid.

> phenomenon – something that I had never seen before. What seemed like literally millions of brilliant streamers of light broke out from the man's head and torso, shooting out perpendicular to the contours of his body. The circumference of the display of light was huge, with light streaming at least four or five feet from his body into the surround. The array was dazzling. There was light beyond light, and color beyond color – every hue, every shade imaginable. The gradations of color were so subtle, in fact, that I was certain I had never seen some of the colors before. Although every aspect of the spectrum appeared to be well represented, the whole array tended to give off a particular hue, with a preponderance of color in the purple or violet range. The radiant light was full of dynamic energy; it seemed very powerful alive. It moved, swirling and changing as the man oriented towards me. The streamers of light just kept flooding out unendingly from somewhere in the depths of his body.[12]

But more sinisterly;

> ...there was something beyond the beauty – something profound that seemed almost sinister. Uncannily, as if written in the explosive ripples of the very light itself, I saw the man's whole history – his like, his private thoughts and experiences, his destiny itself laid bare – and there was something ominous lurking at the core of it. It was an ineffable experience that afterwards I found very difficult to describe or explain. By way of analogy, it seemed as though I had stumbled across the man's private diary and learned some dark, terrible secret that I wasn't meant to know. And the essence of it, all of the information about the man's consciousness, and his life prospects and purposes, was coded in the colors of the light. It was as close as you could come to seeing a lifetime of consciousness quantized in brilliant streams of light....[13]

Michaeleen Maher would consider her experience more than a mere "hallucination", but the second-path is quite happy with the interpretation. In

[12]Maher (1999, pp. 52–53).

[13]Ibid.

agreement with Scott's notion of "derangement", psychiatric and neurological disorders are primarily held responsible.[14] And while there's absolutely no evidence that Maher's aura was the product of brain disorder, many other complex witnessed experiences are fair game for the hallucination label.

Apart from schizophrenia,[15] neurological drivers of complex visual hallucinations include stroke and neurodegenerative diseases such as *Parkinson's disease* and the clinically similar *Lewy body dementia*.[16] More rarely, visual hallucinations can be produced by auto-immune diseases, prion disease, diencephalic lesion and *delirium tremens*.[17]

Regarding disorders of the lower visual system, *Charles Bonnet Syndrome*[18] is perhaps the most celebrated (if that's an appropriate word to use). It is a consequence of macular degeneration; the deterioration of the macula (centre) of the visual field. As might be expected of a degenerative disorder, Charles Bonnet syndrome mostly occurs in the elderly populations and most likely to be reported when patients are immobile and in a darkened environment. These patients report visions of human faces, human figures and part figures (such as a torso without a head) as well as recognisable or bizarre animals, large and small. These vivid images can be stationary or they can be moving along with the background scene,[19] as the following story demonstrates;

> I looked up from where I was sitting across at the south facing window. And in that window, I saw some movement; activity. I could not make it out from where I was sitting but I thought it may be a wasp and went over to interfere with it if it was. It turned out to be an insect, but a long one: if you could imagine a dragonfly with wings at both ends. It was a little like that, which persuaded me that it was a phantom, a Bonnet product. Where we got the idea from, I do not know. Anyway, I went up to the window to get a closer look which is where I saw it had wings at each end,

[14] Kosman and Silbersweig (2018).
[15] Gao et al. (2018).
[16] Lin et al. (2018).
[17] Arnulf (2013).
[18] Mobbs and Watt (2011).
[19] Arnulf (2013).

> and as I did so, it flew towards me and, without thinking, I clapped on it. When I opened my hands, it fell out of them down on to the ground, not freely; quite slowly. And as it did so, it turned somehow towards me. I suppose it must have been a head but I did not see anything clearly like an individual head; not then. And when it came to rest eventually on the ground, it looked towards me rather, very sadly really. It made me feel quite guilty.[20]

Delusional parasitosis is occasionally accompanied by *seeing* the parasites as well as feeling them. This is an example of a *multimodal hallucination*, where two or more sensory modalities are represented (as opposed to unimodal, meaning the hallucination consist of only a single sensory modality).[21] Only a couple of cases have ever been documented, such as this one;

> A 74-year-old woman began to see insects and animals in her food and became very reluctant to eat a meal. She also started seeing insects under her skin and had the tactile sensation of their crawling, burrowing, and biting; this made her believe that because she ate a meal containing the insects' eggs, they had hatched in her body and moved beneath her skin. The patient frequently struck a desk with her arms to crush the insects under her skin, which eventually caused subcutaneous hemorrhages. She had a history of a left-sided meningioma, which had been surgically removed 17 years before, and a right putamen hemorrhage, which was surgically treated 5 years earlier....The patient also saw maggots, centipedes, very tiny horses, dogs, cats, turtles, and insects with human faces beneath the skin of her arms. She often saw many scarabs on the wall and a cat beside her bed, all of which were nonexistent.[22]

Multimodal hallucinations are more convincing to patients than unimodal experiences because seeing a person talk to you while feeling them touch you "has many of the hallmarks of actual, veridical experience".[23]

---

[20] Collerton et al. (2016, pp. 198–199).

[21] Dudley et al. (2018).

[22] Hirakawa et al. (2016, p. 10).

[23] Dudley et al. (2018).

Nevertheless, unimodal visual hallucinations of *very strange* entities can make an equally-lasting impression, as in the next two stories from patients suffering from Lewy body dementia;

> An 85-year-old, widowed, retired farmer from Western Ireland was referred to his local Psychiatry of Old Age service with a history of visual hallucinations over a period of months. The hallucinations consisted of seeing children and small people in his home that were silent and mischievous. The images were vivid. They could appear at any time of the day or night and were present on a daily basis. Despite his efforts to engage them in conversation, they never spoke. Similarly, they did not steal objects, nor did they touch chocolate he left out for them. He originally believed that they appeared after the death of his wife to force him to leave his home. Although he admitted that he had seen them dragging chairs up onto the roof, he denied believing that they could harm him. No other psychotic symptoms were present and his mood was good. He had no past psychiatric history or family history of psychiatric illness.... A diagnosis of Lewy body dementia was made.[24]

Similar hallucinated people were identified as *leprechauns* by an elderly female patient;

> An 82-year-old lady from Western Ireland was referred to our service with a history of visual hallucinations. These consisted of seeing people in her house whom she believed were leprechauns. Although these people did not communicate with her, she blamed them for stealing her money. At the time of her initial assessment, she had ... mild Parkinsonian symptoms suggestive of Lewy body dementia.[25]

Hallucinations that conform to a patient's field of vision have been labelled *content-specific delusion*, although the word "delusion" better refer to incorrect *ideas*, rather than incorrect perceptions. Irrespective of semantics;

---

[24] Foy (2011, pp. 362–363).
[25] Foy (2011, p. 363).

> These content-specific delusions fit perfectly into the experiencer's everyday life and go unnoticed because they are seamless with previous, and subsequent, non-hallucinated states of awareness.[26]

Most importantly, the content also behaves in a very sophisticated manner. The following story from a schizophrenic patient named Ruth is a detailed example of how hallucinated imagery can be "painted over" reality to merge with legitimate perceptual content. In Ruth's case, she persistently witnessed an apparition of her father who was still living (but residing in a distant place), and which resembled in "precise detail" his every feature, down to his clothes, features, movements and expressions. However these were as he *once* was, at a time when Ruth was familiar with him (she hadn't seen him in a long time). Ruth;

> …was able to make some adjustments, such as letting the hair grow longer, but there was nothing that would imply clairvoyance, no external knowledge that was not available to her mind, nothing that she could not have found in her own knowledge and memories of him.[27]

Moreover;

> The apparition appeared to blend perfectly with the setting in which it was seen. It would utilize the furniture that was there, move round objects, seem to lift articles from tables, and so forth.[28]

The apparition would open doors, show his reflection in mirrors, and behave as though he was listening to real people who were talking to Ruth (these people couldn't see the apparition). And he was not solely visual, since Ruth could hear him speak and make other noises, and even smell him and feel his draught when he opened a door.

Naturally, Ruth's physician concluded the father was a hallucination. This is confirmed by some clever experimentation; Ruth reported seeing the solid figure of her father moving between her and a light source, but

[26]Jinks (2016, p. 77).
[27]Evans (1984, p. 189).
[28]Ibid.

her pupils *remained* constricted due to the light (that is, they were not affected by an imaginary figure). And sitting in a dark room but believing the apparition had switched on a light, Ruth was incapable of reading a book (although she truly believed the room was fully illuminated).

Hallucinations are a powerful tool for the second-path. Scientific reality assumes a universe of material objects perceived accurately by the senses. When something is perceived that *doesn't* conform to what's allowable (an apparition isn't allowable), and isn't easily dismissed as an illusion, it must be a hallucination. That applies to *any* paranormal experience, whether it is the vision of a departed soul, an angel or any other extraordinary entity.[29] The second-path asks us to imagine the frozen figure of Tom, a door handle gripped tightly in one hand and a key-card in the other, staring blankly down the hallway. Deep in his frontal cortex, particular memory and judgment mechanisms are activated that overlay a complicated visual image onto an empty environmental scene, which results in a conscious hallucination.[30]

For neuroscience, the conception of Tom staring zombie-like into space as his frontal cortex concocts a ghost in the absence of sensory input is quite reasonable. It happened to Ruth and countless other patients who see extraordinary visions. As a neurological summary, complex visual hallucinations are said to derive from three possible types of disorder; (1) *irritative* (involving the accidental discharge of higher association cortices and this is falsely recognised as coming from an actual sensory stimulus—epilepsy is an example), (2) *release* (as in Charles Bonnet syndrome, where faulty visual input results in mistaken cortical interpretation) and (3) *faulty visual processing* (due to lesions creating incorrect cortical excitation patters from normal visual input; *Lhermitte's peduncular hallucinosis* is given as an example).[31]

[29]Mobbs and Watt (2011).

[30]Whether there is a phenomenological difference between seeing an actual man walking the corridor, and "seeing" the exact-same hallucination, is the subject matter of disjunctivism, but this is a philosophical issue beyond the scope of this book (e.g., Fish 2005; Haddock 2011).

[31]Cancelli et al. (2004).

Then again, Tom hasn't been diagnosed with epilepsy, a stroke, Parkinson's disease, Lewy body dementia, schizophrenia, Charles Bonnet Syndrome or something with the impressive name of "Lhermitte's peduncular hallucinosis". The time-course of such psychiatric hallucinations is also much longer than the typical apparition sighting (they can last for hours),[32] and *most* human-form hallucinations are relatively static[33] not moving like Tom's dark-suited man (so many of the examples I've supplied, like Ruth's father, are quite exceptional).

So while visual hallucinations *do* occur, and are frequently caused by brain dysfunction, it's long been appreciated that paranormal experiences are rarely of this nature. *The Census of Hallucinations* was a project conducted by the SPR in the early 1880s and examined sensory hallucinations during wakefulness; In Tyrrell's words "of the externalised kind which constitutes an apparition".[34] In this era there was recognition of the difference between hallucinated apparitions, which were visual, *and those of the insane*, which were primarily auditory.

Tyrrell argued that apparitions are a special type of hallucination corresponding closely to something legitimately external to the witness but which is not physically present in their local surrounds. These are distinct from the purely subjective hallucinations experienced by the mentally ill and drug-takers. In this context, "voices in the head" are true (insane) hallucinations while visual encounters with ghosts—the type of story the Society sought—were considered a special type of hallucination. In Tyrrell's opinion, these are *telepathic* hallucinations induced directly on the brain that could be independently verified.

For the second-path (and scientific reality), *The Census of Hallucination* is of historical interest only, and the concept of a telepathic hallucination is irrelevant. After all, modern neuroscience has determined that hallucinated imagery can emerge from processes distinct from telepathic induction (of which there is little or no supporting evidence). And yet it must be grudgingly acknowledged that Tyrrell and the SPR have a point; not all hallucinations have an origin in psychiatric or neurological disorder.

[32]Parra (2015).

[33]Collerton et al. (2005).

[34]Tyrell (1953).

A healthy brain can also generate hallucinatory experiences. This requires revisiting consciousness, but from the perspective of how it might *alter* to allow hallucinatory experiences. That's epitomised by the expression *altered states of consciousness*, a concept which becomes the focus of the next chapter.

References

Akinci, E., Oncu, F., & Topcular, B. (2016). Tactile Hallucination and Delusion Following Acute Stroke: A Case Report. *Journal of Psychiatry & Neurological Sciences, 29*(1), 79–84.

Arnulf, I. (2013). Dream Imagery, Rapid Eye Movement Sleep Behavior Disorder, and Hallucinations. *Sleep and Biological Rhythms, 11*(Suppl. 1), 15–20.

Billock, V. A., & Tsou, B. H. (2012). Elementary Visual Hallucinations and Their Relationships to Neural Pattern-Forming Mechanisms. *Psychological Bulletin, 138*(4), 744–774.

Cancelli, I., Marcon, G., & Balestrieri, M. (2004). Factors Associated with Complex Visual Hallucinations During Antidepressant Treatment. *Human Psychopharmacology: Clinical & Experimental, 19*(8), 577–584.

Choong, C., Hunter, M. D., & Woodruff, P. W. (2007). Auditory Hallucinations in Those Populations That Do Not Suffer from Schizophrenia. *Current Psychiatry Reports, 9*(3), 206–212.

Collerton, D., Perry, E., & McKeith, I. (2005). Why People See Things That Are Not There: A Novel Perception and Attention Deficit Model for Recurrent Complex Visual Hallucinations. *Behavioral and Brain Sciences, 28*(6), 737–757.

Collerton, D., Taylor, J.-P., Tsuda, I., Fujii, H., Nara, S., Aihara, K., et al. (2016). How Can We See Things That Are Not There? Current Insights into Complex Visual Hallucinations. *Journal of Consciousness Studies, 23*(7–8), 195–227.

Dowben, J. S., Kowalski, P. C., & Keltner, N. L. (2017). Formication, Tactile Hallucinations, Delusional Parasitosis, and Morgellons: Enough to Make Your Skin Crawl. *Perspectives on Psychiatric Care, 53,* 220–221.

Dudley, R., Aynsworth, C., Cheetham, R., McCarthy-Jones, S., & Collerton, D. (2018). Prevalence and Characteristics of Multi-modal Hallucinations in People with Psychosis Who Experience Visual Hallucinations. *Psychiatry Research, 269,* 25–30.

Evans, H. (1984). *Visions, Apparitions, Alien Visitors.* London: Book Club Associates.
Fish, W. (2005). Disjunctivism and Non-disjunctivism: Making Sense of the Debate. *Proceedings of the Aristotelian Society, 105*(1), 119–127.
Foy, K. (2011). Leprechauns and Lewy Body Disease. *Transcultural Psychiatry, 48*(3), 362–364.
Gao, X., Zhang, W., Yao, L., Xiao, Y., Liu, L., Liu, J., et al. (2018). Association Between Structural and Functional Brain Alterations in Drug-Free Patients with Schizophrenia: A Multimodal Meta-Analysis. *Journal of Psychiatry and Neuroscience, 43*(2), 131–142.
Haddock, A. (2011). The Disjunctive Conception of Perceiving. *Philosophical Explorations, 14*(1), 23–42.
Hirakawa, H., Terao, T., Kanehisa, M., Ninomiya, T., & Ishii, N. (2016). Coexistence of Delusional Parasitosis and Complex Visual Hallucinations with Micropsia. *The Journal of Neuropsychiatry and Clinical Neurosciences, 28*(1), e10–e12.
Inglis, B. (1985/1986). *The Paranormal: An Encyclopedia of Psychic Phenomena.* London: Paladin.
Jinks, A. (2016). *The Disappearing Object Phenomenon: An Investigation.* Jefferson, NC: McFarland.
Kosman, K. A., & Silbersweig, D. A. (2018). Pseudo-Charles Bonnet Syndrome with a Frontal Tumor: Visual Hallucinations, the Brain, and the Two-Hit Hypothesis. *The Journal of Neuropsychiatry and Clinical Neurosciences, 30*(1), 84–86.
Lin, W., Xie, Y.-C., Cheng, P.-Y., Dong, L.-Y., Hung, G.-U., & Chiu, P.-Y. (2018). Association of Visual Hallucinations with Very Mild Degenerative Dementia Due to Dementia with Lewy Bodies. *PLoS ONE, 13*(10), 1–8.
Maher, M. (1999). Riding the Waves in Search of the Particles: A Modern Study of Ghosts and Apparitions. *Journal of Parapsychology, 63*(1), 47–80.
Mobbs, D., & Watt, C. (2011). There Is Nothing Paranormal About Near-Death Experiences: How Neuroscience Can Explain Seeing Bright Lights, Meeting the Dead, or Being Convinced You Are One of Them. *Trends in Cognitive Sciences, 15*(10), 447–449.
Parra, A. (2015). Seeing Rare Things with the Mind's Eye: Visual Imagery Vividness and Paranormal/Anomalous Experiences. *Australian Journal of Parapsychology, 15*(1), 37–51.
Price, B. (2016). Hallucinations: Insights and Supportive First Care. *Nursing Standard, 30*(21), 49–58.
Tyrell, G. N. M. (1953/1963). *Apparitions.* New York: Collier Books.

17

Altered States of Consciousness

The expression *altered state of consciousness* was coined by psychiatrist Arnold Ludwig in 1966, who wrote;

> For the purpose of discussion, I shall regard "altered state(s) of consciousness" [hereafter referred to as ASC(s)] as any mental state(s), induced by various physiological, psychological, or pharmacological manoeuvres or agents, which can be recognized subjectively by the individual himself (or by an objective observer of the individual) as representing a sufficient deviation in subjective experience or psychological functioning from certain general norms for that individual during alert, waking consciousness.[1]

The individual receives a different (and incorrect) phenomenological experience of reality. For this reason, ASCs have more recently been labelled *altered states of phenomenology*[2] (although I'll stick with "ASC" for convenience). In an ASC there are "archaic modes of thought" (also known as *primary process thought*), and a shortfall in reality testing; people lose the

[1] Ludwig (1966, p. 225).

[2] Luke et al. (2012) and Rock and Krippner (2012).

T. Jinks, *Psychological Perspectives on Reality, Consciousness and Paranormal Experience*,
https://doi.org/10.1007/978-3-030-28902-7_17

ability to distinguish between what's real and what isn't, and what is cause and what is effect.[3]

Charles Tart's slightly later definition of an ASC contains useful information about what the state actually *feels like*. In Tart's opinion, ASCs involve;

> ...a qualitative alteration in the overall pattern of mental functioning, such that the experiencer feels his consciousness radically different from the way it functions ordinarily.[4]

This description can be applied to hallucinations stemming from mental illness, hence the two expressions (hallucination and ASC) are often used reciprocally. The advantage of *altered state of consciousness* is that it's a much broader concept that takes into account many non-pathological possibilities. For instance, deep relaxation is an example of a non-pathological state of consciousness that is *altered from the normal*.[5] And in the strictest sense ASCs are *temporary*, unlike psychiatric disease which can be permanent.[6]

Nonetheless there is an unavoidable connection between the hallucinations arising from mental illness and the more general term "ASC", as revealed by Hilary Evans;

> ...hallucinations result not from the illness per se, but from the altered state of consciousness which is induced by the illness[7]

In other words, ASCs produced by illness or brain damage cause pathological hallucinations while ASCs *not* produced by illness (e.g., deep relaxation) create unusual experiences which might also be labelled "hallucinatory". So when considering paranormal experiences like Tom's ghost or Michaeleen Maher's aura, if travelling along the second-path it is more

[3]Glicksohn (1990).
[4]Tart (1972, p. 1203).
[5]Naish (2005).
[6]Kokoszka (1999–2000).
[7]Evans (1984, p. 54).

correct (and courteous) to identify these reports as benign hallucinations stemming from altered consciousness, rather than something psychiatric.[8]

That said, the fact that quite healthy people can experience something akin to a "hallucination" that might take on human form (like a traditional ghost) is not evidence for dismissing the possibility of actual, autonomous "ghosts". This is a fallacy known as *affirming the consequent*, where the legitimate statement "hallucinations can *cause* ghost sightings" is unreasonably reversed and the resulting blanket proposition taken as equally correct; that ghost sightings are *caused* by hallucinations. That's because there are other options to what causes a ghost sighting…including a real ghost![9]

Fair enough, but surely that's being too generous to ghosts? In mainstream psychological literature it's common for *any* anomalous (or even religious) experience to be considered an ASC. Indeed it's common to see the words *paranormal*, *anomalous* and *religious* used interchangeably.[10]

So where is the second-path taking us? Well, in the fashion of deep relaxation, many varieties of ASC are sleep-related, so there's no need to consider paranormal experiences as psychotic! Apart from daydreaming (during the day of course), and sleep at night accompanied by dreaming,[11] there is the *hypnagogic state* (i.e. the transitional state between waking and sleeping) and the *hypnopompic state* (i.e. the transitional state between sleeping and waking). These states are favourite rationalisations for paranormal experiences,[12] particularly the hypnagogic state—which are a favourite "off the shelf" explanation for a ghost seen or its voice heard while nodding off to sleep—this is the favourite *scientific* explanation for haunted bedroom encounters.

Other non-pathological examples of ASCs include the experience induced by sensory deprivation, meditation (linked to relaxation), and of course ingestion of psychedelic drugs.[13] The latter are particularly relevant to the creation of a conscious space where paranormal experience is

---

[8]Parra (2006).

[9]Again, I'd like to thank an anonymous reviewer for reminding me of this straightforward concept.

[10]Moreira-Almeida and Lotufo-Neto (2017).

[11]Glicksohn (1990).

[12]Sherwood (2002).

[13]Blum (2013).

possible, namely reports of astral travel, NDEs, OBEs, meeting entities, mediumistic ability, psychic powers like telepathy and precognition, seeing the human aura, conversations with plants and animals ("interspecies communication") as well as abduction by aliens and even reincarnation memories.[14]

Whatever the potential cause, there's nothing in the term *altered state of consciousness* that requires the unusual perceptions to be inauthentic—and that includes Tom's ghost. "ASC" simply defines a transition from an ordinary state of consciousness to a non-ordinary one. This may allow legitimate veridical information to be perceived in the style of the first-path (the potential importance of "relaxation" in accessing paranormal information in psi was alluded to on page 68). On the other hand, a paranormal experience may be purely subjective (hallucinatory) in the style of the second-path. It is the *how*, *what* and *why* of an ASC that will determine whether the experience is authentic or not. From the perspective of the second-path, that begins with an understanding of the *symptoms* of an ASC, and what brain state might generate their manifestation.

And there are plenty of symptoms to consider. For example, whether the ASC is a daydream, a meditative state or an OBE, some shared characteristics have been noted. These include the distorted perception of sensory information and a reduction of attention to external stimuli coupled with a shift toward observing internally generated stimuli (physiological states and imaginal scenes). Emotions might also operate quite different, such as a heightened sense of empathy, or personal sensation of joy and love.[15]

These similarities (and potential differences) have been systematically mapped by in the 1970s by psychologist Adolph Dittrich, who collected data on the invariants of anomalous experience (having assumed these were the consequence of ASCs) in order to develop a standardized instrument for assessing so-called ASC symptoms. In German this is the *Aussergewohnliche Psychische Zustande* (APZ) or translated into English, the *Altered States of Consciousness questionnaire*.[16]

---

[14]Luke (2011).

[15]Tart (1986) and Modestino (2016).

[16]Kokoszka (1999–2000).

Based on the analysis of participants' responses the instrument identifies three broad categories of ASC symptom. The first is *oceanic boundlessness* (OBN). This is a pleasurable experience of "self-dissolution" or disintegration where the individual feels disconnected from the stable boundaries of their ego, and they are able to achieve what they might describe as unity with, but also transcendence from, space and time. Emotions are positive (ecstatic) and profound knowledge is believed to be gained. This is the facet of an ASC most associated with mystical and spiritual experiences, astral travel, NDEs and OBEs.[17]

The second experiential facet of an ASC is *anxious ego-dissolution* (AED) (or for a more dramatic translation from the German, "dread of ego dissolution"[18]). This represents the negative consequences arising ASC-induced self-dissolution. Rather than OBN bliss, there is anxiety stemming from a loss of control over the body and its behaviour, also known as *ego disintegration.*[19]

AED represents the "bad trip" aspect of an ASC than *may* be induced by psychedelic drugs, or a hellish NDE where the traveller experiences disturbing scenes in the afterlife. AED is also evocative of schizophrenic symptoms such as paranoia, thought disorder and loss of control of thoughts.[20] This encourages the idea that ASC experiences are just as inauthentic as true schizophrenic hallucinations.

The third general symptom of the ASC is *visionary restructuralization* (VIR). Instead of the positive or negative impression contextualised by ORN and AED, the VIR facet describes how what is *visually* seen in an ASC is different to what is normally seen in an ordinary state of consciousness. Synaesthesia is one symptom, as is the propensity to interpret visual scenes and/or objects in an alternative way. That's the ASC equivalent of an illusion—a shadow becomes a man. The visions could be memories and imaginations endowed with abnormally vivid characteristics, such as stronger colours or textures, leading to an impression of objectively reality. On the other hand, a vision could possess completely altered features;

[17] With exceptions, such as negative NDEs.

[18] Kokoszka's translation (1999/2000).

[19] Liechti et al. (2017).

[20] Kokoszka (1999–2000).

micropsias (objects smaller than normal) and the rarer *macropsias* (objects larger than normal) are examples of unusual visual restructures—classified under the broader dysfunctional label *Metamorphopsia* (meaning a deviation in visual patterns). These have established pathological origins, so the symptoms of a visual disorder like Charles Bonnet Syndrome is branded an expression of VIR.[21] Another symptom of VIR is witnessing real objects (including living things) behaving in normal manners but which cannot be independently verified by other percipients. This leaves some to speculate that these correspond to the "visionary experiences" (i.e., hallucinations) extolled by seers and religious figures.[22]

While not able to *explain* why an ASC produces strange visual experiences, if the report is brief, illogical and unexpected—like Tom's ghost—then the VIR facet of an ASC at least provides an initial classification, helping pin down why the altered state took the form it did. Loosely speaking, VIR describes the invariant features of *any* visual paranormal experience. An experience that's primarily VIR might be interpreted as a ghost sighting. However the addition of OBN symptoms, on top of VIR, will become Nadia's astral journal, and if AED features are present, a disagreeable alien abduction.[23]

Having determined the commonality of *primary* APZ dimensions with paranormal events, the next step is to link each of the symptoms to repeatable and temporary changes in cortical state. That allows some closure on the *hows* and *whys* of a paranormal experience. The most detailed research linking phenomenological experience and brain science is from the field of psychopharmacology. Seeing an apparition can be a consequence of psychedelic drug use,[24] and while it's unlikely Tom had ingested mescaline (or something else) during his stay at the hotel, the fact a ghost sighting are able to be induced give credence to the view that what he saw wasn't a deceased person's "energy" but a hallucination produced by an altered state. Tom might not have been on drugs that afternoon, but knowledge

[21] Midena and Vujosevic (2015).

[22] Kokoszka (1999/2000).

[23] In addition to these "Big 3" categories of ASC symptom are two secondary dimensions labelled acoustic alterations and vigilance reduction, and below these two are eleven more homogenous lower-order factors which are really of only specialist interest (Studerus et al. 2011).

[24] Luke (2011).

of how drugs can produce ghost sightings might lead to an understanding of how Tom saw his (hallucinated) ghost.

In the second-path of scientific reality, psychedelic drugs don't act in remarkable ways to shift filters or dial tuners, allowing actual discernment of cryptic environmental information or other-dimensional realms. Instead, through normal biological mechanisms they produce the same changes in cognition, mood and perception that people encounter in everyday dreams, religious exaltation,[25] or when they have flashes of vivid involuntary memory. That's why neologisms like *hallucinogens* and *psychotomimetics* were created; the drugs facilitate so-called "altered states of consciousness" (really, hallucinations and psychoses[26]), which are purely subjective and have no basis in reality.

Interest in the experiential effects of psychedelic drugs like mescaline, LSD, DMT and the psilocybin mushroom has a long history in literature and early psychology. The use of psilocybin and mescaline is humorously discussed in the following passage;

> Hallucinogenic plants first came under the modern spotlight in the mid-1800s as botanists and ethnologists expanded their knowledge. Lewis Carroll's *Alice in Wonderland* (1865), was written by an author who was familiar with English botanist Mordecai Cooke's early text (1860/2012) on psychoactive mushrooms and their effects. Psychiatrist and romance writer S. Weir Mitchell, who penned an account of peyote in 1896, gave some peyote buttons to the godfather of psychology, William James. British psychologist Havelock Ellis wrote articles about his own experiences with peyote, noting in particular 'the more delicate phenomena of light and shade and color' (Ellis, 1898). Ellis, in turn, passed some buttons[27] to William Butler Yeats, but Yeats later reported that he preferred hashish.[28]

Surprisingly, marijuana (unlike opiates and cocaine) has also been shown to have psychedelic effects,[29] specifically the delta-9-tetrahydrocannabinol

[25]Luke et al. (2012, p. 98).

[26]Koethe et al. (2009).

[27]The buttons are what grows after the head of the peyote is cut off and a plate is left—these are removed and dried and consumed in various ways e.g., chewed or powdered.

[28]Hansen (2014, p. 680).

[29]Luke et al. (2012).

(Δ9-THC) ingredient which can produce schizophrenic symptoms in the form of psychotic reactions and cognitive disruption.[30]

Different drugs possess different molecular structures and consequently provoke different effects on the CNS (and the resulting emergent or epiphenomenal stages of consciousness). Therefore, while all psychoactive drugs encourage ASCs, the resulting states will possess divergent characteristics.[31] Currently there's only superficial understanding of how drug type effect phenomenological symptoms,[32] specifically what (alleged) paranormal experience results from the various categories of psychoactive chemical. Very broadly, anthropological investigation of psychedelic drug use by traditional cultures[33] has determined that most psychoactive plants are able to patients suffering from spiritual disease, however there are specifics; contact with the spirit world, and spiritual enlightenment, requires ingestion of psilocybin, mescaline, ibogaine or ayahuasca[34] Ibogaine and ayahuasca also help with gaining advice from the ancestors. Apparently, gods and goddesses can be consulted using artemisia absinthium and ibogaine, while religious and philosophical understanding is best achieved using psilocybin, mescaline and Artemisia absinthium.

Outside of traditional cultures, the anaesthetic ketamine also has a psychedelic reputation;

> In sub-anesthetic doses, it produces a hallucinogenic state in humans that is again in a class of its own and that has been described by some as a state in which mentality takes on an abstract geometric spatial quality that morphs in concert with thoughts, feelings, and perceptions.[35]

---

[30] Koethe et al. (2009).

[31] Sayin (2014).

[32] Luke (2011).

[33] Sayin (2014): From "African native religions; Bwiti Cultures; South American native religions; Amazon Cultures; Central American Cultures; Mexican subcultures; Aztec, Maya and Inca; Wiccan and witch subcultures; Satanists; American Indians; Greek and Hellenistic cultures; Sufis; Hassan Sabbah's Hashissins; Hindu, Indian and Tibetan cultures [through to] some of the Nordic subcultures".

[34] I'll shortly discuss more about ayahuasca.

[35] Presti (2011, p. 36).

And regarding bizarre encounters with non-human creatures, psilocybin mushrooms and straight DMT are apparently the best way to meet with those reptilian (alien) entities, as narrated in this story from Fran;

> I witnessed a giant snake looking close at me, teaching me infinite understanding and charging my soul battery. I also gave birth to a strange thing, a machine made of wood and fabric that would only walk if the wind blew.[36]

In this realm you can also meet insect entities in the form of praying mantises.[37] Here's a detailed but darker story involving an encounter with these sentient, other-dimensional insects;

> …I was camping with three friends at Island Lake near Nevada City, California. A couple of us decided to take "heroic" doses of Psilocybin mushrooms one evening. I chewed down four grams, retired to my tent, closed my eyes, and got horizontal on my air mattress. As the effects of the 'shrooms came on, my inner vision revealed what looked like a dank moss-green hospital emergency waiting room. I seemed to be sitting on a bench in this room, and it occurred to me that it was odd that there were no patients being wheeled in or out. Kinda quiet for an ER. After some time, I noticed a few off-white football-sized larvae floating three or four feet off the ground in various spots. Following one of these with my eyes, I then saw an insectoid entity about the size of a small dog, whose back was facing me. It had a long mosquito-like proboscis that I could only partially see. Suddenly, it turned, and--realizing that I saw it--it made a high-pitched buzzing/shrieking sound. (I got the impression that it was sending out a warning alarm.) The entity then initiated telepathic communication with me, explaining that it was quite surprised that I could see it, as this usually didn't occur. It said that it lived by extracting human thought/emotion. Human thoughts were both the currency of its species, as well as their sustenance/energy source. (The needle-like proboscis was looking less friendly by the minute.) I was given the impression that--as the coin of its kind--different types of thought/emotion were valued differently; those with a more intense energy charge, such as fear or love, were worth more. The entity explained that it existed in another dimension so

[36] https://www.acidmath.me/blogs/acidmath-digest/encounters-with-entities.
[37] Luke (2008).

that it could feed off of human thought unhindered. (I got a feeling that the relationship wasn't symbiotic; perhaps these "thought drainers" somehow suck life energy from humans, along with the mental energy.) It claimed that it was the psychic equivalent of an actual insect that feeds on blood, skin, etc., with regard to the extent of any damage it might do to those on whom it fed. Yet I had a nagging feeling that it might not be telling me the whole truth. Maybe these creatures had some influence on inciting wars or disasters in the human realm? The experience left me feeling unsettled for some time afterwards.[38]

References

Blum, H. (2013). Dissociation and Its Disorders. *Psychoanalytic Inquiry, 33*, 427–438.

Evans, H. (1984). *Visions, Apparitions, Alien Visitors*. London: Book Club Associates.

Glicksohn, J. (1990). Belief in the Paranormal and Subjective Paranormal Experience. *Personality and Individual Differences, 11*(7), 675–683.

Hansen, D. (2014). Heavenly and Hellish—Writers on Hallucinogens. *The Psychologist, 27*(9), 680–682.

Koethe, D., Hoyer, C., & Leweke, F. M. (2009). The Endocannabinoid System as a Target for Modelling Psychosis. *Psychopharmacology, 206*(4), 551–561.

Kokoszka, A. (1999–2000). Altered States of Consciousness: A Comparison of Profoundly and Superficially Altered States. *Imagination, Cognition and Personality, 19*(2), 165–184.

Liechti, M. E., Dolder, P. C., & Schmid, Y. (2017). Alterations of Consciousness and Mystical-Type Experiences After Acute LSD in Humans. *Psychopharmacology, 234*(9–10), 1499–1510.

Ludwig, A. M. (1966, March 4–6). Possession States in Primitive People, from the Symposium Sponsored by R. M. Bucke Society, Montreal. *Archives of General Psychiatry, 15*.

[38]https://www.erowid.org/chemicals/dmt/dmt_article3.shtml.

Luke, D. (2008). Inner Paths to Outer Space: Journeys to Alien Worlds Through Psychedelics and Other Spiritual Technologies (Book Review). *Journal of Scientific Exploration, 22*(4), 564–569.

Luke, D. (2011). Anomalous Phenomena, Psi, and Altered Consciousness. In E. Cardeña & M. Winkelman (Eds.), *Altering Consciousness: Multidisciplinary Perspectives: History, Culture, and the Humanities; Biological and Psychological Perspectives* (Vol. 2, pp. 355–374). Santa Barbara: Praeger.

Luke, D., Zychowicz, K., Richterova, O., Tjurina, I., & Polonnikova, J. (2012). A Sideways Look at the Neurobiology of Psi: Precognition and Circadian Rhythms. *Neuroquantology, 10*(3), 580–590.

Midena, E., & Vujosevic, S. (2015). Metamorphopsia: An Overlooked Visual Symptom. *Ophthalmic Research, 55*(1), 26–36.

Modestino, E. J. (2016). Neurophenomenology of an Altered State of Consciousness: An fMRI Case Study. *Explore: The Journal of Science & Healing, 12*(2), 128–135.

Moreira-Almeida, A., & Lotufo-Neto, F. (2017). Methodological Guidelines to Investigate Altered States of Consciousness and Anomalous Experiences. *International Review of Psychiatry, 29*(3), 283–292.

Naish, P. (2005). Detecting Hypnotically Altered States of Consciousness. *Contemporary Hypnosis, 22*(1), 24–30.

Parra, A. (2006). 'Seeing and Feeling Ghosts': Absorption, Fantasy Proneness, and Healthy Schizotypy as Predictors of Crisis Apparition Experiences. *Journal of Parapsychology, 70*(2), 357–372.

Presti, D. E. (2011). Neurochemistry and Altered Consciousness. In E. Cardeña & M. Winkelman (Eds.), *Altering Consciousness: Multidisciplinary Perspectives: History, Culture, and the Humanities; Biological and Psychological Perspectives* (Vol. 2, pp. 21–41). Santa Barbara: Praeger.

Rock, A. J., & Krippner, S. (2012). States of Consciousness or States of Phenomenology? In A. E. Cavanna & A. Nani (Eds.), *Consciousness: States, Mechanisms and Disorders* (pp. 55–65). Hauppauge: Nova Science Publishers.

Sayin, H. U. (2014). The Consumption of Psychoactive Plants During Religious Rituals: The Roots of Common Symbols and Figures in Religions and Myths. *Neuroquantology, 12*(2), 276–296.

Sherwood, S. J. (2002). Relationship Between the Hypnogogic/Hypnopompic State and Reports of Anomalous Experiences. *Journal of Parapsychology, 66* (2), 127–150.

Studerus, E., Kometer, M., Hasler, F., & Vollenweider, F. X. (2011). Acute, Subacute and Long-Term Subjective Effects of Psilocybin in Healthy Humans:

A Pooled Analysis of Experimental Studies. *Journal of Psychopharmacology, 25*(11), 1434–1452.

Tart, C. T. (1972). States of Consciousness and State-Specific Sciences. *Science, 176*(4040), 1203–1210.

Tart, C. T. (1986). Consciousness, Altered States, and Worlds of Experience. *Journal of Transpersonal Psychology, 18*(2), 159–170.

18

The *How* and *Why* of Psychedelics

So *how* and *why* do psychoactive drugs work to produce exceptional experience, like the story about meeting reptilians and insectoids told at the end of the last chapter? Unfortunately, the neurochemistry is rather complex. Take the following condensed summary as an example;

> Their basic pharmacology is diverse and includes complex agonist and partial agonist/antagonist actions on 5HT2A, 5HT2C, 5HT1A, dopamine D_2, trace amine associate receptors 1 (TAAR), kappa receptors, various transporters (e.g., serotonergic, dopaminergic, norepinephrine), intracellular messengers, effects on gene expression and epigenetic regulators.[1]

Concentrating the more well-studied pathways, psychedelics like *Lysergic acid diethylamide* (LSD), psilocin, *N,N*-Dimethyltryptamine (DMT) and mescaline are agonists (i.e., they facilitate) receptors for serotonin, particularly in cortical pyramidal cells.[2] These are more correctly

[1] Sellers et al. (2018, p. 116).

[2] Vollenweider and Kometer (2010).

T. Jinks, *Psychological Perspectives on Reality, Consciousness and Paranormal Experience*,
https://doi.org/10.1007/978-3-030-28902-7_18

termed 5-HT2A receptors, the 5-HT referring to serotonin's chemical name 5-hydroxytryptamine, and the "2" variety tend to be favoured by psychedelics.

However subtle variations exist in the way psychedelics act on 5-HT2A receptors. For instance, different psychedelics affect the receptor-attached second messenger pathways within neurons in distinct ways (when activated, these pathways allow the serotonergic neuron to "fire" its action potential). Psychedelics also act more broadly on *other* serotonergic receptors to varying degrees. For instance, DMT also stimulates serotonergic 1, 6 and 7 receptors, while LSD excites dopaminergic D1 and D2 receptors.[3] That is why it's not yet possible to strictly match each drug's particular biological operation with their phenomenological outcome.

Regarding DMT, "straight" oral ingestion or inhalation is not effective because the drug is denatured by monoamine oxidase (MAO) acting in the digestive tract[4] and pulmonary vasculature. This enzyme breaks off the amine group from the DMT molecule, thereby rendering it docile. Instead, DMT is most commonly consumed in a special mixture—as ayahuasca "plant tea", or smoked as bufotenine.[5] Both concoctions possess ingredients that act to inhibit MAO. Ayahuasca is commonly made from the leaves of *Psychotria viridis* which provide the DMT component. However supplementary ingredients are equally as important, such as addition of the malpighiaceous vine[6] *Banisteriopsis caapi* which contains β-carboline alkaloids including harmine, harmaline and tetrahydroharmine. These act as shielding ingredients that protect DMT from the rigours of MAO, and allowing it to enter circulation.[7] And that's not all the harmines can do. Although they are secondary ingredients, they can have their own distinct psychedelic properties. For instance, tetrahydro*harmine* also acts as a serotonin agonist in the fashion of DMT (although it's a reuptake inhibitor rather than acting directly on the receptor),[8] and acts to expand the altered state of *experience* following ayahuasca consumption.

[3] Preller and Vollenweider (2018).

[4] Luke et al. (2012).

[5] Blainey (2015).

[6] A family of shrub found in the Americas.

[7] Halberstadt et al. (2012).

[8] Valle et al. (2016).

There's a popular first-path belief that pure DMT creates a "technological"-themed trip when successfully ingested (e.g., in an experimental setting). This characterised by reference to psychedelic researcher Terrence McKenna's infamous "self-transforming machine elves", which are proposed to be one (rigid) expression of a particular setting on the perceptual filter/tuner mechanism. On the other hand, the addition of MAO-inhibiting ingredients such as the harmines will turn the tuner to a different, time-extended setting and create a more "rich, luxurious and organic-style" trip[9] involving plant entities and reptilian or insect-like creatures.

Not so for the second-path. It's simply a lucky coincidence that a traditional brew made from the ayahuasca vine already has this built in MAO protection device. And from a functionalist point of view, that's why ayahuasca just happens to "work" and create a vivid Altered States of Consciousness (ASC) while countless other brews derived from countless other vines don't work. And as to the differences in experience between pure DMT and ayahuasca, that's due to the "extra" impact from secondary constituents like harmaline. The resulting state of consciousness will be altered and the phenomenological experience will vary as a direct response to the novel pattern of activation created across serotonergic (and other) receptor types.

Hence;

> ...it may well be that such differences are relevant to understanding the different subjective signatures of the various classical psychedelics. As more is learned about the molecular actions of these substances, it may be increasingly possible to formulate hypotheses as to how neurochemical differences are related to mental-state effects.[10]

The answer lies in the connection between the serotonin system and how hallucinations are generated. Although primarily a gastrointestinal neurotransmitter, serotonin's lesser role in the brain is popularly regarded as a facilitator of mood—a rather vague concept encapsulating "happiness"

[9]Hancock (2005, p. 433).

[10]Mishor et al. (2011, p. 104).

and "wellbeing". This idea is informed by the link between the 5-HT gene and affective states,[11] and the efficacy of drugs acting as serotonergic enhancers (specifically, serotonergic reuptake inhibitors) to treat anxiety and depression.[12] However 5-HT2A receptor activation by LSD, psilocybin, mescaline and ayahuasca have been shown to provoke *visual* hallucinations, and when drugs that have the opposite actions are administered to patients suffering from Parkinson's disease psychosis (such as the 5 inverse agonist Pimavanserin) these hallucinations are attenuated.[13]

The most suitable explanation for serotonergic hallucinations involves a mismatch of bottom-up and top-down central nervous system (CNS) processing, better known as the *information integration theory of consciousness*.[14] Parkinson's patients exhibit an *over*activity of the serotonergic system(s) in prefrontal regions—something that psychedelic drug agonists mimic. This overactivity reduces inhibition in various other prefrontal "cognitive" regions in close anatomical proximity. These have imposing names like the middle, superior and inferior frontal gyrus, the middle temporal gyrus, the pre-supplementary motor area, the anterior cingulate cortex and the putamen. The reduction of inhibition corresponds to *over*-excitation of these top-down cortical regions, which prevents them from integrating bottom-up perceptual information into seamless visual perception. The resulting conscious experience, and the feelings that are produced because of them, will be "aberrant".[15]

The impact of psychedelic serotonin agonists on these same cortical information processing systems will further prevent the gating or control of internal and external neural inputs. This prevents the brain from selectively attending to (therefore limiting) the information entering consciousness, and it becomes "swamped". Put in neurophysiological language;

[11]Taylor (2011).

[12]Blum et al. (2000).

[13]Schmidt et al. (2018).

[14]Bartocci (2004).

[15]Geyer and Vollenweider (2008).

> ...thalamic gating deficits...result in sensory overload with excessive processing of exteroceptive and interoceptive stimuli leading to a collapse of integrative cortical functions.[16]

Overwhelmed by both external and internal material, cognitive processing and ego-functioning breaks down, and the end-product is hallucination.[17] Indeed, psychedelic experiences produced by an agent such as psilocin are not due to increased brain activity (measured as heightened blood flow), but overall *decreased* activity—particularly in the prefrontal cortex. There is an "uncoupling" of communication between the prefrontal and cingulate cortices which is necessary for ego function and introspection, so the result is a "state of unconstrained cognition"[18] which might appear novel and anomalous to the percipient, but is merely a product of temporary neurophysiological malfunction.

Linking these experiences to the categories of the ASC developed by Adolph Dittrich, a shift from external reality to internally-derived material will generate sensations matching those of oceanic boundlessness (OBN) and *anxious ego-dissolution* (AED). The following sensations reported by this drug-user link the effects of the psychedelic with Dittrich's symptoms;

> I felt like a marionette, I felt isolated from everything and everyone, I was not able to complete a thought, my thought repeatedly became disconnected, I felt as though I were paralyzed.[19]

And with the disruption of external visual coherency, there will be greater reliance on internally-sought imagery corresponding to the visionary restructuralization (VIR) dimension. The same patient;

> ...saw regular patterns in complete darkness or with closed eyes.[20]

---

[16] Vollenweider and Geyer (2001, p. 496).

[17] Bartocci (2004).

[18] Vollenweider and Geyer (2001).

[19] Schmidt et al. (2018, p. 1468).

[20] Schmidt et al. (2018, p. 1468).

How might this come about? Alongside frontal regions, psychedelic drugs have also been shown to act on the visual cortex, an area with abundant 5-HT2A receptors. Electroencephalography (EEG) measurements across the occipital and parietal cortices following ayahuasca and psilocin consumption show a reduction of alpha wave power. Alpha waves are the most robust brain signal detected through EEG, emanating from the occipital lobes during relaxation and with the eyes closed (although the subject is still awake). They are believed to represent inhibition of the visual cortex[21] (or an indication that the cortex is merely in baseline operation), so when alpha rhythms are *reduced* there is a concurrent increase in brain metabolism and blood flow measured using PET and fMRI.[22] That's accompanied subjectively by reports of visual phenomena, which explains psychedelic hallucinations. This is taken as confirmation that psychedelics suppress inhibitory alpha frequencies in the visual cortex. The consequent *arousal* creates the visionary phenomena experienced when there are no external cues present in the environment.[23]

This leads to the second-path conclusion that if psychedelic drugs can accidently generate the various facets of an ASC, then there's nothing paranormal about altered states. For instance, *near death experiences* (NDEs) are argued to be the product of a similar style of neural "hyperfrontality"[24] imitated by the actions of psilocybin, mescaline, DMT and ketamine. These drugs increase glucose metabolism in the frontal and occipital cortices [25] and this corresponds with greater OBN scores (whereas AED is associated with thalamic overactivity and orbitofrontal underactivity). Bear in mind the OBN dimension conforms very closely to NDE reports.

Ketamine can also create the positive emotional traits (and visions) reported during alleged NDEs;

---

[21]Valle et al. (2016).

[22]Schmidt et al. (2018).

[23]Valle et al. (2016, p. 1171).

[24]Mobbs and Watt (2011) rail against "(Moody 1975) and other bestseller books have largely omitted discussion of any physiological basis for these experiences, and instead appear to prefer paranormal explanations over and above scientific enlightenment", hence ignore case studies where false inductions account for individual elements of NDEs using neurophysiology".

[25]Vollenweider and Geyer (2001).

> ...through evoking the rewarding properties of the opioid system and misattribution of events due to disruption of the prefrontal cortex. Such processes may occur naturally and similar systems are evoked when animals are under extreme danger. For example, dopamine and opioid systems become active when an animal is under predatory attack. Thus, these endogenous systems come into play during highly traumatic events and have likely evolved to aid in the survival of the organism.[26]

Ketamine also acts as an *N-methyl-D-aspartate (NMDA) receptor antagonist.*[27] Glutamate is an excitatory neurotransmitter that combines with a variety of receptors including the NDMA-type, named because it is favoured by the amino acid derivative of glutamate, NMDA. Ketamine's action in blocking (antagonising) the NDMA receptor and affecting glutamine function has a similar hallucinogenic affect to the outcome of psychedelics acting on the serotonergic system, which suggests they are pharmacologically related. For instance, there's plenty of evidence that serotonergic psychedelics also increase glutamine activity following stimulation of those 5-HT2A receptors positioned on pyramidal cells in the prefrontal cortex.[28]

Although the first and second-paths are very different with how they tackle paranormal experience, there's overlap when it comes to hallucinations; the drug model shares similarities to the perceptual filter/tuner. After all, the concept of "unconstrained cognition" fits nicely with the idea of psi as a *perceived* experience, whereby psychedelics permit access to knowledge of everything that has happened or is happening, everywhere.[29] Psychiatrist and neuroscientist John Smythies proposed that because cortical glutamate is constantly active, even during sleep,[30] its ability to continually trigger NMDA receptors is what prevents the brain from accidently entering altered states—specifically those resembling NDEs. In this sense,

---

[26]Mobbs and Watt (2011, p. 449).

[27]Reissig et al. (2012); Dextromethorphan (DXM) is a widely available over-the counter cough suppressant (e.g., Robitussin®, Coricidin®) that acts like ketamine.

[28]Vollenweider and Kometer (2010).

[29]Luke et al. (2012).

[30]Compared with more "volatile" noradrenaline, dopamine and serotonin.

the glutamate/NDMA receptor axis is the respectable scientific equivalent to Bergson's mechanism, preventing spontaneous anomalous experiences.

> Therefore, NDEs occur when the NMDA receptors are inactivated brain-wide, either by ketamine or some...ketamine-like substance, or by the cessation of neuronal activity during the natural ND [near death] state.[31]

In other words, NMDA antagonists like ketamine, and serotonergic psychedelics impacting impact on the glutaminergic system, disrupt the filter. In the first-path this is the neurological cause of authentic paranormal experiences.[32] For second-path scientific reality, the unconstrained cognition following the breakdown of a sensory/information filter doesn't imply any *actual* anomalous information, just a lot of overwhelming and temporarily forgotten material. Smythies himself argues that the experiences produced by psychedelics derive from personal imagery, accumulated during a lifetime, rather than from anywhere mystical.[33]

So putting aside the authenticity of the experience, why is DMT—an *exogenous* chemical—able to influence 5-HT receptors in the first place? Is there a *natural* version of DMT in the body? To get to the answer takes quite a journey. In 1951 Smythies and his colleague Humphry Osmond[34] developed the *transmethylation hypothesis*. Smythies noted the strong positive relationship between schizophrenia and stress. Stress (and anxiety) increase adrenaline levels in the body, and adrenaline is chemically similar to mescaline. He was also aware that brain dopamine could be "methylated", which involves the addition of a methyl (-CH3) group to an existing chemical. When this occurs dimethylphenylamine is formed, and this new constituent is psychosis-producing (i.e., it's a psychotogen).[35] By the same argument, if there were a dysfunction of the adrenaline-producing

[31]Smythies (2011, p. 149).
[32]Luke et al. (2012).
[33]Smythies (2011).
[34]Also Huxley's friend.
[35]Mills (2010).

adrenal glands, any adrenaline that is manufactured might also "accidently" become methylated, resulting in a molecule closely resembling mescaline. The actions of this endogenous mescaline in the CNS was argues as the driver of schizophrenic hallucinations.

The transmethylation hypothesis has been expanded in more recent years to include the (also accidental) methylation of compounds other than adrenaline, such as endogenous indolealkylamines which (surprise!) can include N, N-di*methyl*tryptamine—DMT. And rather than act directly on serotonergic receptors like the *drug* DMT, this home-grown version is instead subsumed by the serotonergic neurons that project through the cortex and re-released as a serotonin-mimicking neurotransmitter. So the phenomenological (psychotic) experiences of DMT consumed in an Ayahuasca brew, and schizophrenia-induced DMT, will ultimately be similar.[36] Stretching the idea even further, if the secretion of melatonin from the pineal gland controls *sleep*, the release of endogenous DMT as a derivate of melatonin controls *dreams*.[37]

Schizophrenic symptoms can also arise from cannabis use. The cannabis sativa plant contains a psychoactive constituent delta-9-tetrahydrocannabinol (Δ9-THC), [38] and as its dose increases there is substantial change in phenomenological symptoms. Intoxication occurs at low amounts, but consumption increased signs pass from arousal, amiability and euphoria to anxiety, panic and paranoia and ultimately psychosis. Psychotic experiences comprise the "Big 3" dimensions and sub-categories of Dittrich's Altered States of Consciousness questionnaire's (APZ's), including shifts in body image and ego-dissolution, how time and space is perceived and of course visual and auditory hallucinations.[39] And like natural substances with DMT as a constituent, cannabis contains additional natural ingredients that aren't themselves psychedelic (the phytocannabinoids such as cannabinol and cannabidiol) but which modulate the ASC produced by Δ9-THC ingestion.

---

[36]Grammenos and Barker (2015).

[37]Luke et al. (2012).

[38]Koethe et al. (2009).

[39]Preller and Vollerwielder (2018).

Endogenous *endocannabinoid receptors* activated by Δ9-THC are (controversially) suspected of altering whole-brain "neural synchrony", producing psychotic, schizophrenic-like symptoms. The specifics of this synchrony may relate to the dopamine hypothesis, which causally links irregular increases in frontal dopamine with the manifestation of schizophrenia.[40] As a small component in this process, the actions of "home-grown" cannabinoid-like molecules on endocannabinoid receptors might be a *good* thing; they dampen any overactivity occurring in these frontal dopaminergic pathways and lower the probability of psychosis.[41] Should Δ9-THC be consumed and the synchrony disturbed, the probability of psychosis is increased.

From this brief background of drug neuropharmacology, a second-path treatment of ASCs, hallucinations and paranormal experience can be advanced. It goes like this; the brain has ways of producing visual impressions during ordinary consciousness. These scenes are a product of complex, top-down neural pathways which rely on the predictable combination of many different categories of neurotransmitter, including serotonin and dopamine, with a wide variety of associated receptor types. If there's something chronically wrong with the workings of these systems—for instance a condition labelled "schizophrenia"—then psychosis is likely. There's nothing special about the imagery that's *perceived* in psychosis. The bottom-up sensory elements experienced in healthy waking consciousness are jumbled due to an "inability to integrate", and what the patient eventually receives is equivalent to the dimensions described in Dittrich's APZ; "oceanic boundlessness", ego-dissolution" and "visionary restructuralization". Extrinsic drugs that imitate natural neurotransmitter action will cause similar but acute imbalances to the same image-generating cortical systems. The outcome will be comparable (but brief) hallucinations.

The first-path conviction that drugs like mescaline or DMT alter hypothetical filters and tuners, allowing perception of an expanded universe, is simply too farfetched. The second-path's application of Occam's razor

---

[40] For example, Howes and Kapur (2009).

[41] Koethe et al. (2009).

allows for a much more rational and sensible solution to explain how subjective visions arise from such unusual states of consciousness—a solution supported by decades of empirical evidence.

References

Bartocci, G. (2004). Transcendence Techniques and Psychobiological Mechanisms Underlying Religious Experience. *Mental Health, Religion & Culture, 7*(2), 171–181.

Blainey, M. (2015). Forbidden Therapies: Santo Daime, Ayahuasca, and the Prohibition of Entheogens in Western Society. *Journal of Religion and Health, 54*(1), 287–302.

Blum, H., De Boer, J. A., Bosker, F. J., & Slaap, B. R. (2000). Serotonergic Drugs in the Treatment of Depressive and Anxiety Disorders. *Human Psychopharmacology: Clinical and Experimental, 15*(5), 315–336.

Geyer, M. A., & Vollenweider, F. X. (2008). Serotonin Research: Contributions to Understanding Psychoses. *Trends in Pharmacological Sciences, 29,* 445–453.

Grammenos, D., & Barker, S. (2015). On the Transmethylation Hypothesis: Stress, N, N-Dimethyltryptamine, and Positive Symptoms of Psychosis. *Journal of Neural Transmission, 122*(6), 733–739.

Halberstadt, A. L., Nichols, D. E., & Geyer, M. A. (2012). Behavioral Effects of α, α, β, β Tetradeutero-5-MeO-DMT in Rats: Comparison with-MeO-DMT Administered in Combination with a Monoamine Oxidase Inhibitor. *Psychopharmacology (Berl), 221*(4), 709–718.

Hancock, G. (2005). *Supernatural: Meetings with the Ancient Teachers of Mankind.* London: Century.

Howes, O. D., & Kapur, S. (2009). The Dopamine Hypothesis of Schizophrenia: Version III—The Final Common Pathway. *Schizophrenia Bulletin, 35*(3), 549–562.

Koethe, D., Hoyer, C., & Leweke, F. M. (2009). The Endocannabinoid System as a Target for Modelling Psychosis. *Psychopharmacology, 206*(4), 551–561.

Luke, D., Zychowicz, K., Richterova, O., Tjurina, I., & Polonnikova, J. (2012). A Sideways Look at the Neurobiology of Psi: Precognition and Circadian Rhythms. *Neuroquantology, 10*(3), 580–590.

Mills, J. A. (2010). Hallucinogens as Hard Science: The Adrenochrome Hypothesis for the Biogenesis of Schizophrenia. *History of Psychology, 13*(2), 178–195.

Mishor, Z., McKenna, D. J., & Callaway, J. C. (2011). DMT and Human Consciousness. In E. Cardeña & M. Winkelman (Eds.), *Altering Consciousness: Multidisciplinary Perspectives: History, Culture, and the Humanities; Biological and Psychological Perspectives* (Vol. 2, pp. 85–119). Santa Barbara: Praeger.

Mobbs, D., & Watt, C. (2011). There Is Nothing Paranormal About Near-Death Experiences: How Neuroscience Can Explain Seeing Bright Lights, Meeting the Dead, or Being Convinced You Are One of Them. *Trends in Cognitive Sciences, 15*(10), 447–449.

Moody, R. A. (1975/2001).*Life After Life: The Investigation of a Phenomenon—Survival of Bodily Death.* New York: HarperOne (Original work published 1975).

Preller, K. H., & Vollenweider, F. X. (2018). Phenomenology, Structure, and Dynamic of Psychedelic States. *Current Topics in Behavioral Neurosciences, 36,* 221–256.

Reissig, C., Carter, L., Johnson, M., Mintzer, M., Klinedinst, M., & Griffiths, R. (2012). High Doses of Dextromethorphan, an NMDA Antagonist, Produce Effects Similar to Classic Hallucinogens. *Psychopharmacology, 223*(1), 1–15.

Schmidt, A., Müller, F., Lenz, C., Dolder, P. C., Schmid, Y., Zanchi, D., et al. (2018). Acute LSD Effects on Response Inhibition Neural Networks. *Psycho logical Medicine, 48*(9), 1464–1473.

Sellers, E. M., Romach, M. K., & Leiderman, D. B. (2018). Studies with Psychedelic Drugs in Human Volunteers. *Neuropharmacology, 142,* 116–134.

Smythies, J. (2011). Ketamine, Bergson and NDEs. *Journal of the Society for Psychical Research, 75*(904), 148–150.

Taylor, W. D. (2011). Yes, There Is a "happiness" Gene … but it's Not That Simple; the 5-HTT Gene That Transports Serotonin in the Brain Can Lead to Feelings of Satisfaction in Life. *Duke Medicine Health News, 17*(7), 7.

Valle, M., Maqueda, A. E., Rabella, M., Rodríguez-Pujadas, A., Antonijoan, M., Romero, S., et al. (2016). Inhibition of Alpha Oscillations Through Serotonin-2A Receptor Activation Underlies the Visual Effects of Ayahuasca in Humans. *European Neuropsychopharmacology, 26* (7), 1161–1175.

Vollenweider, F. X., & Geyer, M. A. (2001). A Systems Model of Altered Consciousness: Integrating Natural and Drug-Induced Psychoses. *Brain Research Bulletin, 56,* 495–507.

Vollenweider, F. X., & Kometer, M. (2010). The Neurobiology of psychedelic Drugs: Implications for the Treatment of Mood Disorders. *Nature Reviews Neuroscience, 11,* 642–651.

19

Global Changes to the Altered State

This rational and sensible conclusion arises from the difficulty attributing Tom's vision to a psychiatric disorder, or the influence of psychedelic drugs. He wasn't under the influence of even a headache tablet (or so he says). For this reason, it's disingenuous to identify the cause of a "typical" ghost sighting as mescaline or ayahuasca ingestion. And if schizophrenia is ruled out (with regards Tom at least), *endogenous* N,N-dimethyltryptamine (DMT) (or some other psychedelic neurotransmitter) cannot be held responsible either. There needs to be another way to induce transient disturbances in visual processing that will generate *detailed, witnessed* paranormal encounters; external (or maybe internal) influences that tap into the same image-forming pathways of the cortex, altering consciousness and creating a similar mental disturbance to those reported by schizophrenics or psychedelic drug-users.

Various external inputs have been *behaviourally* documented that create psychedelic effects. In other words, when people indulge in certain behaviour they report experiences similar to the symptoms of Altered states of consciousness (ASCs), including visual hallucinations. A well-documented example is repetitive *noise*, in the form of percussive effects

T. Jinks, *Psychological Perspectives on Reality, Consciousness and Paranormal Experience*,
https://doi.org/10.1007/978-3-030-28902-7_19

from drumming or monotonous prayer or chanting.[1] ASCs can also be induced through the *absence* of input—typified by sensory deprivation facilitated by an enclosed flotation tank,[2] but also fasting and solitude[3];

> …going to one's special place…a riverbank, a desert, a beach, a garden, a mountain, or a pine forest.[4]

Additional but contentious *self*-induction techniques include meditation, "self-hypnosis" and "autogenic training".

When discussing ASC induction techniques that have "traditional", "cultural" or "esoteric" connotations (as opposed to an illness or recreational drug origin), the popular label is *shamanism*. One has to be careful with this word, however, since it is apparently too narrow to describe what is considered a complex phenomenon that also represents the practices of the "medicine man", diviner, witch doctor, medium and healer in very different contexts.[5] It is argued that since the common features among these occupations are only vaguely understood, the word shaman should be applied only to Eurasian and subarctic examples. Instead, *magico-religious practitioner*[6] is a better expression describing an individual belonging to a pre-agrarian tribe, clan or broader culture who has been identified as skilled in spiritual journeying.[7] This variously comes about through a hereditary lineage or some peculiarity in their demeanour or behaviour[8] (nevertheless "shaman" is a much easier word to use than magico-religious practitioner, so I'll stick with it).

Briefly, the shaman gains their exceptional knowledge through two pathways—the advice of existing shaman and/or through a special dream

[1] Lehmann et al. (2001).
[2] Jonsson et al. (2014).
[3] Rioux (1996).
[4] Ibid., p. 62.
[5] Krippner (2002).
[6] Winkelman (1990).
[7] Jinks (2016).
[8] Johnson (2003).

or visionary trance.[9] From the second pathway, the alterations in consciousness that are sought can be generated by both drug and non-drug taking behaviours.[10] The latter include those self-induction techniques like rhythmic drumming and austerities (fasting and starvation, isolation and sleep deprivation) to enter higher or lower (underworld) realms for the purpose of learning "timeless mysteries and secrets" which are of use to their communities on the shaman's return.[11] Examples of gained skills include healing and medicinal wisdom[12] and perhaps the "awakening" of psychic abilities (mediumship).[13] The shaman is frequently guided in their journey by their ancestors[14] or other non-human entities, and from that perspective there are certain parallels between a shamanic experience and *near death experiences* (NDEs), astral journeying, and many alien abduction accounts. That leads to the possibility of a collective neurological "capacity" to enter ASCs, in turn explaining the consistencies of experience across cultures.[15] From there it's a short leap to the popular idea that modern alien abduction is simply the expression of a spontaneous, shamanistic-like ASC. The difference between abductions and shamanism is the *reason* for the experience rather than the method used. Shamans don't indulge in random recreational drug-induced "trips", or spontaneously enter an altered state with no understanding of why it has occurred. People who haven't the cultural support and training of a shaman, but incidentally undertake the same practices, can generate comparable (but to them frightening) consequences.

An instance of a shaman-like practice leading to lucid visions of entities is told by the very sceptical author Michael Shermer. As Shermer tells the story, he was highly fatigued during an endurance bike and encountered aliens;

---

[9] Hume (1995).

[10] Lehmann et al. (2001).

[11] Winkelman (1990).

[12] Eliade (1972) (translated by Trask).

[13] Ring (1989).

[14] Rioux (1996).

[15] Hume (1995); For example, several comments made by those experimenting with a posture called The Olmec Prince, reported feelings of physical transformation from human to animal.

> Early one August morning whilst riding in Nebraska he spied a large craft with bright lights. It overtook him and forced him off the road, before alien beings emerged from the craft and abducted him for 90 minutes. Next thing he knew, he was riding back along the road with no memory of what happened during that time. He recounted the incident to a film crew not long afterwards, citing it as a perfect example of a complex hallucination brought about by the psychological disturbance of sleep deprivation.[16]

Luckily for Shermer, his scepticism provided him with the precise second-path answer; the event was caused by a brain chemical imbalance brought about by sleep deprivation—described in cognitive terms as the attenuation of those executive control regions in the prefrontal and parietal cortex which hold in place the *ordinary* state of consciousness and provide the sense of self to waking consciousness.[17] This is the same process that occurs during the stage of entering sleep and the beginning of dreaming. The fatigue-generated ASC is therefore identical to a hypnogogic state, where bizarre and unfamiliar imagery is released from a brain memory library[18] and interpreted by the *contemporary* individual as paranormal. Without shamanic training, the novice will report whatever cultural interpretation they can think of—grey space aliens being one of the most prevalent.[19]

Tom hasn't brain damage or dementia, but did he see a ghost because he was fatigued? Or very, *very* hungry? Or was he listening to percussive music through his headphones (perhaps be forgot to mention this trivial fact)? Or did he just fall into a spontaneous (and temporary) ASC based on a combination of these ingredients and others unknown. It really doesn't matter, because the second-path has a smoking gun to explain the ghost in the corridor. People like Tom travel the first-path and report phenomenological experiences that they label paranormal. Other people who also report highly unusual encounters (with aliens or ghosts or any number of other things), but they travel the second-path. They *don't* assume what they see (or sense in general) is paranormal because they are aware

[16]Shermer (2005, p. 34).
[17]Fosse et al. (2004).
[18]Jinks (2016).
[19]McNally and Clancy (2005).

of concurrent, physically unusual circumstances that distort waking consciousness. They know they've taken *Lysergic acid diethylamide* (LSD) or they know they're exhausted and hungry, hence might be open to false experiences. And enough is known of *how* and *why* these states are produced (garnered from research into drug effects on forebrain substrates involving serotonin and other neurotransmitters) to be confident that a comparable process gives rise to the *hows* and *whys* of everyday paranormal sightings—even if it's not possible to pin down the initiator of every incident that's reported, including Tom's. People like Tom aren't able to arrive at the same second-path conclusion because they lack an appreciation of scientific reality.

Examined a different way, the second-path's treatment of consciousness maps ASCs along a rough perception-hallucination continuum.[20] The mild stuff includes those transient conditions associated with sleeping and daydreaming. More serious psychedelic drug consumption progresses the ASC towards ego-dissolution, based on a complex interaction amongst dosage, route of administration, personality traits, the type of imagery that's released, and both suggestibility and the expectancy of the drug-taker.[21] Physical austerities can result in the same mid-range effects. And at the far end of the spectrum, serious chronic and psychotic disorders such as schizophrenia or degenerative dementia correspond to profound phenomenological outcomes. These causes might all be distinct form one another, however their actions are grounded in the same central nervous system (CNS) substrates, leading to comparable impacts on consciousness.

When there's topographical "overlap" between or amongst these various causes there will be shared invariant symptoms identified by scales including the Altered States of Consciousness questionnaire (APZ)—so daydreaming and schizophrenia can possess commonalities, like the visionary restructuralization (VIR) dimension. There's nothing mystical about this *model psychosis*, where disorders of thought, perceptual disturbances and in extreme cases hallucinations can be experimentally or recreationally induced.[22] That is, the symptoms of alleged paranormal and mystical

[20]This isn't necessarily correct, as I'll discuss later with Tart's discrete states.

[21]Metzinger (2009).

[22]Koethe et al. (2009).

experiences can be "artificially" generated by stress,[23] psychedelics, or in a laboratory setting through sensory deprivation[24] (e.g., Ganzfeld and psi abilities),[25] and by this argument the results equally apply to ecstatic shamanic, mystical or religious experiences.[26] Case closed, according to the second-path.

Perhaps the only thing missing is a stronger typology of *experience*—not just the symptoms mapped by psychological instruments such as the APZ. Why do some ASCs result in seeing ghosts, and others alien abductions, if there are shared origins? For the second-path, conscious states derive from brain activity and are fully explained by brain activity, so it's logical to take the correlational evidence relating serotonin and alpha waves (and everything else) to extraordinary experiences, and conclude that changes in occipital serotonin and alpha waves cause oceanic boundlessness (OBN), *anxious ego-dissolution* (AED), VIR (and all the other symptoms) in an altered state. But what are the mechanisms underlying this restructuralization? Luckily the second-path is comfortable discussing states of consciousness independently of neurophysiology, so it can provide answers to why the various experiences emerge.

Running with the notion of an elementary conscious state that people "possess" and which is comprised of *elements*, there's a useful metaphor which describes consciousness as a container with objects or materials positioned inside. Allowing some artistic licence, whole consciousness can be thought of as a *snow-globe*—a glass (or plastic) container within which the elements of reality are held in place. These elements of consciousness make up phenomenological experience and represented by the countless tiny white flakes that flutter around inside the globe. This implies two aspects of functional consciousness—the *container* and the *constituents*. The container and the constituents aren't necessarily different things, as they're not actually "things" like snow-globes or artificial snow flakes. However for the purpose of this discussion they can be discussed as separate independent mental entities.

23 Irwin (2018).

24 Kelly (1978).

25 Parker et al. (1997).

26 Wright (1994).

Alterations in the underlying brain state, such as acute or chronic changes in serotonin activity, will affect the container (the snow-globe) and/or the constituents (the flakes) and this is reflected in the various ASC symptoms. Supplementing the "Big 3" treatment of ASCs, the snow-globe and snow flake metaphor provides commentary on how and why ASCs can be experientially different, which in turn gives greater clarity on the corresponding paranormal experiences.

The snow-globe itself represents the elementary nature of global consciousness[27] and the individual's sense of embodied self. Nothing exists outside of the snow-globe's boundaries that is recognisable or meaningful. If this directly experienced state of being is translated by the individual as *unusual* (i.e., the shape of the container "feels" distorted) then the experience conforms to an ASC. Such a change in global state might be described by this story from Mitchell;

> I was 12 years old and at school. Everything appeared to be in its right place when I was sitting in the classroom. All the students were in the room and the teacher was out front, and the desks and the posters on the wall and the door were where they always are. But it didn't feel right. It didn't feel like it was real, and it could all fall apart in an instant. It seemed held together, but deep down I knew it wasn't, and I was a bit scared about what would happen if it did fall apart, and what I'd see behind the classroom if you get what I mean. I also had this horrible sense of dread. like I was going to die or it was the end of the world. This type of thing happened to me a few times when I was young, and later on I learned that this might have been a type of epileptic seizure that only happened when I was growing up. I've never had them again.[28]

Mitchell is describing an alteration in his *fundamental milieu*, which is difficult for him to describe other than "it doesn't feel right", or to put words in his mouth, the glass snow-globe that represents his consciousness had distorted to the point of shattering (accounting for his description of falling apart). All Mitchell could do was admit that he felt scared despite

[27] Polito et al. (2010).

[28] Mitchell, personal communication, February 2004.

everything around him looked perfectly normal. The constituents of his reality were unaltered, but his container was *ab*normal.

He ultimately decided on a medical attribution that is entirely appropriate when travelling the second-path. After all, it's well established that *temporal lobe epilepsy* can produce ASCs.[29] Temporal lobe epilepsy (TLE) is characterised by seizures focussed in the cortical region beneath the temples (the lateral temporal lobes) and limbic areas underlying these (the medial temporal lobes) which have a role in memory storage and retrieval.[30] Sufferers frequently report non-visual sensory hallucinations like phantom smells and tastes,[31] but in terms of Mitchell's experience of a global change of state, one TLE characteristics is of particular interest, involving the disruption of *vestibular* processing.

The vestibular cortex integrates information about bodily acceleration arriving from the inner ear's detection organs, with sensory information coming from elsewhere—like visual movement. This allows the maintenance of steady eye position (known as *ocular reflexes*) and sustained postural balance and gait.[32] It is clinical practice for patients with TLE to have the activity of their temporal lobes monitored with recording electrodes to identify the location of seizures. If these electrodes have stimulating capacity, those same areas can receive directed current and the phenomenological experience narrated by a vigilant patient. The right angular gyrus is proximal to the vestibular cortex, and low currents over this gyrus in the order of 2–3 milliamps (mA) lead patients to report "sinking feelings"—presumably due to artificial activation of vestibular neural pathways. When current is increased over 3mA the patient can experience an external-centred, rather than viewer-centred model of "self" where they "see" and "feel" themselves in a disembodied state.[33] The sense of presence is also described.[34] For instance, following low current experimental stimulation a patient reports;

[29] Wright (1994).

[30] Maillard et al. (2004).

[31] Since the temporal lobes are distant from the occipital lobes, visual hallucinations are quite rare.

[32] Gu (2018).

[33] Bensley (2003).

[34] Mobbs and Watt (2011).

> I see myself lying in bed, from above, but I only see my legs and lower trunk.[35]

And when current is increased even further, this same patient reports this newly disconnected sense-of-self floating above the bed and touching the ceiling. Stimulation (or abnormal activation) of the temporal lobes' right superior temporal gyrus also impacts the vestibular cortex, and engenders *mental diplopia*. In this state, a patient feels as though they're in two different places at the same time.[36] That's the elegant second-path solution for most reports of out-of-body experience (OBE).

Interference with a person's vestibular integration therefore affects how they perceive their global conscious state, represented as the normal sense of embodiment. While Mitchell's narrative doesn't conform mental diplopia brought about by a temporal lobe seizure (or experimental induction of seizure activity in this region), however Tom's ghost *might* be diplopia; what he saw in the corridor was his own self from a different perspective! That's a tough case to argue as there was no accompanying sense of alteration in his sense-of-self or any evidence to suggest he suffered from TLE, but for the second-path it's a more rational solution than a discarnate spirit.

Returning to Mitchell's experience, a more careful explanation would be that he suffered from *dissociation*, in that he felt *dissociated* from his normal global state (or to trivialise his experience, he no longer feels like his normal, snow-globe self!). From the standpoint of scientific reality, the "qualia" of the body or the sense of being contained "within" the container of a body is manufactured by complicated neurophysiological processes.[37] When these processes are disrupted the experience is one of dissociation. The vestibular symptoms of TLE are only one example of how dissociation can arise—there are many other causes, some of which will be covered in the upcoming chapter.

---

[35] Blanke et al. (2002, p. 269).

[36] Tong (2003).

[37] Rabeyron and Caussie (2016).

References

Bensley, D. A. (2003). Can Minds Leave Bodies? *Skeptical Inquirer, 27*(4), 34–40.

Blanke, O., Ortigue, S., Landis, T., & Seeck, M. (2002). Stimulating Illusory Own-Body Perceptions. *Nature, 419,* 269–270.

Eliade, M. (1972). *Shamanism: Archaic Techniques of Ecstasy.* Princeton, NJ: Princeton University Press.

Fosse, R., Stickgold, R., & Hobson, A. (2004). Thinking and Hallucinating: Reciprocal Changes in Sleep. *Psychophysiology, 41,* 298–305.

Gu, Y. (2018). Vestibular Signals in Primate Cortex for Self-Motion Perception. *Current Opinion in Neurobiology, 52,* 10–17.

Hume, L. (1995). By Knot of One the Spell's Begun, By Knot of Two it Cometh True… Magic, Altered States, and Multiple Realities. *Social Alternatives, 14*(4), 23–27.

Irwin, H. J. (2018). Stress Sensitivity and Minimal-Self Dysfunction as Predictors of Anomalous Experiences and Paranormal Attribution. *Journal of the Society for Psychical Research, 82*(1), 1–14.

Jinks, A. (2016). *The Disappearing Object Phenomenon: An Investigation.* Jefferson, NC: McFarland.

Johnson, T. (2003). Shamans, Medicine Men, and Other Healers. *Native American Medicine, 2003,* 20–32.

Jonsson, K., Grim, K., & Kjellgren, A. (2014). Do Highly Sensitive Persons Experience More Nonordinary States of Consciousness During Sensory Isolation? *Social Behavior and Personality, 42*(9), 1495–1506.

Kelly, M. T. (1978). Sensory Deprivation and Research on Telepathy. *Parapsychology Review, 9*(4), 24–26.

Koethe, D., Hoyer, C., & Leweke, F. M. (2009). The Endocannabinoid System as a Target for Modelling Psychosis. *Psychopharmacology, 206*(4), 551–561.

Krippner, S. C. (2002). Conflicting Perspectives on Shamans and Shamanism: Points and Counterpoints. *American Psychologist, 57*(11), 3–66.

Lehmann, D., Faber, P. L., Achermann, P., Jeanmonod, D., Gianotti, L. R., & Pizzagalli, D. (2001). Brain Sources of EEG Gamma Frequency During Volitionally Meditation-Induced, Altered States of Consciousness, and Experience of the Self. *Psychiatry Research, 108*(2), 111–121.

Maillard, L., Vignal, J.-P., Gavaret, M., Guye, M., Biraben, A., McGonigal, A., et al. (2004). Semiologic and Electrophysiologic Correlations in Temporal Lobe Seizure Subtypes. *Epilepsia, 45*(12), 1590–1599.

McNally, R. J., & Clancy, S. A. (2005). Sleep Paralysis, Sexual Abuse and Space Alien Abduction. *Transcultural Psychiatry, 42*(1), 113–122.
Metzinger, T. (2009). Why Are Out-of-Body Experiences Interesting for Philosophers? The Theoretical Relevance of OBE Research. *Cortex, 45*(2), 256–258.
Mobbs, D., & Watt, C. (2011). There Is Nothing Paranormal About Near-Death Experiences: How Neuroscience Can Explain Seeing Bright Lights, Meeting the Dead, or Being Convinced You Are One of Them. *Trends in Cognitive Sciences, 15*(10), 447–449.
Parker, A., Frederiksen, A., & Johansson, H. (1997). Towards Specifying the Recipe for Success with the Ganzfeld: Replication of the Ganzfeld Findings Using a Manual Ganzfeld with Subjects Reporting Prior Paranormal Experiences. *European Journal of Parapsychology, 13,* 15–27.
Polito, V., Langdon, R., & Brown, J. (2010). The Experience of Altered States of Consciousness in Shamanic Ritual: The Role of Pre-existing Beliefs and Affective Factors. *Consciousness and Cognition: An International Journal, 19*(4), 918–925.
Rabeyron, T., & Caussie, S. (2016). Clinical Aspects of Out-of-Body Experiences: Trauma, Reflexivity and Symbolisation. *L'Évolution Psychiatrique, 81*(4), e53–e71.
Ring, K. (1989). Near-Death and UFO Encounters as Shamanic Initiations. *ReVision, 11*(3), 1–22.
Rioux, D. (1996). Shamanic Healing Techniques: Toward Holistic Addiction Counseling. *Alcoholism Treatment Quarterly, 14*(1), 59–69.
Shermer, M. (2005). Abducted. *Scientific American, 292*(2), 34.
Tong, F. (2003). Out-of-body Experiences: From Penfield to Present. *Trends in Cognitive Science, 7*(3), 104–106.
Winkelman, M. J. (1990). Shamans and Other 'Magico-Religious' Healers: A Cross-Cultural Study of Their Origins, Nature, and Social Transformations. *Ethos, 18*(3), 308–352.
Wright, P. A. (1994). A Psychobiological Approach to Shamanic Altered States of Consciousness. *ReVision, 16*(4), 164–172.

20
Dissociation

For the purpose of understanding how dissociation can generate paranormal experiences, it helps that Mitchell is self-aware of the change in his global condition since he's able to provide a verbal summary of what it feels like to be in a dissociated state (if indeed he *has* undergone dissociation), just as self-awareness helps to characterise what *ordinary* consciousness feels like (qualia).

Dissociation is a form of ASC involving a *lessening* of contact with the personal body (dubbed *asomatic consciousness*), body ownership and selfhood.[1] This arises from aberrations integrating all the multisensory information (e.g., visual, tactile, somatosensory, and vestibular modalities)[2] normally required for maintaining the stable sense of embodied self.[3] Using more technical language, there is disruption of the *neuromatrix*, a massively intricate network of interconnected neurons (and other

[1] Braithwaite et al. (2017).

[2] Rabeyron and Caussie (2016, p. 58).

[3] Braithwaite et al. (2017).

T. Jinks, *Psychological Perspectives on Reality, Consciousness and Paranormal Experience*,
https://doi.org/10.1007/978-3-030-28902-7_20

support cells) which assesses afferent proprioceptive data and creates stable body image.[4]

In its most benign form, pretend play and, daydreaming are dissociative states,[5] and since these are quite normal behaviours, dissociation is not necessarily pathological.[6] *Hypnosis* is another example of a mild dissociative ASC with strong links to daydreaming. It is generally defined as a form of extreme relaxation where an individual's focus of attention shifts from external environmental stimuli to internal material, like private memories.[7] Hypnosis is also linked to paranormal experiences, since people in deep hypnosis sometimes report impressions that resemble symptoms of NDEs and OBEs. This includes a sense of euphoria, floating out of body, and the perception of a bright light.[8] However there's an accusation that hypnosis creates false memories (pseudomemories)[9] and this potential distortion increases as a function of desire to recall, while simultaneously giving the hypnotic subject the wrong impression that these pseudomemories are highly accurate. Therefore, the NDE and OBE experiences are attributed to erroneous recall of mildly unusual internal sensations gleaned while the subject is in a state of deep relaxation.

Furthermore, it is argued that subjects cannot "do things" or "see things" under hypnosis that are beyond their normal abilities, and the hypnotic state doesn't provide these people with unique therapeutic powers or allow them to alter the actions of their autonomic nervous system (the part of the CNS not under voluntary control). One extreme view is that hypnosis is merely an attempt by the subject to give the *appearance* of being in a trance.[10] From that perspective hypnosis isn't special; it indicates extreme conformity rather than being an *actual* ASC—a view that's nonetheless hotly contested.[11]

---

[4]Rabeyron and Caussie (2016, p. 58).

[5]Castillo (2003).

[6]Moreira-Almeidaa and Lotufo-Neto (2017).

[7]Rogo (1979).

[8]Cardeña and Krippner (2010).

[9]Lynn et al. (1991) and Wagstaff et al. (2008).

[10]Spanos (1996).

[11]Naish (2005); who says that hypnosis is closely linked to changes in the reality monitoring process.

Dissociation can also be induced by psilocybin, DMT and ketamine, and encouraged by stress and post-traumatic stress disorder, schizophrenia, brain lesions and TLE,[12] personality traits of dissociation as well as merely possessing the "happy schizotypal" profile characterised by a marked tendency to have anomalous experiences *not* considered pathological.[13]

More benignly, dissociative states can also be provoked in the absence of sensory information,[14] or when the CNS is confronted with conflicting information. For instance, sensory deprivation can cause the data-starved brain to create a different interpretation of the self, based on whatever incomplete or incorrect environmental clues it can draw on. That might be interpreted in consciousness as a general impression of "weirdness", although the ASC that's produced is mild—such as the state achieved during periods in a those floatation tanks I mentioned previously (called REST or *restricted environmental stimulation technique*). This consists of;

> ...weightlessness, altered perception of time...changes in body perception, and emotional expression [with] attention and thoughts ... more directed towards mindfulness or the intuitively and visually "here-and-now" state.[15]

The dissociation caused by conflicting sensory information is shown in this simple exercise that frequently results in an uncomfortable alteration in a subject's somatic boundaries—the early stages of a global alteration of the sense-of-self. Two people sit behind one another on chairs, close enough so that the person in the rear—the subject of the demonstration—can touch the person in the front's nose with their right index finger. The subject closes their eyes. A third person stands to the side of the sitting subject and strokes the subject's nose in a continual but irregular pattern while holding their right hand by the wrist (index finger outstretched) and performing the same stroking pattern on the front person's nose. Sound confusing? If you were the subject with eyes closed this is what you would experience; the information you are primarily receiving is somatosensory; the feeling of being *touched*, stoked and tapped on the side of the right

[12]Rabeyron and Caussie (2016, p. 58).

[13]Ibid.

[14]Ibid.

[15]Jonsson et al. (2014, p. 1496).

nostril. However there's a concurrent feeling of you also *touching* a right-side nostril at an outstretched, arm's-length distance ahead. Although you are aware this is no more than an amusing demonstration, without visual cues to override the touch sensations you will begin to experience an unconscious mental conflict. Your top-down brain receives information that the nose is being stroked, but simultaneously you are felt to be stroking something resembling a nose with your index finger at a distance of about 2 feet further into space. A few seconds might pass, and there's a good chance you'll let out a gasp. When asked why, you exclaim that the impression is that your nose feels two-foot long in space! Once you open your eyes, or the stroking pattern is mismatched between the two nostrils, the "illusion" will immediately disappear.[16]

Strictly speaking, your transient two-foot long nose *is* an illusion, since sensory information is being misinterpreted and a new (but incorrect) judgment made. Yet there is a hallucinatory quality to the demonstration, because a sensation of something that *doesn't exist* has been produced. This will impact on how consciousness construes the physical body in space. It doesn't take much effort to provide enough conflicting data to the brain to upset the "normal" sense of self and produce a simple but effective illustration of self-dissociation.

Dissociative states like Mitchell's are more serious, as there is a;

> …loss of self-regulation, self-cohesion, and self-continuity. The familiar sense of self is lacking and the change is usually experienced as distressful.[17]

These match the debilitating features of an ASC described by questionnaires such as Dittrich's APZ, for example the self-dissolution, alteration of ego-boundaries and heightened mood of OBN couple with the negative themes of AED; disintegration of coherent self, the separation of self from the world, loss of self-control over autonomy, feelings of estrangement and anxiety or panic. However dissociation can also be measured using tailored scales, like the 28 item self-report *Dissociative Experiences*

---

[16]E.g., Ramachandran and Blakeslee (1998).

[17]Blum (2013, p. 428).

Scale.[18] Should a high score on these instruments be coupled with traits like absorption, magical ideation and "unconventional" (paranormal and anomalous) *beliefs*, then a propensity to report fantasy paranormal *experiences* is to be expected.

Linking the outcome to the cause, a transient quirk of Mitchell's developing teenaged brain found him unable to effectively integrate somatosensory (and other environmental) information in the classroom, and as a consequence his global state was only just being sustained by an over-worked nervous system attempting to prevent his embodied self "falling apart". That accounts for at least some of the expressions he used to describe the condition.

Had a different pattern of dysfunction struck Mitchell's CNS, he might have reported an alternative style of global breakdown; a *heautoscopic hallucination*. These hallucinations differ from mental diplopia in that the patient witnesses their own body double without a sense of embodiment in *either* form. They are also more severe than the mildly dissociative sense of presence where the body-double is only *sensed* rather than observed visually.[19]

Luckily for Mitchell he avoided the most severe expressions of dissociation—a complete dissipation of the embodied self and a complete loss of self-identity (extending the analogy, his snow-globe is shattered into countless pieces). These are true pathological conditions where self-awareness is absent. Such disorders certainly don't apply to Mitchell's classroom oddity, and certainly not Tom's ghost sighting, but they are a (controversial) style of ASC that have strong paranormal attributions.

Dissociative states lacking awareness have been recognised as far back as the late 1800s by Pierre Janet,[20] and when profound and dysfunctional they are termed dissociative *disorders*. Labelling a dissociative disorder as another type of ASC is somewhat controversial; while dissociative disorders are characterised by some as altered (and divided) states of consciousness,[21]

[18] Bernstein and Putnam (1986).

[19] Braithwaite et al. (2017).

[20] Bühler and Heim (2011).

[21] Blum (2013).

they lack the temporary and non-psychiatric features that many claim are the basis of true ASCs.[22]

It's not possible to contemplate how it actually *feels* to be in this completely dissociative condition. Unlike Mitchell's narrative, there's no first-hand account of the subject's phenomenological experience in the absence of self-awareness (assuming it's even possible to speak of experience without awareness). All that's quantifiable are variations in the subject's overt behaviour—mannerisms and expressions that are open to independent observation.

The component of complete identity loss is found in the *dissociative fugue*, classed as alterations in self-consciousness, emotional processing and executive (cognitive) function[23] that results in a loss of personal memory and impaired recall that's far more significant than mere forgetfulness.[24] Often in the case of fugue, the loss of personal identity is accompanied by "sudden enigmatic travel" as in this example;

> The person at the center of this report is a 54-year-old married woman with two adult children who was seen through a university psychology clinic. Ms. X's initial appearance at a mental health service was shortly after she experienced the sudden onset of retrograde amnesia for the previous 33 years *and for her own current identity [emphasis added]*. She reportedly was unable to recall memories from the age of 21 years. She was found by police driving near a railway crossing in the rural community in which she lived, apparently searching for her childhood home, which was actually in another community several hundred kilometers away. According to reports at the time, she did not appear to recognize either her husband or her children, nor did she recognize her own photo on her driver's license, and she believed she was 18.[25]

The most extreme manifestation is *dissociative identity disorder* (DID, formally multiple personality disorder or MPD). While considered by some an iatrogenic phenomenon (meaning the therapists or other influential

[22] Kokoszka (1999/2000).
[23] Staniloiu et al. (2010).
[24] Helmes et al. (2015).
[25] Helmes et al. (2015, p. 102).

people project these "multiple personality" symptoms onto a suggestible patient, who then act-out the disorder in either an aware, or non-aware, state of consciousness),[26] DID is nevertheless listed in the 5th Edition of the American Psychiatric Association's Diagnostic manual (DSM IV). Therefore, the concept has considerable mainstream support. In terms of origin, DID is linked to traumatic childhood events coupled with a patient's predisposition to methods of coping. The result is some expression of dissociation. That produces amnesia, auditory hallucinations (particularly voices), the sense of being controlled by outside agents, and of course changes in personal identity. This doesn't necessarily manifest as gross behavioural traits—the alteration might be expressed as different styles of handwriting in a diary.[27]

Ultimately there's a purpose for the patient entering this extreme dissociative state, since they require;

> …the presence of at least two distinct identities or alters who (a) have a relatively stable personality and way of relating to the world and (b) take over or control the psychobiological function, which results in time loss for the main identity.[28]

DID shares similarities with *Possession trance* (PT), which might be viewed as a "shamanic" (or mediumistic) version of DID, but induced by the subject in a cultural context through sensory deprivation, dance or drugs. This leads to them being controlled or even replace by an external identity.[29] While of anthropological importance, the social significance of PT has no place in scientific reality and the second-path, which classifies PT as just another *trance and possession disorder*—as named in the World Health Organization's ICD-10 and the DSM-IV.[30] The fact that affected people don't tend to believe in the psychiatric diagnosis, and treatment by neuroleptics (anti-psychotic drugs) is ineffective,[31] is not of any great concern.

---

[26]Scroppo et al. (1998).

[27]Zeligman et al. (2017).

[28]Ibid., pp. 65–66.

[29]Wright (1994).

[30]von Lucadou and Wald (2014).

[31]E.g., Tucker and Neppe (1988).

The proposed aetiology for the fugue component of dissociative disorders focusses on the inhibition of memory retrieval[32] and the neurological breakdown in the continuity of self. Once again top-down brain areas responsible for memory and executive functions are implicated, such as the prefrontal cortex and the autobiographical-housing (and retrieving) medial temporal lobes and posterior parietal cortex.[33] In the case of dissociation, these brain substrate show *lowered* neural activity.

Beyond this neurophysiological musing, the *reason* for dissociation is unavoidably psychological rather than a mechanistic. It's not too fringe a proposal to state that memories can have both a "threatening" and also a "painful" quality. Although these memories are only *components* of consciousness, when recognised they might be powerful enough to jeopardize the stability of the global state of consciousness—the snow-globe. For this reason bad recollections must be expunged from awareness,[34] and the most extreme (erroneous) manifestation of this process is the removal of *everything* from consciousness, which accounts for the pathological amnesia. The suggestion here is that serious dissociation involves not so much a distortion of the snow-globe container but a complete divorce the container and the constituents as a form of "protection" of global consciousness. Simplistically, the snow-globe loses awareness of the snowflakes as it must be protected at all costs. Alternately, the mental constituents may be segregated and compartmentalised into coexisting but orthogonal "psychic organizations" within the container of global consciousness.[35] Presumably that would result in two (or more) conscious identities emerging that might have no mutual recognition, which is a hallmark of DID.

---

[32] Henning-Fast et al. (2008).

[33] Ibid.

[34] Ibid.

[35] Blum (2013).

References

Bernstein, E. M., & Putnam, F. W. (1986). Development, Reliability, and Validity of a Dissociation Scale. *Journal of Nervous and Mental Disease, 174*(12), 727–735.

Blum, H. (2013). Dissociation and Its Disorders. *Psychoanalytic Inquiry, 33*, 427–438.

Braithwaite, J. J., Watson, D. G., & Dewe, H. (2017). Predisposition to Out-of-Body Experience (OBE) Is Associated with Aberrations in Multisensory Integration: Psychophysiological Support from a "Rubber Hand Illusion" Study. *Journal of Experimental Psychology: Human Perception and Performance, 43*(6), 1125–1143.

Bühler, K.-E., & Heim, G. (2011). Etiology, Pathogenesis, and Therapy According to Pierre Janet Concerning Conversion Disorders and Dissociative Disorders. *American Journal of Psychotherapy, 65*(4), 281–309.

Cardeña, E., & Krippner, S. (2010). The Cultural Context of Hypnosis. In S. J. Lynn, J. W. Rhue, & I. Kirsch (Eds.), *Handbook of Clinical Hypnosis* (2nd ed., pp. 743–771). Washington, DC: American Psychological Association.

Castillo, R. J. (2003). Trance, Functional Psychosis, and Culture. *Psychiatry, 66*(1), 9–21.

Helmes, E., Brown, J.-M., & Elliott, L. (2015). A Case of Dissociative Fugue and General Amnesia with an 11-Year Follow-Up. *Journal of Trauma and Dissociation, 16*(1), 100–113.

Hennig-Fast, K., Meister, F., Frodl, T., Beraldi, A., Padberg, F., Engel, R. R., et al. (2008). A Case of Persistent Retrograde Amnesia Following a Dissociative Fugue: Neuropsychological and Neurofunctional Underpinnings of Loss of Autobiographical Memory and Self-Awareness. *Neuropsychologia, 46*(12), 2993–3005.

Jonsson, K., Grim, K., & Kjellgren, A. (2014). Do Highly Sensitive Persons Experience More Nonordinary States of Consciousness During Sensory Isolation? *Social Behavior and Personality, 42*(9), 1495–1506.

Kokoszka, A. (1999–2000). Altered States of Consciousness: A Comparison of Profoundly and Superficially Altered States. *Imagination, Cognition and Personality, 19*(2), 165–184.

Lynn, S. J., Milano, M., & Weekes, J. R. (1991). Hypnosis and Pseudomemories: The Effects of Prehypnotic Expectancies. *Journal of Personality and Social Psychology, 60*(2), 318–326.

Moreira-Almeida, A., & Lotufo-Neto, F. (2017). Methodological Guidelines to Investigate Altered States of Consciousness and Anomalous Experiences. *International Review of Psychiatry, 29*(3), 283–292.
Naish, P. (2005). Detecting Hypnotically Altered States of Consciousness. *Contemporary Hypnosis, 22*(1), 24–30.
Rabeyron, T., & Caussie, S. (2016). Clinical Aspects of Out-of-Body Experiences: Trauma, Reflexivity and Symbolisation. *L'Évolution Psychiatrique, 81*(4), e53–e71.
Ramachandran, V. S., & Blakeslee, S. (1998). *Phantoms in the Brain.* London: Fourth Estate.
Rogo, D. S. (1979). *The Poltergeist Experience: Investigations into Ghostly Phenomena.* New York: Penguin.
Scroppo, J. C., Drob, S. L., Weinberger, J. L., & Eagle, P. (1998). Identifying Dissociative Identity Disorder: A Self-Report and Projective Study. *Journal of Abnormal Psychology, 107*(2), 272–284.
Spanos, N. P. (1996). Hypnosis: Mythology Versus Reality. In N. P. Spanos (Ed.), *Multiple Identities & False Memories: A Sociocognitive Perspective* (pp. 17–28). Washington, DC: American Psychological Association.
Staniloiu, A., Markowitsch, H. J., & Brand, M. (2010). Psychogenic Amnesia—A Malady of the Constricted Self. *Consciousness and Cognition, 19,* 778–801.
Tucker, G. J., & Neppe, V. M. (1988). Neurology and Psychiatry. *General Hospital Psychiatry, 10*(1), 24–33.
von Lucadou, W., & Wald, F. (2014). Extraordinary Experiences in Its Cultural and Theoretical Context. *International Review of Psychiatry, 26*(3), 324–334.
Wagstaff, G. F., Cole, J., Wheatcroft, J., Anderton, A., & Madden, H. (2008). Reducing and Reversing Pseudomemories with Hypnosis. *Contemporary Hypnosis, 25*(3/4), 178–191.
Wright, P. A. (1994). A Psychobiological Approach to Shamanic Altered States of Consciousness. *ReVision, 16*(4), 164–172.
Zeligman, M., Greene, J. H., Hundley, G., Graham, J. M., Spann, S., Bickley, E., et al. (2017). Lived Experiences of Men With Dissociative Identity Disorder. *Adultspan Journal, 16*(2), 65–79.

21

Component Changes to the Altered State

Global changes of state like DID are all very interesting to deliberate, but such clinical illustrations really don't correspond to a singular ghost encounter like Tom's. The configuration of an ASC better suited to ghost sightings involves a stable global consciousness with *constituents* that appear partly or wholly uncharacteristic. Or to continue the analogy, the individual is comfortable *being* their snow-globe, but the floating flakes they perceive within this glass dome are not normal. This is a more useful conceptualisation of the conscious experience of ASCs, as it is more common for self-aware people to describe changes in the *components* of objective reality, rather than changes in the fabric (which explains Mitchell's difficulty in narrating his experience).

And in the same way that Mitchell (and others) become aware of the fabric of their consciousness only when it *alters* (everyday reality simply *is*, and there's no need to monitor it), people might only become aware of the *components* of consciousness as they change. An example of a alteration in a physical component would be your awareness of a sharp and unexpected stomach pain. In these circumstances, you know what your stomach should "feel like" but you rarely attend to this resting state. You

T. Jinks, *Psychological Perspectives on Reality, Consciousness and Paranormal Experience*,
https://doi.org/10.1007/978-3-030-28902-7_21

also realise that this new, painful experience is not normal, but in the case of stomach pain it is not necessarily unique. Stomach aches might not be common for most people, but they are to be anticipated at some stage throughout life—even sharp ones.

Occasionally your awareness might be drawn to specific *mental* elements of consciousness that are also unexpected—perhaps even unique—and conclude that consciousness is not the way it should be and that the components are no longer grounded in present perception (i.e., the here-and-now). This expression of an altered state of can be identified as either *partial* or *complete*, and within each of these categories the components can be either *non-sensory* or *sensory.*

Non-sensory partial component changes refer to situations where only one (or a few) of non-sensory elements of consciousness are altered, while the percipient's global state remains constant alongside most mental constituents.

Defining a single non-sensory element of consciousness is problematic, but let's assume it is the mental equivalent of a single physical impression like stomach pain. An example might be the recognition of singular novel thought, affect, motivation or interest that an individual considers unusual in comparison to what they believe to be a normal component of their mental space. The sense of presence is one such example, since it is an impression that cannot be even remotely interpreted as part of everyday experience.

A popular (albeit controversial) rationalisation for the sense of presence is as a consequence of *hemispheric intrusion*. Anatomically and functionally the two sides of the brain are pretty much identical, although there are specific differences in temporal lobe function such as the discovery by Norman Geschwind and colleagues at Harvard Medical School in the 1960s of a much larger, superior aspect of the temporal lobe (planum temporale) on the left side in humans. This was soon correlated with left hemisphere language speciality, which accounts for the mainstream belief that the human left hemisphere has a much stronger involvement in language production than the right hemisphere.[1]

[1]Purves et al. (2001).

Since the ability for language leads to self-concept ("I" and "me"), the left hemisphere has long been identified as the seat of the sense-of-self and self-awareness.[2] Typically the equivalent area of the *right* hemisphere lacks language capability, instead specialising in spatial awareness and perhaps non-language traits like intuition and emotion.[3] Since humans are so language-dependent, there is a one-directional dominance of the left hemisphere over the right. This is where TLE returns to the discussion. If a person suffers a small temporal lobe seizure that upsets the unilateral (left-to-right) channel of information that travels along the hemispheric bridges of the corpus callosum and anterior commissure, there might be some rare "back flow" from the right hemisphere to the left—given the ominous name of *intrusion*.[4] This change in neural influence would be processed in global consciousness as a highly unusual *constituent* experience requiring an explanation. For the top-down brain, there's no framework for easily integrating excessive right hemispheric temporal-spatial (or intuitive, or emotional) material into normal waking consciousness mediated by the left hemisphere. According to neuroscientist Michael Persinger's vectorial hemisphericity hypothesis, popular in the 1980s and 1990s, a mechanism to deal with intrusion is to *project* or *externalise* the sensation, which becomes an invisible and personal "presence" but which in fact represents the neural operation of the right temporal lobes.[5]

It is argued that these *ego-alien intrusions* explain encounters with guardian angels, spirit guides, demons or ET aliens,[6] depending on circumstance and cultural background of the percipient. Speculating further, if the CNS subsequently attempts a visual depiction of this presence, what is generated is a doppelgänger or tulpa. This neurological description might sound a little farfetched (and has its detractors), but for the second-path it is surely preferable to a "real" psychic projection, or a "real" invisible spirit.

---

[2] Booth et al. (2008).

[3] Fallik and Eliot (1986).

[4] Persinger et al. (1994).

[5] Ibid.

[6] Jinks (2012).

Another example of an anomalous non-sensory component of consciousness is the impression of déjà vu. Déjà vu is defined as a variety of paranormal experience involving irrational familiarity;

> On the bus with my Uni friends and felt like I had been there before, had the same conversation and people had the same positions, although I don't remember seeing the faces before but when I was on the bus in that position the faces I saw fitted the vision perfectly.[7]

Modern academic definitions emphasise that this familiarity must be accompanied by the knowledge that such an experience is *false*.[8] It's perfectly natural to have these sensations, but if the experiencer believes them to be legitimate, then the déjà vu is diagnosed as pathological.[9]

The earliest (and most popular) neuropsychological explanations for déjà vu involved a *dual-pathway* of perception, where the two hemispheres processing the same information do not proffer this imagery to consciousness at exactly the same time, and this is the basis of the sense of familiarity. The content of the second stream of information has already been processed through the *first* stream, but the brain still views this second-arriving material as fresh. The simplest rationalisation the brain makes is that what is now seen must have been witnessed at some time in the (more distant) past.

These ideas no longer conform to modern understanding of CNS information processing.[10]

Contemporary theories are more likely to associate temporal lobes dysfunction with déjà vu. As I've mentioned, this region is a storehouse of autobiographical memories. In the present moment, waking consciousness can draw on the imagery, familiarity and emotional components of autobiographical memories as a useful addition to mental life. Most of the time the retrieval of life memories, and their familiarity, are heavily assimilated.[11] If this balance is broken—perhaps through a microseizure

---

[7] Moulin (2018, p. 281).

[8] Cleary and Claxton (2018).

[9] Ibid.

[10] Moulin (2018).

[11] Ibid.

affecting the operation of the neuromatrix—then memories will not trigger familiarity and the experiencer will be inclined to search for an anomalous interpretation. The most obvious would be a reincarnation memory. Alternately, if the sense of familiarity is *not* accompanied by a memory, and can only be attached to a here-and-now (novel) perceptual scene, then the impression will arise of having previously been there.[12] This sensation might be compounded by a feeling of attachment (emotional valence) to the experience of place.[13]

A further proposal is that a location might be visited, or a person witnessed, and this automatically triggers an autobiographical memory of an earlier encounter with this *same* scene or individual. That's the normal process of recall. However in neurological déjà vu, the *knowledge* component of this earlier time is accidently withheld from the recollection. Explained from a cognitive perspective;

> …déjà vu arises because we forget the source of the information, or perhaps the source of the information was never consciously known. If processing at encoding is sufficiently shallow, or the interval between encoding and retrieval sufficiently long, we may find information highly familiar without being able to remember why.[14]

Temporal lobe seizures and déjà vu have also been experimentally correlated,[15] most likely because of transient and reduced activity in the perirhinal and entorhinal regions of the medial temporal lobes important in autobiographical memory and the discrimination of familiarity.[16] And perhaps if drowsiness (rather than TLE) causes a similar fleeting brain mistake in a healthy individual,[17] then a benign explanation for the déjà vu experience has been found.

[12] Spatt (2002).
[13] Moulin (2018).
[14] Ibid., p. 62.
[15] Bancaud et al. (1994).
[16] Guedj et al. (2010).
[17] Spatt (2002).

Another non-sensory component change in consciousness is the *sense of dread* often induced by *sleep paralysis*. This is an experience that has strong paranormal overtones, as the description from this sufferer suggests;

> Some nights he would kick, thrash, or scream, but not necessarily awaken. Sometimes he would suddenly wake up with a sense of dread, a racing heartbeat, and profuse sweating.[18]

While very common (occurring in as many as 40–50% of people at some stage in life[19]), sleep paralysis is more complex than a singular hypnagogic or hypnopompic hallucination. It's a disorder of rapid-eye movement (REM) sleep and displays a particular range of EEG behaviour. If you recall, Alpha waves are the type of brain activity recorded when you're awake and in a relaxed state. They are of a consistent shape and medium frequency. Beta waves are recorded when your brain is alert and active, and have an irregular pattern and higher amplitude and frequency to Alpha waves. Then there are low frequency and synchronous Theta waves which appear on an EEG when the patient is falling asleep, known as *Stage 1 sleep*. A few minutes into sleep, Stage 2 activity is recorded by EEG equipment, with a return to a Theta-like irregular pattern alongside the appearance of spikey, high amplitude K complexes. Then a little later there is Stage 3 sleep, with characteristic high amplitude, low frequency delta waves. Stage 3 sleep is also known as *slow-wave sleep*.

Finally, about 90 minutes after entering sleep a Stage 4 pattern is observed. This is a combination of irregular Theta and Beta activity which corresponds to the attenuation of external information processing by the sleeper's brain, and an increasing engagement with endogenous imagery (the concept of dreaming). Stage 4 is behaviourally represented by the rapid movement of the eyeballs behind closed eyelids (REM)[20] and physiological changes typified by lowered heart rate and respiration, sensory

[18]Neubauer (2009, p. 16).
[19]Sherwood (2002).
[20]Nishino et al. (2000).

gating[21] and skeletal muscle atonia (a loss of muscle strength, presumably an evolutionary advantage to ensure your dreams aren't "acted out").[22]

When it comes to sleep paralysis, occasionally there might be a direct entry into the REM state from the earliest stages of entering sleep, avoiding the 90 or so minutes of lead-up through the various stages. This is said to occur as a consequence of dissonant activity of neurons originating in a brain stem region called the *pons* that project to the cortex and drive the shift from wakefulness to sleep, and vice versa. The REM-on cells might be overactive, or the REM-off cells might be underactive when first falling asleep, speeding up the process and causing the sleeper to enter dreaming far more quickly than normal.

Most importantly, when REM errors occur there's no straightforward transition into the ASC corresponding to sleep and dreaming (or *out of* the sleeping state if the individual is waking up). An ordinary state of consciousness is maintained, but with many unusual and intrusive features. These features are registered and integrated into consciousness and construed as meaningful. Hence *reasons* are sought for the REMs and the uncomfortable feeling of paralysis caused by muscle atonia, as well as other features that typically occur in REM sleep (but which the sleeper isn't normally aware of), such as random muscle twitches and auditory impressions from middle-ear activity,[23] as well as weak but noticeable vestibular-related sensations of floating and sinking.[24]

In the fashion of vectorial hemisphericity, these REM-generated percepts are unavoidably projected onto the environment as the sleeper's brain assumes that's where the origin of these symptoms lies. That might result in the sense of presence[25] which is a manifestation of the *intruder experience*.[26] And when REM-on neurons situated in the pons send their sleep signals to the cortex via the relay centre of the thalamus, there can be concurrent stimulation of the *amygdala* which normally moderates behaviours

---

[21] Cheyne et al. (1999).

[22] Mahowald et al. (1998).

[23] De Gennaro et al. (2000).

[24] Comparelli et al. (2000).

[25] Ibid.

[26] Girard and Cheyne (2004).

in response to fear-inducing stimuli.[27] In this manner, the sense of dread is triggered by amygdaloid activation and since the individual is still self-aware, this odd feeling manifests as the negative component of the wider sleep paralysis experience—the "presence" in the room that's preventing the sleeper from moving happens to be the *cause* of this intense fear. This presence, coupled with breathing difficulties (due to sleep-related muscular paralysis) and the middle-ear noises interpreted as panting or whispering, leads to a neat explanation for the "Old Hag" (or "Ag Rog") phenomenon studied in the late 1970s. This witch-like entity was blamed for night-time attacks on dozens of people residing on in northeast coastal Newfoundland;

> In a typical attack the victim suddenly awakes feeling unable to move and experiences a heavy weight pressing on his chest. The figure of an animal or a human is sometimes seen astride the chest. The victim, although aware of other people in the household, is unable to overcome his feeling of paralysis until someone touches him, shakes him, or calls his name.[28]

Since the victims show no physical or psychological differences to the general public, something temporary and non-pathological must be responsible, and sleep paralysis is a neat solution.[29] Indeed sleep paralysis has been the stock-standard explanation to bedroom entity encounters ever since.

Briefly, another non-sensory, partial component change in consciousness is the impression gained through a psi event (or more correctly in the second-path, the *impression* that one has experienced a psi event). This state of consciousness *feels* unusual because there is a component of knowledge or understanding that arrives spontaneously.[30]

Regarding ASCs involving non-sensory component changes, a distinction can be made between hallucinations that seamlessly fill a *part* of consciousness, and those that contribute to the *entire* experience of consciousness. The latter can be labelled *complete* component changes.

---

[27] Amaral (2003).

[28] Ness (1978, p. 96).

[29] Ibid.

[30] Braud (1975).

The degree to which components vary from what's expected in ordinary consciousness, or simplistically, the "number" of components that change, will determine the intensity of this style of ASC. Where there is significant alteration of content a *profoundly altered state of consciousness* (PASC) will arise.[31] In a PASC, the individual remains embodied, but the content and modalities of the impression of reality is all-encompassing and very different to everyday "baseline experiences". Returning to the snow-globe analogy, all the flurrying snowflakes have disappeared, and a very different set of objects have taken their place.

Conceptualising gross alterations in non-sensory cognitive and/or affective states is difficult, although symptoms would closely resemble the APZ dimension of OBN. In this profound condition—perhaps initiated by certain types of meditation[32]—there might be recognition of the embodied state *as a whole* but a significant reduction or even complete loss of reality testing (corresponding to a loss of ego functioning[33]) leading to impressions of time distortion coupled with unexpected achievement of noesis and a profound feeling harmony with personal self, the universe and the divine.[34] This has been labelled *clairvoyant reality*, where a psychic practitioner receives immediate and profound non-sensory understanding of the interrelationships amongst the objects and events in the world around them and the illusory nature of time.[35]

It is much easier to comprehend changes in the *sensory* components of consciousness. An ASC consisting of a *sensory partial* component change retains an individual's global fabric and their ordinary non-sensory impressions (specifically, the embodied self, their normal cognitions and emotions). However, facets of a *perceived* scene are atypical. These correspond to the visual, auditory and tactile hallucinations I've mentioned previously (and presumably olfactory and gustatory ones too). They are primarily unimodal, meaning a change in vision, or sound, or touch *only*—although the experience is nevertheless more complex than the most basic single component change, which refers to a specific pain, or a single hue, or

[31] Naish (2005).

[32] For example, *contemplative meditation.*

[33] Hanes (2012).

[34] Ibid.

[35] Braud (1975).

direction of movement. In terms of novelty, the components might have no relationship with the perceived environment. For instance synaesthetic processes could conceivably translate bodily perceptions into specific visual images overlayed on a "normal" scene.[36]

Partial component changes are most commonly associated with neuropathology, for instance Charles Bonnet Syndrome, but as I've suggested visual hallucinations can be produced spontaneously; Michaeleen Maher's vision of an intricate human aura on an otherwise normal day is a sensory partial component change. Tom's ghost is also a spontaneous, unimodal hallucination, although it might be expected that the cause of these two experiences are very different.

Then there are sensory *complete* component changes, where most or all of the elements of perception are transformed, and derive from something other than objective reality. Psychoactive drugs (and endogenous chemicals[37]) are most commonly linked to complete component (visual) changes. The story of meeting insects in a "dank moss-green hospital emergency waiting room" following consumption of psilocin involves a vision that takes up all of conscious awareness. In the same vein, an alien abduction storyline like the Allagash kidnapping—complete with the inside of a spacecraft and an operating table—corresponds to a complete component ASC, hence the popular linking of alien abduction with acute changes in brain neurochemistry.[38]

A less dramatic example of the complete component change ASC is spontaneous and non-pathological *lucid dreaming*, where the dreamer remains grounded in their sense of identity, and their existential framework appears intact, but they know the storyline they are experiencing is dream material. Normal dreaming is accompanied by impaired capacity of control, but in lucid dreaming there is complete access to cognitive resources such as autobiographical memories, and the dreamer has the ability to govern the direction of the dream's plot.[39]

---

[36] Rabeyron and Caussie (2016).

[37] E.g., Strassman (2005).

[38] Luke (2008).

[39] Appel et al. (2018).

In addition, the characters in the lucid dream—which are often reported by the dreamer to be as real or "realer" than those encountered in waking life—can possess autonomous behaviour and intentionality.[40] In the following example, a man named Jon meets his long-dead father and dead aunt in a lucid dream;

> I walked over to them and understood that they came from the other side. I decided to shake hands with them, actually to test the texture, thinking about the paranormality of the situation. Their hands felt physical, but not as dense as normal matter. They were smiling amicably at me, and warm feelings were flowing both ways. No words were exchanged.[41]

This feeling isn't necessarily different from the "disembodied embodiment" of the parasomatic state[42] although at the current level of understanding, it's not possible to pin down the stage where an orthodox, embodied sense-of-self transitions to an altered global consciousness.

Occasionally, the complete component sensory ASC *lacks* the "realer than real" feeling of a lucid dream or an OBE—as this next story from Jo attests, experienced when she was wide awake;

> I woke up at about 8am on Sunday morning and everything was different. The way the room was arranged was different, my bed, and where the door was. The door was different. I remember thinking straight away that I'd come home and everything was normal then, and I had definitely come home to the right address. I also knew I wasn't dreaming, because it still felt like I was home, and I actually didn't feel upset at all. I didn't bother getting out of bed, but I did have a drink of water from a cup that was next to my bed. That cup was about the only thing that was normal that I could see. I could feel everything like normal, and I saw everything normally. I guess I must have been hallucinating, but it was a pretty convincing hallucination. Either that or I was in another dimension. I figured the best way to get out

[40] Although they do differ from OBE experiences, which apparently possess greater volitional involvement (e.g., Blackmore 1986).

[41] Puhle and Parker (2017, p. 155).

[42] Rabeyron and Caussie (2016).

> of it was to go back to sleep. When I woke again a few hours later the room was back to normal. This made me start believing in the paranormal.[43]

In terms of why a complete component change occurs, the serotonergic "filter" hypothesis for hallucinations provides valuable insight. Should serotonin imbalance create heightened responsivity to overwhelming amounts of visual stimuli, then when those (same) objects are attended to, they might be interpreted as novel or more intense than any regular visual experience.

Also of interest are the emotions and beliefs that *emerge* from experience. These are seemingly derived from the storyboard of new (or seen-as-new) sensory components. In Jo's story, she mentions not feeling upset, rationally analysed her surrounds in a typical way so as to draw conclusions about what might be happening to her. This implies the neural processes behind the creation of the sensory "hallucination" are efficient in tapping into, and controlling, the stable *global* state while producing conforming non-sensory responses to an exceptional and challenging sensory experience (I'll return to the implications of this statement later in the book).

How would a *combination* of global and component changes be experienced? By definition the outcome must involve the alterations in personal identity, the embodied self and global consciousness coupled with changes in what people are "seeing" and "feeling".

Examples of these would be NDEs, *some* reports of OBEs where *different* embodiment is reported and Nadia's astral journey. Here is another story from a recreational ketamine user who reports floating in a disembodied state while experiencing highly unusual imagery;

> … when I closed my eyes a lot of information started to happen. Colors, patterns, cross-connections in sensory perception. Like sound and inner visions sort of got confused. I got deeper and deeper into this state of realization, until at one point the world disappeared. I was no longer in my body. I didn't have a body. And I reached a point at which I knew I was going to die …. .And what incredible feelings that evoked! … …I just yielded. And then I entered a space in which …there aren't any words ….I mean, at-one-with-the-universe, recognizing-your-godhead ….The feeling

[43] Jo, Personal communication 20 June 2011.

> was: I was home. That's really the feeling of it. And I didn't want to go anywhere, and I didn't need to go anywhere. It was a bliss state. Of a kind I had never experienced before. I hung out there awhile, and then I came back. I didn't want to come back. I guess in the deep state it was no longer than half an hour.[44]

According to this view (and like dissociative disorders and psychoses), NDEs, OBEs and astral journeys emerge when overwrought brain processes "in extreme crisis"[45] (and their psychological counterparts) undermine the embodied self—while concurrently the mental void becomes occupied with "retrospective fantasies"[46] made real through intense absorption.

Invariant component changes linked to these profound ASCs can be identified and explained using shrewd neurophysiological language. The NDE is typified by "seeing" a dark tunnel ending in a world of light, and this effect can be matched to the experiences of pilots subjected to substantial G-forces and who *sometimes* experience *hypotensive syncope* (in other words, fainting because their blood pressure has dropped significantly) and *retinal ischemia* (where the visual system is oxygen deprived), and who subsequently report tunnel-like impressions as they quickly lose access to peripheral vision.[47] There's enough similarity between the symptoms of fainting, and tunnel-experiences, for second-path researchers to suggest that this both features are due to straightforward alterations in brain function, even if these states are initiated in quite different ways.

References

Amaral, D. G. (2003). The Amygdala, Social Behavior, and Danger Detection. *Annals of the New York Academy of Sciences, 1000*(1), 337–347.

---

[44] Ingram et al. (2018, p. 31).

[45] Stafford Betty (2004).

[46] Schnaper and Panitz (1990).

[47] Mobbs and Watt (2011).

Appel, K., Pipa, G., & Dresler, M. (2018). Investigating Consciousness in the Sleep Laboratory—An Interdisciplinary Perspective on Lucid Dreaming. *Interdisciplinary Science Reviews, 43*(2), 192–207.

Bancaud, J., Brunet-Bourgin, F., Cahauvel, P., & Halgren, E. (1994). Anatomical Origin of Déjà Vu and Vivid 'Memories' in Human Temporal Lobe Epilepsy. *Brain, 117*(1), 71–90.

Blackmore, S. J. (1986). Where Am I? Perspectives in Imagery, Memory and the OBE. In D. H. Weiner & D. I. Radin (Eds.), *Research in Parapsychology* (pp. 108–111). Metuchen, NJ: Scarecrow Press.

Booth, J. N., Koren, S. A., & Persinger, M. A. (2008). Increased Theta Activity in Quantitative Electroencephalographic (QEEG) Measurements During Exposure to Complex Weak Magnetic Fields. *Electromagnetic Biology and Medicine, 27*(4), 426–436.

Braud, W. G. (1975). Psi-Conducive States. *Journal of Communication, 25*(1), 142–152.

Cheyne, J. A., Newby-Clarke, I. R., & Rueffer, S. D. (1999). Relations Among Hypnagogic and Hypnopompic Experiences Associated with Sleep Paralysis. *Journal of Sleep Research, 8,* 313–317.

Cleary, A. M., & Claxton, A. B. (2018). Déjà Vu: An Illusion of Prediction. *Psychological Science, 29*(4), 635.

Comparelli, A., De Carolis, A., Kotzalidis, G. D., Masillo, A., Ferracuti, S., & Conesa, J. (2000). Geomagnetic, Cross-Cultural and Occupational Faces of Sleep Paralysis: An Ecological Perspective. *Sleep and Hypnosis, 2*(3), 105–111.

De Gennaro, L., Ferrara, M., Urbani, L., Bertini, M., De Gennaro, L., Ferrara, M., et al. (2000). Complementary Relationship Between Wake and REM Sleep in the Auditory System: A Pre-sleep Increase of Middle-Ear Muscle Activity (MEMA) Causes a Decrease of MEMA During Sleep. *Experimental Brain Research, 130*(1), 105–112.

Fallik, B., & Eliot, J. (1986). An Examination of Possible Age Differences in Interrelationships Between Intuition, Cognitive Style, and Hemispheric Preference Variables. *Perceptual and Motor Skills, 63*(3), 1251–1257.

Girard, T. A., & Cheyne, J. A. (2004). Individual Differences in Lateralization of Hallucinations Associated with Sleep Paralysis. *Brain and Cognition, 9*(1), 93–111.

Guedj, E., Aubert, S., McGonigal, A., Mundler, O., & Bartolomei, F. (2010). Déjà-vu in Temporal Lobe Epilepsy: Metabolic Pattern of Cortical Involvement in Patients with Normal Brain MRI. *Neuropsychologia, 48*(7), 2174–2181.

Hanes, K. (2012). Unusual Phenomena Associated with a Transcendent Human Experience: A Case Study. *Journal of Transpersonal Psychology, 44*(1), 26–47.

Ingram, R., Kang, H., Lightman, S., Jane, D., Bortolotto, Z. A., Collingridge, G. L., et al. (2018). Some Distorted Thoughts About Ketamine as a Psychedelic and a Novel Hypothesis Based on NMDA Receptor-Mediated Synaptic Plasticity. *Neuropharmacology, 142*, 30–34.

Jinks, A. (2012). *An Introduction to the Psychology of Paranormal Belief and Experience.* Jefferson, NC: McFarland.

Luke, D. (2008). Inner Paths to Outer Space: Journeys to Alien Worlds Through Psychedelics and Other Spiritual Technologies (Book Review). *Journal of Scientific Exploration, 22*(4), 564–569.

Mahowald, M. W., Woods, S. R., & Schenk, C. H. (1998). Sleeping Dreams, Waking Hallucinations, and the Central Nervous System. *Dreaming, 8*(2), 89–102.

Mobbs, D., & Watt, C. (2011). There Is Nothing Paranormal About Near-Death Experiences: How Neuroscience Can Explain Seeing Bright Lights, Meeting the Dead, or Being Convinced You Are One of Them. *Trends in Cognitive Sciences, 15*(10), 447–449.

Moulin, C. J. A. (2018). *The Cognitive Neuropsychology of Déjà Vu.* London: Routledge.

Naish, P. (2005). Detecting Hypnotically Altered States of Consciousness. *Contemporary Hypnosis, 22*(1), 24–30.

Ness, R. C. (1978). The Old Hag Phenomenon as Sleep Paralysis: A Biocultural Interpretation. *Culture, Medicine and Psychiatry, 2*(1), 15–39.

Neubauer, D. N. (2009). Fearful Sleeping and Waking. *Primary Psychiatry, 16*(11), 16–18.

Nishino, S., Riehl, J., Hong, J., Kwan, M., Reid, M., & Mignot, E. (2000). Is Narcolepsy a REM Sleep Disorder? Analysis of Sleep Abnormalities in Narcoleptic Dobermans. *Neuroscience Research, 38*, 437–446.

Persinger, M. A., Bureau, Y. R. J., Peredery, O. P., & Richards, P. M. (1994). The Sensed Presence as Right Hemispheric Intrusions into the Left Hemispheric Awareness of Self: An Illustrative Case Study. *Perceptual and Motor Skills, 78*(3), 999–1009.

Puhle, A., & Parker, A. (2017). An Exploratory Study of Lucid Dreams Concerning Deceased Persons. *Journal of the Society for Psychical Research, 81*(3), 145–160.

Purves, D., Augustine, G. J., Fitzpatrick, D., Katz, L. C., LaMantia, A.-S., McNamara, J. O., et al. (Eds.). (2001). *Neuroscience.* Sunderland: Sinauer Associates.

Rabeyron, T., & Caussie, S. (2016). Clinical Aspects of Out-of-Body Experiences: Trauma, Reflexivity and Symbolisation. *L'Évolution Psychiatrique, 81*(4), e53–e71.

Schnaper, N., & Panitz, H. L. (1990). Near-Death Experiences: Perception Is Reality. *Journal of Near-Death Studies, 9*(2), 97–104.

Sherwood, S. J. (2002). Relationship Between the Hypnogogic/Hypnopompic State and Reports of Anomalous Experiences. *Journal of Parapsychology, 66*(2), 127–150.

Spatt, J. (2002). Déjà Vu: Possible Parahippocampal Mechanisms. *Journal of Neuropsychiatry and Clinical Neurosciences, 4*(1), 6–10.

Stafford Betty, L. (2004). Mind, Paranormal Experience, and the Inadequacy of Materialism. *International Philosophical Quarterly, 44*(3), 373–392.

Strassman, R. (2005). Hallucinogens. In M. Earleywine (Ed.), *Mind-Altering Drugs: The Science of Subjective Experience* (pp. 49–85). Oxford: Oxford University Press.

22

Mental Imagery and the Jigsaw

> Specific content of the experience could be determined by the symbolic expression of the needs of the subject.[1]

Snow-globes and snow-flake metaphors help classify the many different types of ASC, but an assessment of the personalisation of experience remains elusive; we still need an answer to the earlier puzzle of why certain styles of ASC result in seeing the snippet of a ghost, while others create a full-blown alien abduction trauma. Or more fittingly, why do different people experience different kinds of *mental imagery* made real by whatever brain disturbance they're inflicted with?

The second-path solution is that people possesses unique beliefs, memories, and expectations which are stored in various cortical areas. If these regions are affected by *any* of the potential initiators of an (hallucinatory) ASC, then particular phenomenological states will be reported. Take the example of a partial component visual hallucination. This may be brought

[1] Rabeyron and Caussie (2016, p. 58).

T. Jinks, *Psychological Perspectives on Reality, Consciousness and Paranormal Experience*,
https://doi.org/10.1007/978-3-030-28902-7_22

about by fatigue. Tom was very likely fatigued after a long day at the conference, and even though he might protest that what he saw was real, it must be accepted that the impact of weariness on the occipital lobe caused the release of a brief cartoon-like image of a man who wasn't there. But why a man and not a donkey?

That's explained by *Seepage theories.* The content of hallucinations is due to a seepage (or distillation) of imagery from pre-conscious to conscious awareness.[2] This material displays two characteristics. The first is fixed; stable content made up of *physiological constants* that form the basis of the symptoms of NDEs and OBEs, the Old Hag phenomenon derived from sleep paralysis, ghostly apparitions, as well as poltergeist activity and psi experiences such as telepathy, premonition, and clairvoyance.[3] These are reported by all types of people, and are culturally independent. There's simply neural hard-wiring assisting experience of a white tunnel of light during brain trauma, or a sleep paralysis episode when the neurological variables are just right (for instance, a combination of acute metabolic disturbance, fatigue and lying on the back rather than the side when falling asleep).

However there are individual differences in each of these paranormal reports, and these derive from the second characteristic of imagery—variations in interpretation which are culturally *dependent.* And they're based on probability. It is more likely that if we are fatigued, and stored visual imagery seeps from our deeper consciousness, that this will be of people and places (and objects and events) that we're most familiar with. We will see people in our partial component hallucination, not donkeys. We don't think of donkeys very often. That's why Tom saw a walking man—he sees them all the time.

A clever way of examining how this nebulous material (both fixed and personal) can affect mental imagery was *not* developed to support scientific reality; for the second-path, mere reference to those "beliefs, memories, and expectations" is enough to thoroughly explain the visual content of ASC. Nevertheless, if stronger insight into the personal nature of ASCs and paranormal experiences that involve partial and complete component

[2]López Rodrigo et al. (1997).

[3]von Lucadou and Wald (2014).

changes then Charles Tart's analogy of the *Jigsaw* is terrific. I'll borrow much from Tart's work in the following discussion.[4]

In his analogy, the CNS is a builder of a jigsaw puzzle of many pieces. In an ordinary state of consciousness there is a complete "rectangular" representation of reality, with none of the regular pieces missing. The shape of the rectangle is equivalent to our intact snow-globe, and the pieces are the snow-flakes. Tart's analogy is actually more useful, since the puzzle pieces can be visualized as physical objects of numerous shapes and sizes. Like Titchener's elements, they are primarily constructed from external sensory information infused with fluctuating secondary states such as non-sensory emotions, physiological drives (like hunger or thirst), mental images and other possibilities—essentially the beliefs, memories and expectations of the second-path. The closer in shape and size these hypothetical blocks are, the more *related* they are, like the nodes of a neural network model (which to a certain extent they resemble).

Overall, the intact rectangular jigsaw puzzle possesses a comfortable "feel" to it, and represents the established global state. Tart calls this a *baseline* consciousness, but that only applies to the most common expression of ordinary mind. It's the way people usually perceive themselves when they go about their everyday business.

Tart argues there are multiple examples of ordinary consciousness that are different to the baseline state, and different to each other, but they are still *ordinary*. This includes being drunk, or dreaming during sleep.[5] Being drunk or dreaming represent what is known as a *discrete state of consciousness*. The components might closely resemble those present in baseline consciousness, although unexpected emotions and cognitions can emerge during drunkenness, or unusual imagery during dreaming. This corresponds to a *very mild* partial component ASC.

However discrete states are *not* systematically modified as they progress from mild to severe. The effect of progressing from left to right along a continuum[6] might help to separate benign ASCs (like dreaming) from more serious ASCs (like sleep paralysis) and very debilitating examples

[4]Tart (2008).

[5]So these might be altered states of consciousness, but they are only *ordinary* altered states!

[6]A term I've earlier used to describe the various expressions of ASCs (including non-ordinary states).

(like the state induced by schizophrenia). However in terms of discrete expressions of the ordinary state, drunkenness *does not* seamlessly merge into dreaming (or anything else). These are completely different (discrete) types of consciousness with their own special physiognomies. They also possess very specific methods of transition *into*, and *out of*, them from resting consciousness. So for drunkenness there needs to be the drinking of alcohol, and for dreaming there needs to be entry into REM sleep from earlier stages of sleep (presuming there is no transition dysfunction, as attributed to sleep paralysis).

For the resting state to be maintained for long periods of time, or until the next state of consciousness is entered into—drunkenness or dreaming or any other similar example, a person requires responsible, hard-working, brain-based stabilisation mechanisms. These need to be adaptable to constant changes in the internal milieu (that is, our physiology), the external environment and the person's responsive behaviour. This will ensure the components of ordinary consciousness and the normal impression it affords do *not* change too radically. It is very unhelpful for our survival if we are prone to enter into an ASC spontaneously (unless perhaps you are a shaman and the community is supporting you). Of course it happens—Tom's ghost might be an example of a spontaneous ASC as are any of the examples I've provided so far—but they generally don't last long enough to be detrimental to health. Realistically there are no predators ready to pounce and eat us as we privately indulge in a non-ordinary conscious spectacle.

As Tart says;

> …we are so used to doing this work [maintaining ordinary consciousness] that we normally never notice it.[7]

It also helps that the hypothetical jigsaw pieces of consciousness form a tight cluster in their rectangular configuration, so it's quite hard to destabilise them.

[7]Tart (2008, p. 139).

> A low pile of actual physical blocks like this could handle a lot of shaking and pushing - environmental or bodily change - while still maintaining their relationship with each other.[8]

That's the everyday condition of the baseline state. In terms of *how* this state is altered, there is a three step process that describes not so much the neurophysiological cause, but *how* the material gets there and most importantly *where* it comes from (no *why* is available at this stage, as this method is co-opting Tart's work to suit the second-path of scientific reality). Although the model most properly refers to entering a discrete state of ordinary consciousness from the resting state, for the sake of this discussion it is also applicable to partial and total content ASCs that have hallucinatory connotations since these are definitely *not* examples of discrete states of ordinary consciousness!

The first step in moving from the resting state to any other state is the disruption of stabilisation that holds the baseline jigsaw pieces in place. The contents of consciousness are never static—after all, everyday thinking is a dynamic procedure. Nonetheless these jostling components fall within an envelope of possibilities that we understand as being *common, normal, regular.* There are numerous theoretical devices that keep the contents of consciousness within these boundaries, so the disruption needs to be extensive. Tart uses the example of bolts holding a structure together. A sufficient number of bolts need to be removed to jeopardise the structure's integrity. Therefore, entering an ASC is not as simple as having a single unsettling stimulus like a startling loud noise enter consciousness. That's equivalent to removing a single bolt—the structure might be weakened but won't collapse. There might be momentary mental turmoil as a result of the loud sound, but this will quickly resolve and the contents of consciousness will swiftly settle back to normal as the individual recovers from their initial shock.

Of course, destabilisation can be quick and effective if the correct tools are available to remove the most crucial bolts. These bolts are brain systems like those familiar cortical feedback loops important for functioning consciousness, and the tools that are effective in locating these bolts and

[8] Ibid.

straining them to their limits (and beyond) are psychedelics and all the other induction methods we're acquainted with—severe fatigue, the sensory deprivation of sensory input, or rhythmic continuity of sensory input (e.g., dancing and drumming).

During this phase, imagine the jigsaw puzzle of the baseline state being shaken to the point where the puzzle pieces begin to break free from one another, or at least lose their coherent relationships. If they all went their separate ways the outcome would be a total fracturing of consciousness, perhaps resulting in *no* qualia—*nothing*. That corresponds to a true fugue state. However, that won't transpire if Tart's theoretical *patterning forces* are at play following the short disruptive period. These forces;

> .."push" or "pull" disrupted psychological functioning and subsystems toward the new pattern of the desired ASC.[9]

The result is a new arrangement of jigsaw pieces fitting together in a different way—a different picture, so to speak. And forcing the analogy, an intact rectangular perimeter safeguards the global state and sense of self despite the component changes (a few or many). The components of consciousness are also *maintained* by the patterning forces, which prevents the ordinary baseline product from re-exerting itself. In this way, the forces that initiate the ASC can be both disruptive *and* patterning.

This seems quite ridiculous, as a real jigsaw puzzle with actual coherent picture (e.g., of lighthouse or a landscape) can't be rearranged to form anything but a jumbled mess. However the jigsaw analogy is just that, a loose analogy. The idea is that re-patterned consciousness *can* be experienced as something that's lucid, but *altered* from normal. The more extensive the re-patterning, the more unusual the experience. Meanwhile the jigsaw puzzle perimeter might bulge and flex such that it is no longer a rectangular conformation, which would suggest an impact on the sense of normal embodiment. This would be the most extreme outcome, equivalent to a change in both the global and component states of consciousness.

[9]Tart (2008, p. 142).

During re-patterning, some baseline jigsaw puzzle pieces might remain in place and retain their proper appearance in consciousness. Nevertheless there might be a change in the relationship amongst these and other pieces, where some emerge more prominently than their counterparts in the new phase. In the fashion of a Bergson-style filter that's been *loosened*, re-patterning might involve the entry of external stimuli normally unable to access the mixing-palette of consciousness, while preventing other, regular stimuli from admittance. Examples include unattended background scenery taking on greater significance to the observer, or hearing a person speaking in the vicinity who would normally be ignored. The following case concerns a 17-year-old man from Switzerland suffering from a temporal lobe tumour. The impact of this tumour appears to have relaxed his putative Bergson-style filter, with the added (rare) jigsaw piece of a gustatory hallucination accompanied by the loss of temporal awareness.

This particular patient had suffered from epileptic seizures due to the tumour since he was 15, but the sensation was personally pleasurable. He experienced the sensation of time having "stopped", allowing a profound understanding of *everything* happening in his vicinity; a situation, event or conversation. For the short time he was in this state (he believed they lasted up to 40 seconds, but observers agreed only between 1 and 3 seconds) he perceived and understood the multiple, simultaneously streams of information efficiently and effortlessly, as though he had removed an attentional bottleneck. This allowed him to determine potential future outcomes with ease, such as a subsequent topic in a conversation, although he didn't feel this corresponded to precognitive ability. And his "ability" was often accompanied by a gustatory hallucination—a completely novel but agreeable taste.[10]

In the young man's story, the reorganisation of consciousness allows clearness of meaning to *appear* to be achieved from what might normally be a low-information or incoherent situation. Hence perceiving this situation (or object or event) during the re-patterning process will provide a temporary noesis or an unexpected feeling or pleasure not possible in the baseline state.

[10]Picard (2013, p. 2495).

Perhaps of greater interest are the hypothetical puzzle pieces that "exist" (so to speak) *beyond* the rectangular boundary that is the finished jigsaw of baseline consciousness. Tart refers to these pieces as *latent psychological potentials* which are rarely present in, or available to, baseline consciousness. Since I'm using this analogy to support the second-path, these pieces need not be esoteric, but are basic physiological events as trivial as conjunctivitis, minor skin sensations, a dry mouth, tachycardia or even a full bladder![11] Then there are stored *mental* possibilities that have difficulty ever reaching awareness. In this situation, regular baseline pieces are thrown out during ASC induction and replaced by unanticipated fragments from the dynamic *un*conscious. Perhaps these are "lost" memories stowed in a diffuse neural array (our "neuromatrix"), and when triggered by a novel disruptive force they exert themselves as realistic mental image in reconfigured consciousness.

Stabilisation lasts for as long as the new pattern can withstand the external and internal changes impacting on consciousness. And yet through experience or training, a person can adapt to the forces and facilitate the temporal length of the ASC. For example, when going from the baseline state to the discrete state of hypnosis,[12] there might be little initial success without "situational props" to assist the disruption and patterning.[13] These could be *quiet* and *comfort*. Should the sessions be repeated, less reliance will be placed on these circumstances to achieve, and hold, the hypnotic state—even in adverse environments.

Returning to ordinary consciousness involves a reverse "de-induction" procedure, where new disruptive forces successfully interfere with re-patterned consciousness, a short transition period takes place as the various bits-and-pieces reform, and the individual returns to the baseline state. While de-induction can be complicated, it is not necessarily a difficult process, and much quicker than inducing an ASC. That's because getting back to baseline is, in Tart's words, incredibly *overlearned*.

[11] Preller and Vollenweider (2018).

[12] Assuming hypnosis is a legitimate ASC.

[13] Tart (2008).

References

López Rodrigo, A. M., Paíno Piñeiro, M. M., Martínez Suárez, P. C., Caro, M. I., & Lemos Giráldez, S. (1997). Hallucinations in a Normal Population: Imagery and Personality Influences. *Psychology in Spain, 1*(1), 10–16.

Picard, F. (2013). State of Belief, Subjective Certainty and Bliss as a Product of Cortical Dysfunction. *Cortex, 49*(9), 2494–2500.

Preller, K. H., & Vollenweider, F. X. (2018). Phenomenology, Structure, and Dynamic of Psychedelic States. *Current Topics in Behavioral Neurosciences, 36,* 221–256.

Rabeyron, T., & Caussie, S. (2016). Clinical Aspects of Out-of-Body Experiences: Trauma, Reflexivity and Symbolisation. *L'Évolution Psychiatrique, 81*(4), e53–e71.

Tart, C. T. (2008). Accessing State-Specific Transpersonal Knowledge: Inducing Altered States. *Journal of Transpersonal Psychology, 40*(2), 137–154.

von Lucadou, W., & Wald, F. (2014). Extraordinary Experiences in Its Cultural and Theoretical Context. *International Review of Psychiatry, 26*(3), 324–334.

23

The Enigma of Unconscious Imagery

Having briefly summarised the jigsaw analogy of altered states, it's now possible to ponder how any new "pieces" of a reconfigured consciousness might enable unusual (paranormal) experiences. I've previously mentioned "lost" memories as a potential source of this mental imagery, and the likelihood that Tom would see the elusive figure of a man rather than a donkey in the hotel corridor, since Tom has perceived more humans in his life than donkeys. Then again, Tom believed he saw an actual person, not a memory. That means the end-product of re-patterning must create the same sense of reality as veridical perception. How does this happen?

While they are personally experienced in very different ways, *true* vision and *mental imagery* are strongly related. Both seem to "occur" in our consciousness and they possess similar phenomenal nature.[1] There's a further distinction between *regular* mental imagery of the kind you have when you're imagining tonight's dinner, and the mental imagery associated with visual hallucinations. The latter seem a lot closer to true vision than the former.

[1] Nanay (2016).

T. Jinks, *Psychological Perspectives on Reality, Consciousness and Paranormal Experience*,
https://doi.org/10.1007/978-3-030-28902-7_23

The distinction between mental imagery and hallucination may be teased out using three sets of dual orthogonal dimensions; *non-egocentric* versus *egocentric*, *conscious* versus *unconscious* and *voluntary* versus *involuntary*.[2] The first two dimensions apply to regular mental imagery; these images occur either in some distant and unconnected "abstract visualized space", or in "egocentric space"—meaning they are positioned in close physical proximity to you. Mental imagery is also most commonly thought of as conscious, but there is an unconscious facet which is behaviourally demonstrated by the automatic decisions people make when automatically manipulating mental imagery to achieve a visuo-spatial solution to a problem. By way of example, put yourself in the position of seeing a tipped ornament on a coffee table in a stranger's house. You lift it and immediately place it in the correct position. Presumably there's been an unconscious, reflexive analysis of the mental image of the object that allows you to perform the more appropriate behaviour. You don't need to practice with the actual, novel ornament which you've never seen before to meet the goal of returning it to its rightful position on the table.

It is the third dimension, voluntary versus involuntary, where hallucinations detach from regular mental imagery. You are able to actively summon and control regular mental imagery, like picturing tonight's dinner. When mental imagery is *passive*—meaning it emerges without volition—then *true* hallucinatory experiences will occur.

There's also the fourth more subtle aspect to mental visual imagery—the *feeling of presence* (not to be mistaken with the sense of presence). This feeling provides the additional ingredient guaranteeing that what the imagery *looks like it is real*. Try visualising a spoon lying on a dining room table. What is provided to your consciousness is not the tangible perception of an actual spoon. It's not even a very stable or accurate image. It's more like an *impression* or *concept* of a spoon than something seen with the eyes, but that's the nature of a normal mental image. However, if you unexpectedly see a spoon rise off a dining room table and float in the air in front of you, and it looks as though it is really there….well that's the feeling of presence that signifies a hallucination. It is the same for states of journeying (OBEs and NDEs) that are indistinguishable from actual

[2] Ibid.

perception,[3] or following psychedelic drug-use the reported "bump" from everyday reality to a complete altered state experience that seems entirely beyond (and not a part of) typical imagination. These events are perceived as real, and *remembered* as real, implying that the supporting mechanisms of conventional perception are enacted, including memory formation.

The feeling of presence is a manifestation of *reality monitoring*. When reality monitoring fails, mental imagery can't be distinguished easily from visual experience. This breakdown is facilitated by a cognitive trait called *transliminality*, meaning "across the line". The line represents the separation of conscious from unconscious material, and like many psychological traits transliminality is measured along a continuum (because everyone possesses to some degree the capacity to draw material from the unconscious).[4] For instance, assuming it is the afternoon and there's a prompt for you to remember what you had for breakfast—if indeed you had breakfast—then you are demonstrating movement of information from one source (vaguely referred to as the pre-conscious or unconscious) to another (conscious awareness).

The "line" in consciousness, very similar in many ways to a filter or tuner, is in delicate flux but;

> ...in a healthy adult, the threshold separating [conscious and unconscious] self has become largely stable and only changes punctually and in a predictable manner, as illustrated by the daily cycle of waking, dreaming, and deep sleep.[5]

Where the stability is affected and a *hardening* of the line occurs then a reduction of material emanating from the supraliminal results, and psychological disorders manifest, such as *conversion disorders*, involving;

> ... someone's consciousness not fully manifesting out of the [unconscious] and partly desynchronizing with its physical counterpart (the corresponding body), leading to a loss of awareness of what the body is doing.[6]

---

[3]Bensley (2003).

[4]Thalbourne and Houran (2000).

[5]Brabant (2016, p. 353).

[6]Ibid., p. 351.

Alternately, if the movement of content is very efficient—perhaps *too* efficient—there can be a state akin to spontaneous re-patterning of ordinary consciousness. Beliefs and expectations will more easily cross into awareness, which works well as a method of clarifying paranormal sightings as illusions. Ambiguous environmental stimuli are unclear and often short-lived. An individual expressing a high degree of transliminality might be confronted with a readily available interpretation in consciousness. As there's been no conscious exertion on the individual's part, accepting a ghost or UFO solution will be effortless. Indeed alien abductions[7] and other styles of entity encounters[8] have been attributed to heightened transliminality (expressed by the "victim").

A similar trait that can impact on mental imagery is *boundary deficit*, which refers to a *reduced* ability to distinguish between the sense-of-self and anything outside of, or beyond, the self. The boundary between the two concepts in permeable, and this results in a boundary-deficient individual mistaking their own "self-generated"[9] mental imagery (or thoughts, or other internally-derived stimuli) for external reality.[10] Alternately, they might confuse their own inner speech with an externally-spoken voice, accounting for many cases of auditory hallucination.

In summary, the impression of imaginative thought is neither locally positioned, nor does it possess a feeling of presence. For this reason, normal mental images are never confused with real-life things. However when mental imagery is locally fixed (egocentric) and accompanied by a feeling of presence, and emerges without volition (perhaps encouraged by traits such as transliminality and boundary deficit) then it can be confused with true perception.[11]

As for a *purpose*, the second-path argues regular mental imagery possesses a useful evolutionary function (taking a hard-line view that mental imagery doesn't occur, or is a meaningless accident of neurophysiology, is not helpful). Presumably mental imagery provides an adaptive advantage for important cognitive tasks, such as mentally manipulating objects

[7]Apelle (1995).
[8]Houran et al. (2003).
[9]Aynsworth et. al. (2017).
[10]López Rodrigo et al. (1997).
[11]Nanay (2016).

or scenes to determine future possibilities.[12] And because neural activity underpins normal (visual) perception, such activity must also be responsible for the production of mental imagery. They both emerge from the same brain.

There's considerable support for mental imagery as an offshoot of visual processing. In terms of responsible brain regions, there's an overlap between the striatal and extrastriatal activity that occurs when you witness a scene, and your imagination of that same scene. This is taken as evidence that;

> ...the mental processes that make perception possible are the very same mental processes that make mental imagery possible.[13]

An overlap does *not* mean the two states are indistinguishable, which would account for why mental imagery is not mistaken for real perception when the brain is functioning normally. Mental images arise from degraded cortical patterns derived from the actual visual perceptual pattern. The nature of this degradation (and why there should be variations) is not widely discussed, although it makes sense that healthy mental imagery should be a less complete neural pattern than authentic perception to assist reality monitoring. This prevents the feeling of presence, and triggers our understanding that a mental image has been voluntarily constructed.

Exploring the neurophysiological evidence further, it turns out that authentic perception and mental imagery share similarities in eye movement and pupil dilation. Eyes need to be in constant motion (saccading), for a person to experience visual stimuli *at all*. If an image falling on the retina remains stable, then visual experience comes to an abrupt stop. That's also the case for mental imagery, since;

> ...if subjects are asked not to move their eyes during visualizing, they have difficulties imagining the scene, and if they can do so, they attribute only very rudimentary features to the imagined object.[14]

---

[12] Zhao and Sala (2018).

[13] Nanay (2016, p. 73).

[14] Nanay (2016, p. 77).

This is a shared property of *determinacy* for both authentic perception and mental imagery. When looking at a real object, or attending to a mental image of that same object (and assuming similar cortical activity), examining a single property such as shape or colour will use identical eye movements to make this feature more "determinate" compared to other properties.

But here's the important difference. In authentic perception the information about an object arrives through bottom-up channels and the determinacy is an integral characteristic of the real stimuli. That's different to voluntarily or spontaneously-generated mental imagery where there's *no* stimulus, so the determinacy must come from somewhere else. In terms of the *cortical release phenomenon*[15] (also termed perceptual release theory[16]) these are cognitive and affective constructs; our beliefs, memories and expectations, that supplement the "skeleton" of an imaginary percept.[17] These top-down expectations are mediated by the prefrontal cortex which accesses the workings of lower sensory target cell networks in the posterior sensory cortex to produce;

> ...specific combinatorics determined by the binding codes of the dispositional space, and that consist of memory substitutes for non-remembered perceptions, with the "feeling of reality" being reinforced by the reduced control of prefrontal centers.[18]

What on earth does that mean? Simplistically, we experience the former (imagination) in the absence of the latter (photons of visible light). Without receptive organs *perceiving* something that's transduced by way of a functioning retina there can be no visual (or any other) sensory experience.[19] That gives top-down processes the opportunity to force a pattern of imagery onto the empty canvas of the bottom-up neural assemblies

[15]Arnulf (2013).

[16]Kosman and Silbersweig 2018.

[17]Naish (2005).

[18]Goetzmann et al. (2018, p. 25).

[19]Greene (1999).

that normally provide *real* visual perceptions to consciousness (but in this case, there's nothing "real" to display).[20]

This is believed to be the difference between actually seeing something "out there" in the world, and imagining something in your mind's-eye. For normal mental imagery, there's distortion and interference from countless cognitions. When trying to imagine that spoon on a table, the word "fork" might unavoidably enter your mind through semantic association and it becomes hard *not* to visualise a fork instead. You accept this as annoying and unfortunate interference, unless through personal discipline and practice your meddlesome beliefs, memories and expectations can be held in check.

A non-volitional mental image—most likely a memory of something you've once seen like a man in a dark suit walking left-to-right across your field of vision—is merely the *seed*. With the feeling of presence erroneously activated through failure of the *discrimination reality process*,[21] the nature of the top-down cognitive interference will cause this jigsaw piece entering re-patterned consciousness and allow creation of your very own ghost sighting. And the more unexpected the image, the more credible it is as a foreign (exogenous) product. If the apparition is of a person who comes and goes without trace, and like Tom you can investigate your surrounds to confirm the spectre wasn't physical, then you're not liable to hold something labelled the "discrimination reality processes" responsible and admit you've hallucinated. You'll assume you've seen a *real* ghost.

There are plenty of other paranormal experiences that fit this model of mistaken perception. The brain determines the model of reality best suited to a situation based on incoming sensory information. When there's insufficient stimulation (or perhaps none at all), determinancy from top-down storehouses of memory replaces the missing sensory data. The OBE or NDE is therefore a disorder of top-down cognitive processes where the alleged traveller enters into;

[20]There are problems with this viewpoint, as will be mentioned shortly.

[21]Slade and Bentall (1988).

> ...the inner spaces of dreams and visions...they simply mistake these inner journeys for actual excursions outside the body... misconstruing the subjective contents of their imaginations for the objective properties of the external world.[22]

But why is this dream-like experience accompanied by an out-of-body impression? The answer is elegant. Many memories are stored in the form of a "bird's eye" perspective, meaning they are external to, and above, the individual. People with a propensity for this type of memory representation, and who have entered an ASC, will experience novel jigsaw pieces of mental imagery from this extrapersonal vantage point.[23]

While the second-path cannot supply direct evidence that paranormal experience is the product of the insertion into a mental space of stored (but forgotten) imagery and other sensations, this conjecture is parallel to what's known about hallucinations arising from brain pathology, so there's a certain logic to the claim. In Charles Bonnet Syndrome there is an absence of bottom-up visual information[24] which leaves a visual "blank slate" to be manipulated by higher cognitions. And for hallucinations stemming from Lewy body dementia and Parkinson's disease, the idea that fundamental visual processing areas are not providing the bottom-up material to accurately process is supported by measurements of *hypo*perfusion and *hypo*metabolism of the occipital cortex, and temporal and parietal lobes.[25] If real visual sensations are prevented from entering the brain, it makes sense to replace the missing input with something you already have—memories.

This does *not* mean that for us to "see" (or hear, or feel, taste or smell) something that's not perceived through our senses requires a neural activity pattern to be induced in our cortex identical to an *actual* perception. In terms of conventional mental imagery the cerebral activity recorded in experimental participants who self-report imaginative thoughts is at odds

---

[22] Greene (1999, p. 155).

[23] Blackmore (1992).

[24] Abbott et al. (2007).

[25] Furuya et al. (2006).

with what is expected when they witness a true visual stimulus.[26] Likewise, *hallucinated* imagery doesn't match the pattern of cortical activity typical of everyday vision. In one clinical case, a 66-year-old man with Parkinson's disease reported recurring visions of different kinds of people and animals; threatening tribesmen, chimpanzees swinging in trees, men wearing Civil War-era uniforms and Catholic nuns. As partial component hallucinations (or content specific delusions), they seamlessly conformed to his wider (normal) visual experience. Although the patient was aware that these characters were not authentic (he recognised them as pseudohallucinations), their realism left him profoundly disturbed.

Fortuitously, he suffered sixteen of these pseudohallucinations while undergoing fMRI in a laboratory setting. At the time of these experiences, increased neural activity was recorded in regions not typically associated with visual imagery. These include the cingulate gyrus and insula, and the medial frontal gyrus, postcentral gyrus, thalamus and brainstem.[27] Conversely, *less* activity was recorded in regions corresponding to normal visual system activity—specifically various "gyri" in the occipital, cingulate and temporal lobes.[28] Although these findings came from a single patient, perhaps all hallucinations are caused by a shift from analysing actual visual cortical activity in the occipital (rear) area of the brain, to processing activity in frontal brain regions corresponding to top-down influence. That's supported by the researchers who argue;

> …a misalignment of occipital-frontal circuits leading to imposition of an aberrant "top-down" processing of the visual system that gives rise to interpreted visual perceptions.[29]

Thus far, the primary difference between mental imagery and hallucinated imagery is the vague concept of *feeling of presence*. The more "feeling" you have, the more tangible the imagery.

---

[26] Bridge et al. (2012).

[27] Goetz et al. (2014).

[28] Ibid.

[29] Ibid., p. 116.

One category of imagery that has a very strong feeling of presence is *eidetic* (photographic) imagery.[30] When you perceive a stimulus there will be a residual trace remaining in consciousness for a short time once the actual source has departed. This *after-sensation*, may be a visual image, or in another sensory capacity (like touch) the prolonged impression of pressure on the skin.[31] An early pioneers of eidetic imagery, Erich Jaensch, considered eidetic states to be somewhere between an actual sensory stimulus and an afterimage, with properties of;

> …a clear and lifelike inner experience of stimuli that just occurred or *that are long-standing*.[32]

While eidetic imagery is very short-lived for most people, in some it can manifest for unusual lengths of time. Psychology Professor Joseph Gengerelli refers to a female subject he studied in 1930 who was able to mentally project a circle she had recently examined visually (but now out-of-sight) onto a square target square, and then perform a detailed mental analysis of the two overlayed shapes for goodness-of-fit;

> It was an experience…to watch [her] at work. She often squinted her eyes, sometimes closing one, then the other, much like a person comparing two objects physically present. She rubbed her eyes from time to time as if they had been strained, and after the experiment she complained of a slight headache. She expressed amazement….that everybody could not project simple line figures onto walls or smooth surface, 'I put the circles on the square and notice what happens, if they stick out of the sides of the square of don't reach it'.[33]

This eidetic imagery is able to be held in active consciousness for a considerable time, meaning the imagery must be somehow stored in a strong and persistent neural form. It's not difficult to argue that these are how some normally-inaccessible memories are stored in *everyone*, and when

[30] Sayin (2014).

[31] Feiman (2017).

[32] Feiman (2017, p. 132).

[33] Gengerelli (1976, pp. 220–221).

they are released following ASC induction they join the new patterning as a puzzle piece in a partial component sensory experience. They consist of a short and *vivid* neural program that can be immediately and effortlessly harnessed by top-down processes to become a comfortable ingredient of the new, altered conscious experience.

It's believe the release of eidetic imagery might be assisted by certain psychological traits including schizotypy, dissociation, transliminality and boundary thinness.[34] Combined, these are the personality characteristics of an "anomaly-prone" individual, so any eidetic imagery emerging in a re-patterned consciousness will be accompanied by the impression that what's witnessed is autonomous to the individual, made more substantial by the feeling of presence. That's why anomaly-prone *types* are so open to seeing "real-life" ghosts, or indulging in spontaneous and unexpected OBEs.

References

Abbott, E. J., Connor, G. B., Artes, P. H., & Abadi, R. V. (2007). Visual Loss and Visual Hallucinations in Patients with Age-Related Macular Degeneration (Charles Bonnet Syndrome). *Investigative Ophthalmology and Visual Science, 48*(3), 1416–1423.

Apelle, S. (1995). Point Counterpoint. *Mercury, 24*(1), 9.

Arnulf, I. (2013). Dream Imagery, Rapid Eye Movement Sleep Behavior Disorder, and Hallucinations. *Sleep and Biological Rhythms, 11*(Suppl. 1), 15–20.

Aynsworth, C., Nemat, N., Collerton, D., Smailes, D., & Dudley, R. (2017). Reality Monitoring Performance and the Role of Visual Imagery in Visual Hallucinations. *Behaviour Research and Therapy, 97*, 115–122.

Bensley, D. A. (2003). Can Minds Leave Bodies? *Skeptical Inquirer, 27*(4), 34–40.

Blackmore, S. J. (1992). *Beyond the Body: An Investigation of Out-of-the-Body Experiences.* Chicago: Academy Chicago Publishers.

Brabant, O. (2016). More Than Meets the Eye: Towards a Post-materialist Model of Consciousness. *Explore, 12*(5), 347–354.

Bridge, H., Harrold, S., Holmes, E., Stokes, M., & Kennard, C. (2012). Vivid Visual Mental Imagery in the Absence of the Primary Visual Cortex. *Journal of Neurology, 259*(6), 1062–1070.

[34]Thalbourne and Maltby (2008).

Feiman, G. (2017). Eidetic Memory and School Age. *Journal of Russian & East European Psychology, 54*(2), 130–191.
Furuya, H., Ikezoe, K., Ohyagi, Y., Miyoshi, T., & Fujii, N. (2006). A Case of Progressive Posterior Cortical Atrophy (PCA) with Vivid Hallucination: Are Some Ghost Tales Vivid Hallucinations in Normal People? *Journal of Neurology, Neurosurgery and Psychiatry, 77*(3), 424–425.
Gengerelli, J. A. (1976). Eidetic Imagery in Two Subjects After 46 Years. *Journal of General Psychology, 95*(2), 219–225.
Goetz, C. G., Vaughan, C. L., Goldman, J. G., & Stebbins, G. T. (2014). I Finally See What You See: Parkinson's Disease Visual Hallucinations Captured with Functional Neuroimaging. *Movement Disorders: Official Journal of the Movement Disorder Society, 29*(1), 115–117.
Goetzmann, L., Ruettner, B., & Siegel, A. (2018). Fantasy, Dream, Vision, and Hallucination: Approaches from a Parallactic Neuro-Psychoanalytic Perspective. *Neuropsychoanalysis, 20*(1), 15–31.
Greene, F. G. (1999). A Projection Geometry for Separation Experiences. *Journal of Near-Death Studies, 17*(3), 151–191.
Houran, J., Ashe, D. D., & Thalbourne, M. A. (2003). Encounter Experiences in the Context of Mental Boundaries and Bilaterality. *Journal of the Society for Psychical Research, 67*(873), 260–280.
López Rodrigo, A. M., Paíno Piñeiro, M. M., Martínez Suárez, P. C., Caro, M. I., & Lemos Giráldez, S. (1997). Hallucinations in a Normal Population: Imagery and Personality Influences. *Psychology in Spain, 1*(1), 10–16.
Naish, P. (2005). Detecting Hypnotically Altered States of Consciousness. *Contemporary Hypnosis, 22*(1), 24–30.
Nanay, B. (2016). Hallucination as Mental Imagery. *Journal of Consciousness Studies, 23*(7–8), 65–81.
Sayin, H. U. (2014). Does the Nervous System Have an Intrinsic Archaic Language? Entoptic Images and Phosphenes. *Neuroquantology, 12*(3), 427–445.
Slade, P. D., & Bentall, R. P. (1988). *Sensory Deception: A Scientific Analysis of Hallucination.* Baltimore: Johns Hopkins University Press.
Thalbourne, M. A., & Houran, J. (2000). Transliminality, the Mental Experiences Inventory, and Tolerance of Ambiguity. *Personality and Individual Differences, 28*, 853–863.
Thalbourne, M. A., & Maltby, J. (2008). Transliminality, Thin Boundaries, Unusual Experiences, and Temporal Lobe Lability. *Personality and Individual Differences, 44*(7), 1617–1623.
Zhao, B., & Sala, S. D. (2018). Different Representations and Strategies in Mental Rotation. *Quarterly Journal of Experimental Psychology, 71*(7), 1574–1583.

24
Cultural Wiring

As alluded to in the preceding chapter, the second-path identifies top-down CNS systems as the origin of mental imagery, whereby neural activity associated with these networks corresponds to the phenomenological experience of personal memories. However, there are more options to choose from than isolated individual recollections of autobiographical events—perhaps equivalent to Tom's "walking man". Memories are moulded by *cultural wiring*,[1] or to use another analogy, strained through established systems of attitudes, beliefs and expectations to form the components of active thought. In turn, these systems are configured by social and cultural *forces*, meaning the content of imagery is often shared amongst many people and builds on the background experiences generated by physiological constants.

This idea is covered by the word *psychosocial* or *psychocultural*, where;

> ...all forms of human expression insofar as they have been affected by cultural influence. There can be no sharp line of demarcation between the

[1] Howard (2011).

T. Jinks, *Psychological Perspectives on Reality, Consciousness and Paranormal Experience*,
https://doi.org/10.1007/978-3-030-28902-7_24

> sphere of culture on the one hand, and that of organic and psychological phenomena on the other. Culture is an attribute of human behavior and is therefore to be studied as an integral part of human behaviour.[2]

This provides yet another layer to the phenomenological substance of an ASC. "Psychosocial" is a word implying learned social/cultural information assist the psychological interpretation of internal and external stimuli, derived from a massive warehouse of stored experience derived from movies, TV shows, and pictures from fictional illustrations in books and comics.[3] And most importantly, people with relatively similar cultural backgrounds will have comparable jigsaw pieces potentially available should consciousness be re-patterned during ASC induction.

Psychosocial theory has been long utilised when accounting for the intricate details of witnessed paranormal experiences—particularly alien abductions. One of the best-researched examples centres on the exotic (even for UFOs) 1989 case of Linda Napolitano, investigated at that time by the artist and author Budd Hopkins. As the story goes, Napolitano lived in an apartment building in Manhattan and claimed she was floated out of her 12th floor window and into a waiting UFO. This extraordinary event was witnessed by three men in a car two-blocks away, one being the United Nations Secretary General Javier Perez de Cuellar and the other two his security agents. Following the abduction, Linda went through a series of earthy ordeals, including being kidnapped by one of these very security agents (who appeared to be infatuated with her).

However, as sceptical investigators Joseph Stefula, Richard Butler and George Hansen point out, a novel by Garfield Reeves-Stevens called *Nighteyes* was published not long before the alleged abduction, in April 1989, and the similarities are startling. These include a female character in the story being abducted from a high-rise apartment in NYC, security guards on an early morning stakeout seeing the UFO abduction, her kidnapping to a safe house at a beach by the security agents, the hospitalisation of one of the agents for trauma, and much more, leading Stefula, Butler and Hansen to argue;

---

[2]Bidney (1949, p. 32).

[3]Sparks et al. (1998).

> The parallels are sufficiently numerous to lead us to suspect that the novel served as the basis for Linda's story. We want to emphasize that the parallels are with discrete elements of the case and not with the story line itself.[4]

That's not implying that Linda was deliberately lying. Storylines contained in all facets of life become personal memories, but if we lose the conceptual "tags" that identify these memories as fictional events (or even authentic events experienced by other people and publicised in the media) then they may be mistaken as real memories.

Traditional cultures are not spared from the psychosocial hypothesis. The desert-dwelling Warlpiri of central Australia provide extensive reports of seeing UFOs, and while difficult to establish just for how long they've been seen;

> ...witnesses uniformly described UFOs as similar to most television and film portrayals...large, disc-shaped, brightly lit objects that fly through the sky...piloted by intelligent beings from outer space.[5]

These UFOs are frequently sighted around water supplies—creeks, rain clouds, and waterholes. Apparently that's because the aliens require water. These locations are also the home of Warnayarra—giant multi-coloured rainbow serpents, and yet there is no interaction between the UFOs and the Warnayarra. The suggestion is that the UFO material is gleaned from watching representations of UFOs on television and video (e.g. X-files, Stargate, Star Trek and Independence Day) and this fuses modern mythology with traditional mythology. The themes being passively watched on TV (often repeatedly) are incorporated into a local framework—in this example, water-seeking. The anthropological reason for the UFO sightings is to symbolise identical close encounters between Aboriginal and non-Aboriginal people, seen in the fact that only non-aboriginals are the only ones who actually get abducted by the UFOs.

The strength of psychosocial forces is also temporal. Within a society the content and interpretation of storylines will eventually change over time, hence the popular notion that where people once saw fairies in the

[4] Stefula et al. (1993, p. 14).

[5] Saethre (2007, p. 223).

fields, they now see aliens who alight from silver spaceships.[6] Indeed, the typical alien is;

> ...hairless, small and slender entity with a greyish skin tone, degenerate facial features and large black eyes.[7]

It's been argued that visage derives from Christian demonology, where the modern grey is a high-tech demon.[8] The reptilian-style aliens, however, are (in free-wheeling, sociological language);

> ...ultimately a reactionary and conservative icon, one not allowing for the freeplay of differences, and its seems to represent a future characterized by the "fifth discontinuity", in which a superior species enslaves and perhaps destroys humanity. Hence, the reptoid can be read as an emblem of dystopian warning about limit transgression.[9]

And in terms of being abducted by these creatures, the narrative is said to have evolved from the popular storyline of captivity and *transculturalization* leading to tensions in personal identity extensively found in popular literature. An example is the tale of settlers being captured and imprisoned by American Indians,[10] where the mythical power of the frontier is strong enough to drive a *consistent* and particular experiential storylines.

The influence of psychosocial factors is also prolifically discussed in the field of sleep paralysis. Cultural content doesn't determine its *occurrence*, which is driven by universal physiological constants, but its *specifics*, such as the appearance of the Old Hag. Similarly, shared religious experience is alleged to originate from ASCs induced by shamanic-like practice. Forensic scientist H. Ümit Sayin takes the popular anthropological view that psychotropic plant ingestion has been practiced since at least Neolithic times, and due to thousands of years of common cultural wiring;

[6] Bartholomew et al. (1991).
[7] Kottmeyer (2001).
[8] Partridge (2004).
[9] Lewis and Kahn (2005, p. 62).
[10] Sturma (2002).

> …Most of the mythological figures and characters that exist in tribal religions, in ancient religions and also in the institutionalized modern religions, such as demons, angels, Satan, semi-human animals, gods, goddesses, spirits, ghosts and many others [ghosts, angels, nymphs, spirits, demons, monsters, mythical creatures, supra-natural creatures, Satan, gods and goddesses] originate from the illusions and hallucinations of ancient "ritual people" in Psychedelic Religious Rites (PRR) who had used psychoactive plants during their religious rituals and who had been very 'high' on psychedelic plants……Thus, many factors in the old and new religions, which are a continuation of the old pagan traditions, could have been created by the human imagination and altered perception during religious rituals.[11]

Although not himself convinced, author Grahame Hancock acknowledges;

> …I suppose one possibility that would not entirely contradict the scientific materialist paradigm would be that some hitherto undiscovered library of extraordinarily detailed mental images, sensations and information is…wired into our brains – same for all and accessed via ASC.[12]

There's not really that much more to be said for psychosocial influence over the subject matter of a paranormal experience, other than examining cultural material on a case-by-case basis to determine potential sources of later reports; such as the role played by science fiction (going back to the 1930s) in determining the shape of the alien abduction phenomenon that emerged in the mid-1980s.[13] The best that can be said is that complicated paranormal experiences are "culture bound" constructs of the mind,[14] where

> The commonality of these experiences would suggest that mind and body both play a part in the type of metaphysical experience one undergoes… and

[11] Sayin (2014, p. 428).
[12] Hancock (2005, p. 335).
[13] Kottmeyer (2001).
[14] Hume (1995).

that a specific cultural ontology would supply the details and subsequent meanings of that experience.[15]

Of course, it's fair to ask why mental re-patterning with visual imagery should occur in the first place—why belief in (and experience of) supernatural entities is so universal when other beliefs and experiences also assisting individual and societal longevity wouldn't need the profound separation from reality that ASCs engender.

The easiest answer is that there is *no* evolutionary advantage, or even meaningful reason, to enter an ASCs and experience a culturally-influenced smorgasbord of novel vivid imagery. People simply have the inadvertent capability to do so. Nevertheless we all express individual differences, so due to quirks in neurophysiology and personality some are able to do enter ASCs spontaneously—for example by overproducing endogenous DMT or expressing boundary deficits that allow psychosocial content to leak more efficiently into their mental space.

For the second-path, the strongest argument for the experiences generated by that ASC as *not* paranormal stems from the irrationality of apparitions, OBEs, NDEs, astral journeys or anything else. People who are self-aware while disembodied might *claim* they are able to influence the physical environment. However they are not physical in a physical state, so informed by the dualism conundrum they cannot actually touch objects, let alone move them.[16] The same applies to manifesting as an apparition—there are no material ingredients that would allow such an entity to propagate, and of course there are no physical eyes (or other senses) that would allow this entity to interact with environmental energy. For this reason apparitions, or the concept of astral journeying or bilocating, are against all conceptions of scientific reality, and indeed the very notion of disembodiment is incomprehensible since there cannot be location in time and space if a boundary no longer exists between the brain/body and the environment.[17] Since the same phenomenological experiences[18]

[15]Ibid.

[16]Greene (1999).

[17]Ibid.

[18]Mobbs and Watt (2011).

can be achieved from known aetiologies (like psychedelics), paranormal experiences are simply an alternative expression of the same process.

No sensible mainstream journal has published demonstration of how a traveller has moved an object during an OBE, or brought back perplexing information from an NDE or astral journey (*state-specific knowledge* only accessible in an ASC).[19] When attempting to authenticate this paranormal knowledge, the traveller finds they mistaken;

> For example, they might seem to float up through the ceiling of a building housing their physical body and see a chimney on the roof that they later discover is not there.[20]

That's even presuming these travellers are accurately recalling their experiences after they re-establish an ordinary state of consciousness. After all, human cognitions can be grossly affected by "crisis, injury or novelty".[21]

Regarding the ability of psychedelic drugs to access normally unattainable information, early research on the effect of LSD in changing a user's *philosophical, epistemological* and *ontological* perspectives[22] suggested to Humphrey Osmond (and others) the drug had anomalous capabilities. However by the mid-1960s it was demonstrated that administration of LSD does not miraculously enhance personal creativity or any other skill, although it might assist access to remote ideas and associations[23] (e.g. through improved semantic priming)—an idea affiliated with enhanced transliminality and the fashioning of porous borders between waking consciousness and unconsciousness encouraged by an ASC.

If that's the case;

---

[19] Tart (2008).

[20] Greene (1999, p. 155).

[21] Hanes (2012); although this author claims "the evidence supporting the reality and validity of such events is impressive".

[22] Dyck and Farrel (2018).

[23] Zegans et al. (1967).

Creative individuals may sometimes be able to transform these experiences into a creative work, but it is unlikely that creativity is enhanced by serotonergic hallucinogens in individuals who do not ordinarily show creative behaviour.[24]

References

Bartholomew, R. E., Basterfield, K., & Howard, G. S. (1991). UFO Abductees and Contactees: Psychopathology or Fantasy Proneness? *Professional Psychology: Research and Practice, 22*(3), 215–222.

Bidney, D. (1949). Towards a Psychocultural Definition of the Concept of Personality. In S. S. Sargent & M. W. Smith (Eds.), *Culture and Personality* (pp. 31–55). New York: Viking Fund.

Dyck, E., & Farrel, P. (2018). Psychedelics and Psychotherapy in Canada: Humphry Osmond and Aldous Huxley. *History of Psychology, 21*(3), 240–253.

Greene, F. G. (1999). A Projection Geometry for Separation Experiences. *Journal of Near-Death Studies, 17*(3), 151–191.

Hancock, G. (2005). *Supernatural: Meetings with the Ancient Teachers of Mankind.* London: Century.

Hanes, K. (2012). Unusual Phenomena Associated with a Transcendent Human Experience: A Case Study. *Journal of Transpersonal Psychology, 44*(1), 26–47.

Howard, R. G. (2011). If Paranormal Phenomena Are Information Received by the Conscious Psyche, Then What Is the Conscious Psyche? *Journal of Spirituality and Paranormal Studies, 34*(2), 62–72.

Hume, L. (1995). By Knot of One the Spell's Begun, By Knot of Two It Cometh True… Magic, Altered States, and Multiple Realities. *Social Alternatives, 14*(4), 23–27.

Kottmeyer, M. S. (2001). Grays. In R. D. Story (Ed.), *The Mammoth Encyclopedia of Extraterrestrial Encounters* (pp. 273–281). London: Constable and Robinson Ltd.

Lewis, T., & Kahn, R. (2005). The Reptoid Hypothesis: Utopian and Dystopian Representational Motifs in David Icke's Alien Conspiracy Theory. *Utopian Studies, 16*(1), 45–74.

[24]Preller and Vollenweider (2018, p. 240).

Mobbs, D., & Watt, C. (2011). There Is Nothing Paranormal About Near-Death Experiences: How Neuroscience Can Explain Seeing Bright Lights, Meeting the Dead, or Being Convinced You Are One of Them. *Trends in Cognitive Sciences, 15*(10), 447–449.

Partridge, C. (2004). Alien Demonology: The Christian Roots of the Malevolent Extraterrestrial in UFO Religions and Abduction Spiritualities. *Religion, 34,* 163–189.

Preller, K. H., & Vollenweider, F. X. (2018). Phenomenology, Structure, and Dynamic of Psychedelic States. *Current Topics in Behavioral Neurosciences, 36,* 221–256.

Saethre, E. (2007). UFOs, Otherness, and Belonging: Identity in Remote Aboriginal Australia. *Social Identities, 13*(2), 217–233.

Sayin, H. U. (2014). The Consumption of Psychoactive Plants During Religious Rituals: The Roots of Common Symbols and Figures in Religions and Myths. *Neuroquantology, 12*(2), 276–296.

Sparks, G., Pellechia, M., & Irvine, C. (1998). Does Television About UFOs Affect Viewers' UFO Beliefs? An Experimental Investigation. *Communication Quarterly, 46*(3), 19.

Stefula, J. J., Butler, R. D., & Hansen, G. P. (1993). *A Critique of Budd Hopkins' Case of the UFO Abduction of Linda Napolitano.* Retrieved November 20, 2018, from http://tricksterbook.com/ArticlesOnline/LindaCase.pdf.

Sturma, M. (2002). Aliens and Indians: A Comparison of Abduction and Captivity Narratives. *Journal of Popular Culture, 36*(2), 318–334.

Tart, C. T. (2008). Accessing State-Specific Transpersonal Knowledge: Inducing Altered States. *Journal of Transpersonal Psychology, 40*(2), 137–154.

Zegans, L. S., Pollard, J. C., & Brown, D. (1967). The Effects of LSD-25 on Creativity and Tolerance to Regression. *Archives of General Psychiatry, 16*(6), 740–749.

25

Forcing the Boundaries of the Second-Path: A Purposeful Unconscious

The hefty content presented in the last few chapters provides a sensible, second-path explanation for *how* people see ghosts and perhaps other weird things. Hard science passed much of this *how* question to cognitive psychology to work out, such as *how* mental imagery is distorted by top-down processes, or *how* consciousness is disrupted and re-patterned in an altered state. Anthropology and sociology were also able to provide some input into *why* the imagery might take the form it does, through the psychosocial hypothesis.

To repeat, *why* did Tom see the "ghost" of a man and not a donkey? That comes down to life experience and cultural wiring—he certainly hasn't a strong history of contact with donkeys. Presuming he was in an induced, partial component sensory ASC in the hotel corridor, and re-patterning let in a piece of imagery (eidetic or otherwise) from a top-down storehouse join the novel but transient jigsaw of his waking consciousness, then *probabilistically* the projected imagery will be something more common—a man, not an animal.

If pressed further about the *significance* of the human apparition, the second-path cognitive psychologist should reiterate the randomness of

T. Jinks, *Psychological Perspectives on Reality, Consciousness and Paranormal Experience*,
https://doi.org/10.1007/978-3-030-28902-7_25

hallucinated imagery. However, they might admit that an answer to the question of *why* can be found in a more controversial field—*psychodynamics*.

This urge to overstep scientific reality and ponder the deeper structures of consciousness, thus giving purpose to an apparition sighting doesn't sit comfortably within the parameters of the second-path. And yet it is not *necessary* to pursue these ideas, since the psychosocial hypothesis is enough to provide closure to the storyline of a paranormal experience. Not *necessary*, but possible…and entertaining.

For psychodynamics, the structures of the mind are more than a self-aware consciousness drawing from a formless unconscious storehouse (that's been taken for granted so far). Although far from taking the second-path, Frederick Myers' psychical perspective is helpful. He wrote of the *supra*liminal (conscious or ego-based) mind as balanced above the "line", and the *sub*liminal mind, or minds, below, and which are accessible only in styles of ASC such as dreams, trance, ecstasy and trauma[1] (although Myers naturally didn't use the ASC expression). Hence;

> The information allegedly received from discarnate spirits could, according to Myers, derive instead from the subliminal selves of the living participants of the séance communicating telepathically with the subliminal self of the medium.[2]

The second-path can easily dismiss the telepathic aspect of Myer's work, but might graciously concede that the unconscious is something other than a messy catalogue of potential jigsaw pieces chosen at random. That means granting the unconscious some degree of autonomy over the expression of the database. Information travels backward and forwards across the liminal space in a *dynamic* manner, and this more influential unconscious has a significant ability to sway the here-and-now of mental life (and ultimately, affect behaviour).

Tom sees the ghost of a younger, pallid man because this particular clip was selected over the ghost of a ruddy-faced old man, or a grey-haired

[1] Poller (2015).

[2] Ibid., p. 251.

woman, or a crawling child; any of these human entities are more readily available than the apparition of a donkey, but the *younger man* might have a particular (unknown) purpose that encourages its selection as the apparition of choice.

This does mean a return to dualistic language, and opens the gate to discussion of unconscious primal forces akin to those revealed by Sigmund Freud and later psychoanalysts. This is an anathema to the second-path, but fun to touch on. In this approach a *conservation of energy* premise is applied to the conscious/unconscious nexus. The brain is a battery of energy, and this energy infuses the thoughts, images and other pieces that make up baseline consciousness. If these pieces are discarded from the jigsaw puzzle—something as simple as being forgotten—they do not *dissipate*. Instead they *constellate* in a deeper mind/brain realm. And when those thoughts or images (or any other content of the psyche) are unpleasant or unacceptable to waking consciousness, then they might be held or *repressed* in this space while their energy will be expressed in other forms, such as in the substance of dreams, behaviours such as projection, and illnesses identified as *hysteria*.[3]

For a dynamic unconsciousness to be acceptable to the fringes of the second-path, the label *dual-aspect monism* might be applied.[4] Psychodynamics approaches, including psychoanalysis, focus on subjective experiences in the form of dreams and other mental states as the *actualization of unconscious complexes*. Simultaneously, these experiences are peculiar events brought about by a complex arrangement of cortical neural activity, using professional words such as "neuromatrix".[5] The latter is most *correct*, but the former can be deliberated with assurance because it has no impact on the *reality* of neurophysiology.

Toying with the possibility that a dynamic unconscious governs which jigsaw puzzle pieces that are "seen" during an ASC, a map has been developed that lays out the increasing sophistication of mental imagery in the altered state. This is somewhat like Dittrich's APZ typology, but provides a *purpose* for the different stages. Specific to psychedelic drug use,

[3] Hall (1959).
[4] Solms and Turnbull (2002).
[5] Goetzmann et al. (2018).

the first stage involves those straightforward modifications we've already encountered—impressions of vivid colour and movement in basic geometric shapes, and also mild transformations of environmental scenes (distortions of landscape and architecture), as well as impressions of people and animals.[6] The visual stimuli are believed to derive from unconscious storage—building blocks of autobiographical memory tinged with sociocultural features. Most importantly, they emerge in a random fashion, without any meaning or purpose. That fits well with a grounded second-path approach to ASC. If that's all hallucinations are, then Tom's ghost was an accidental aberration.

Luckily, there is a second stage beyond the release of haphazard imagery, the *recollective–psychodynamic* level. In this stage of the ASC, the experiencer reports heightened arousal and the visual imagery they encounter takes on a very personal flavour. As the boundaries between consciousness and unconsciousness weaken they encounter more energetic unconscious material infused with emotional valence. Such events have a degree of *purpose* because of underlying unconscious motives such as guilt or desire.[7] These scenarios are less common than the components released during the first stage, since a significant breakdown of the liminal space needs to occur before they are accessed. Put simply, the mental illness needs to be severe, or the drug ingestion considerable![8]

A third stage of the ASC surpasses the personal unconscious with its emotion-charged repressed content. This *affective and symbolic* phase involves temporary dissipation of ego-boundaries that signify the breakdown of the global state. Despite these challenges, the experiencer remains aware of the ambiguous situation they find themselves in.[9] Most significantly, the components of this advanced ASC have symbolic meaning,[10] presented in the format of a "symbolic drama". These display;

> …mythological and ritualistic overtones, and subjects may identify features of their own existence in legendary historical figures, fairy tales,

[6] Preller and Vollenweider (2018).

[7] Kosman and Silbersweig (2018).

[8] Relative to the individual.

[9] Preller and Vollenweider (2018).

[10] Ibid.

> and...themes or other symbols and play out their personal drama on these allegoric terms.[11]

These storylines could be construed as invariant "cultural" content shaped by psychosocial processes. And of particular psychodynamic interest;

> When the subjects encounter and struggle with these dramas, they can achieve a "solution," e.g., by imagination, ideation, sensations, and affective or kinesthetic involvement. This can result in a quiet but powerful emotional response and tension release that appears to be transformative and beneficial to the person.[12]

Entry into a component and global ASC consisting of extra-personal material allows you to begin a journey towards *peak experience* where ego function no longer dominates. And as ego dissolution (and boundary porosity) progresses, the fourth stage of the ASC becomes this peak; a deep transcendence of self, replaced by "oneness" with the universe. This is a state of;

> ...pure consciousness [where] there is little phenomenological content, and an absence of dualistic perception and sense of self. Nondual states are characterized by pure awareness, free from fragmentation into dualistic thinking or experience, such as the sense of separateness between self and other.[13]

This can only be achieved in people with highly developed personalities, who have had recurrent psychedelic experiences and who have some understanding of the psychodynamic nature of consciousness;

> When subjects have their eyes open, they retrospectively describe this state as an intense emotionally charged new and unfamiliar perspective, almost like a direct encounter with the "ultimate" reality, which can inspire feelings of awe, sacredness, and eternity. This novel experience is also characterized by a pervasive sense of deep insight into the nature and structure of the

---

[11] Ibid., p. 231.

[12] Ibid.

[13] Wahbeh et al. (2018, p. 19).

> universe that is far beyond the person's usual mode of thinking.… However, when subjects have their eyes closed and turn their attention inward, a state of internal absorption may unfold, with subjects witnessing a vast internal space of objectless infinity that lacks not only the sense of the self, but also all sensory experiences and distracting thoughts. …..may reflect what has been termed as the "pure consciousness event".[14]

Here there is a *positive* purpose to the ASC, regardless of whether the individual recognises it or not. That's applied to the psychoanalytic view of religious experience, where the individual avoids ego-dissolution but is;

> …subjectively invested with supernatural or *preternatural* meaning (interpreting the experience as connected with God, spirits, the devil, or with magical influences) resulting often in a positive personal and social transformation.[15]

On the other hand, altered state experiences can also exhibit a psychodynamic outcome with no necessary transformation. An example is *entity possession*, referring to a dissociative state which appears as though the individual's personality and outward identity (their baseline consciousness) has been "taken over" by a malevolent entity. However this entity is in fact another aspect of the "self".[16] Possession occurs as a function of the *hypnoid phenomenon*, a style of self-induced trance which is a defence mechanism initiated by the individual's unconscious to counter "unbearable psychic pain and trauma".[17]

The depersonalization doesn't necessarily have to involve an "alien" intruder.[18] The self can also be externalised—or *estranged*—to achieve the same method of protection. From this perspective, OBEs are a protective mechanism of escape-from-self;

[14] Preller and Vollenweider (2018, p. 241).
[15] Prince (1994, p. 17) quoted in Bartocci (2004).
[16] Blum (2013).
[17] Ibid., p. 431.
[18] Blum (2013).

> .... if the subject-self is not dissociated from the scene it is experiencing, the psyche can be overwhelmed by the traumatic experience and retains traces of it. BUT If it cuts itself off completely from surrounding reality, its chances of survival decrease – so it needs to OBE ....the body is no longer "inhabitable", is no longer"comfortable" enough, and the psyche looks for an environmental habitat. The OBE appears from this point of view as an extra-corporeal state in which the fundamental structure of the psyche escapes from the body.[19]

Does Tom's one-off apparition sighting display psychodynamic purpose? An answer might lie in the distinction between a psychodynamic *hallucination* and a *vision.*[20]

A psychoanalytic hallucination of the *psychiatric* kind is the variety that possesses a true sense of reality, and these (allegedly) have their origins in repressed childhood sexual knowledge. For instance, Freud's patient Sergeï Pankejeff (the "Wolf Man") experienced the childhood hallucination of a missing digit on his hand, and Freud's explanation was that this scene creation represented his repression of knowledge of seeing his sister abused by his father—apparently the finger had to be cut off (to Pankejeff's awareness) otherwise it would "betray this outrage"....or was it because it symbolised sex?[21] or even his "desire for castration" by his father?[22] Post-Freud, different psychoanalysts have come up with all sorts of different "explanations". This is surely too psychopathological for Tom's ghost.

Not so psychodynamic visions. These are different from everyday dreams, fantasies and daydreams in that they possess a strong feeling of presence hence have the hallmark of a real impression, but they are not real. Instead visions are a repressed memory in embodied form and infused with emotions ranging from hatred and fury, through fear, disgust and shame, to resignation and finally euphoria and exaltation.[23] These emotions are conveyed *figuratively*—meaning they are not confined to an inner mental

[19] Rabeyron and Caussie (2016, p. 64).
[20] Goetzmann et al. (2018).
[21] Roudinesco (2016, p. 198).
[22] Dale (2015, p. 157).
[23] Goetzmann et al. (2018, p. 24).

space; due to their emotional power they are projected onto the world. They appear as;

> Ghosts, angels, and dead or monstrous people are experienced, or perceived visually or with other sensory modes: a stench rises, voices are heard, etc.[24]

Visions might be vivid, they are never taken as *completely* real by the witness—they are the pseudohallucinations of scientific reality, roughly half-way between imaginative thought and actual hallucinations comprising of long-term memories freed from the "control" and "discrimination" of the prefrontal cortex regions. This again leads to the feeling of presence, hence visions can effectively resemble a waking, lucid dream.

That's Tom's ghost explained by the second-path, although psychodynamics—and especially the claims of psychoanalysis, are quite extreme additions to the journey. There's no sensible, measureable and falsifiable *why* for Tom's ghost sighting. Psychodynamics also reeks of dualism. It's far safer to conclude that Tom experienced that accidental release of mental imagery during an ASC brought on by natural causes. These are unable to be determined so long after the event, but they *could* have been identified had measuring instruments been available in place that afternoon. Heart rate, blood pressure, alpha and beta brain waves and more—a constellation of data providing evidence that he was perhaps in a state of deep relaxation, or maybe great stress (we can confidently rule out psychedelic drug use, or macula or CNS damage). The patterning forces holding together ordinary consciousness were momentarily disrupted and a rapid "clip" of a walking man was accidently inserted deftly into self-awareness. The vision has all the peculiarities of eidetic imagery, and happens so quickly that there's no time for Tom to assess whether it's apparent autonomy, and the feeling of presence, is legitimate. In short, it *is* a hallucination, but it is *seen* as *real*.

[24]Goetzmann et al. (2018, p. 24).

References

Bartocci, G. (2004). Transcendence Techniques and Psychobiological Mechanisms Underlying Religious Experience. *Mental Health, Religion & Culture, 7*(2), 171–181.

Blum, H. (2013). Dissociation and Its Disorders. *Psychoanalytic Inquiry, 33*, 427–438.

Dale, I. (2015). *Responses to Self Harm: An Historical Analysis of Medical, Religious, Military and Psychological Perspectives.* Jefferson, NC: McFarland.

Goetzmann, L., Ruettner, B., & Siegel, A. (2018). Fantasy, Dream, Vision, and Hallucination: Approaches from a Parallactic Neuro-Psychoanalytic Perspective. *Neuropsychoanalysis, 20*(1), 15–31.

Hall, C. S. (1959). *A Primer of Freudian Psychology.* New York: Mentor Books.

Kosman, K. A., & Silbersweig, D. A. (2018). Pseudo-Charles Bonnet Syndrome with a Frontal Tumor: Visual Hallucinations, the Brain, and the Two-Hit Hypothesis. *The Journal of Neuropsychiatry and Clinical Neurosciences, 30*(1), 84–86.

Poller, J. (2015). Beyond the Subliminal Mind: Psychical Research in the Work of Aldous Huxley. *Aries, 15*(2), 247–266.

Preller, K. H., & Vollenweider, F. X. (2018). Phenomenology, Structure, and Dynamic of Psychedelic States. *Current Topics in Behavioral Neurosciences, 36*, 221–256.

Rabeyron, T., & Caussie, S. (2016). Clinical Aspects of Out-of-Body Experiences: Trauma, Reflexivity and Symbolisation. *L'Évolution Psychiatrique, 81*(4), e53–e71.

Roudinesco, E. (2016). *Freud: In His Time and Ours* (C. Porter, Trans.). Cambridge: Harvard University Press.

Solms, M., & Turnbull, O. (2002). *The Brain and the Inner World. An Introduction to the Neurosciences of Subjective Experiences.* New York: Other Press.

Wahbeh, H., Sagher, A., Back, W., Pundhir, P., & Travis, F. (2018). A Systematic Review of Transcendent States Across Meditation and Contemplative Traditions Explore. *The Journal of Science and Healing, 14*(1), 19–35.

26

The Problem of the Producer

As an explanation for ghostly hallucinations, the second-path covers all bases. However there's one niggling issue that needs to be confronted....who is the *producer* of the vision? To understand just how important this question is, let's reconsider the dark-suited man from Tom's hallucination. As suggested, it's very likely that at some stage in his life Tom has seen a similar, real man walk across his field of vision from left to right. It's highly probably he sees them all the time! This short visual storyline has been processed and stored as a particular stand-alone configuration of neural activation. When ordinary consciousness is altered in some fashion, this particular pattern happens to be very close to the liminal space between unconscious storage and self-awareness. The man in the business suit (a characteristic of a typical working day in Tom's life) is what Tom sees as a hallucinated ghost. He won't see a donkey.

The same applies to anything else witnessed during your lifetime; even a life-sized reptilian alien you once saw in a horror movie. This would also be stored as an extrastriate activity pattern representing a bipedal reptile, and like Tom's dark-suited man it is a narrative that you can witness as a consequence of your socio-cultural conditioning. It might not be as

T. Jinks, *Psychological Perspectives on Reality, Consciousness and Paranormal Experience*,
https://doi.org/10.1007/978-3-030-28902-7_26

strong as the image of a *human* figure, but will be very likely to appear in visionary form if you are "hooked" on UFOs and the paranormal, or when your ASC is extreme (psychedelic drug use, for example) and able to access slightly deeper levels of unconscious material.

And in these circumstances, the psychosocial hypothesis might not even be a necessary framework for the content of the vision. It has been suggested that under the influence of psychedelics (to take one example);

> The human psyche and consciousness can create many non-existing new forms and formats of creatures, beings, "realities" and also different ways of thinking (illogical or "paralogical") through active transformation and distortion of the original imagery or cognitions.[1]

Everyone's seen an ordinary lizard, and the mind in an altered state can play around with this raw ingredient until, lo-and-behold, the terrifying, human-sized lizard-alien.

Creating novel imagery not the only complicated task performed by the brain in an ASC. There's also the inadvertent process of incorporating separate hallucinated images into an unbroken visual experience, with steady motion and a coherent storyline. The unavoidable conclusion is that this process is coordinated by "something" acting as a "producer" of such a storyline. This "something" is (or are) the force behind the re-patterning of consciousness, arranging available material intelligibly and creating an experiential sequence seen by the witness as "real". The producer is a *who* in the 5W1H method, but not synonymous with the person-as-a-whole, named Tom or anyone else.

The role of this alleged producer is hard to ignore, despite being as dualistic a concept as anything psychodynamics has to offer. "They" are referred to obtusely in the second-path, neuroscientific treatment of visual hallucinations, since these are apparently;

> ...assembled in the "image space" of associative cortex, based on memories according to the *dramatic instructions* of the dispositional space.[2] (emphasis added)

[1]Sayin (2014, p. 293).

[2]Goetzmann et al. (2018, p. 20).

This identifies the producer as a *dramatic instruction-giver.*

What is the monist solution to the producer problem? To make sense of *who* the producer might be (and *how* and *why* "they" behave the way they do), it's worth revisiting a style of altered consciousness that I've briefly touched on, and which is intrinsically linked to the concept of a producer—dreaming.

Earlier I mentioned that regular (non-lucid) dreaming is a straightforward category of ASC involving the involuntary production of a vivid, imagery-based narrative (often fantastical)[3] accompanied by reduced self-reflection, remembered knowledge, and somatic sensations normally present in the awake state.[4] Although these ego components might be attenuated, there is usually *some* feeling of presence—so in this regards dreams are similar to certain types of hallucination.[5] There's also enough self-awareness and self-integration present in non-lucid dreams, and enough meaningfulness in the subject matter, for them to be remembered and reported upon wakening.[6]

In terms of *why* people dream, the common second-path approach is that dreaming has evolved;

> …as a virtual reality mechanism to simulate and rehearse threats and social interactions in order to better cope with them while awake.[7]

For the present purposes, we're not so much looking at *why*, but the *how* and *who* in the production process.

To begin, the building blocks of dreams are the equivalent of closed-eye mental images in the waking state, and recognition occurs through matching these percepts with stored representations.[8] The question is, how can dream imagery be fashioned into a story in the absence of real stimuli? That requires consideration of neural network interactions at

---

[3]Kraehenmann et al. (2017).

[4]Clement (2007/2008).

[5]Nanay (2016).

[6]Clement (2007/2008); although "memory for even vivid dreams is evanescent and tends to fade quickly unless special steps are taken to retain it" (Hobson et al. 2000, p. 795).

[7]Stumbrys and Erlacher (2017, p. 40).

[8]Clement (2007/2008).

subcortical and cortical levels of the brain, and the similarity between psychedelic states and dream states is helpful in this regard. For example, serotonergic-acting psychedelics like LSD create hallucinations in waking people with phenomenological similarities to ordinary dreaming.[9] These include all the typical symptoms we're familiar with in an ASC—visual imagery coupled with changes in thinking style, disinhibition of emotions, motivations as well as alterations of control of the sense-of-self (also known as *depersonalisation*).[10] There's also improved access to autobiographical memories and mechanisms are enacted that allow fantasising about the future.

That's knowledge based on *personal* reports of the psychedelic state. Physiological measurements in participants who have consumed LSD show saccadic eye movements and REM features characteristic of dreaming. Concurrent cortical recordings show increased activity in the temporal lobes, and bottom-up activation of the primary visual cortex leading to basic *spontaneous* entopic imagery (like geometric figures) that might potentially develop into to more complex "scenic hallucinations" as the signal travels up the visual hierarchy (although no stronger mechanism of how exactly that occurs is forthcoming).[11] That said, this imagery *still* wouldn't be of any significant complexity if its origins are from bottom-up visual entry-points. The theory with LSD use, and applied to dreaming, is that the novelty and plot arises is through the creativity of top-down substrates simultaneously in operation, as these indulge in;

> …divergent, nonlinear thinking, which is defined as a thought process where many responses to a single question or problem are generated and where the responses are often novel or lead to a broader perspective than in normal thought.[12]

Hence the cogent producer in this model of hallucination is not an additional "thing" in the brain (like a dualistic and autonomous unconscious) but the multiple states of possibilities which arise from any singular small

[9]Kraehenmann et al. (2017).

[10]Ibid.

[11]Ibid.

[12]Kraehenmann et al. (2017, p. 2039).

event of lower origin (basic imagery generated by psychedelic drugs, or the seed of dream idea) which are automatically scaffolded as the content rises up through cortical pathways until a unique scenario ends up the "winner" of the drug-imagery/dream contest.

Perhaps the most detailed assessment of the passive dream construction mechanism (without the requirement of a producer) is found in the *activation synthesis theory*, where the origins of dreams are bottom-up and sub-cortical in origin.[13] The source for the initiation of dreaming will be familiar—the pons, and it's worth going into a little more detail about this region's function not simply in generating REM sleep, but dreaming.

The pons is a resident of a functional brain stem region known as the reticular formation, with its "driving centre" the reticular activating system (RAS). This has a strong association with keeping the brain "aroused"—shorthand for saying it provides the signal (or energy, or some special input) that maintains steady consciousness. The RAS projects from brainstem to the forebrain and sends neural offshoots to the thalamus and hypothalamus releasing the neurotransmitters as varied as serotonin, noradrenaline, dopamine, acetylcholine and glutamate.[14] These all stimulate higher brain centres, hence the role of the RAS in helping us "remaining aware".

On the other hand, certain regions within the RAS also appear to have the *opposite* role to generating arousal. The *pons* is situated in the lower RAS, and probing this location with stimulating electrode generates an EEG pattern similar to REM (something we've briefly touched on when discussing sleep paralysis).[15] In particular, two sub-regions appear to be the source of projections to the forebrain, and take responsibility for sleep generation. The first is the r*aphe nucleus*, which releases serotonin, and the second is the *locus coeruleus*, which releases NAd.

However, there is a region even lower in the brain stem, in the medial pontine reticular formation known as the *gigantocellular tegmental field*. This substrate projects ACh fibres to the forebrain, and experimental destruction of these neurons inhibits REM sleep. This REM reduction could also be driven by collateral damage to an area near the GTF called

[13]Kramer (2002).
[14]Edlow et al. (2012).
[15]Kovalzon (2016).

the *peribrachial region*. In this model, peribrachial cells activate GTF cells, but also project directly to forebrain. Hence REM sleep is still able to be generated when the peribrachial is intact, but the GTF has been destroyed by lesion.[16]

So how do the raphe nucleus and the locus coeruleus fit into the model? Well, the raphe and locus coeruleus both project to the peribrachial region, and hence REM sleep consists of reciprocal interaction. Simply put, the raphe and locus coerulus are active during waking, and suppress the REM "driving" function of peribrachial cells. When their activity declines at the very beginning of the sleep cycle, peribrachial cells are "freed" to exert their influence over the forebrain, and generate REM sleep through the GTF pathway.

While these events are taking place, *ponto-geniculo-occipital* (PGO) waves travel up the brainstem, through the mid-brain and bursting into the occipital cortex. As the region tasked with fundamental visual processing, activity in the occipital cortex *seems* to coincide with the closed rapid eye movements associated with REM sleep, which has the hallucinatory nature of a psychedelic incident (the peculiar imagery, extended narrative, decreased self-reflection and self-control typical of the ASC).[17] These sweeping signals projecting to the frontal and occipital cortices during REM sleep do not contain any theme. However, mechanisms of matching and recognition are initiated by the *passive* cortex in automatic response to the ascending activity—not planning a storyline like a producer, but merely "elaborating" random content that arrives. Put succinctly, "making the best of a bad job".[18] Dream organisation is assisted by the *inferior parietal cortex* which is speculated to provide spatial integration of the various arbitrary signals thrown together in the occipital lobe. Other mid-brain *non-visual* areas such as limbic system (amygdala), anterior cingulate and parahippocampal cortex and hippocampus then bestow on the dream the characteristics of emotion and meaningfulness.[19]

---

[16]Wickens (2009, p. 268).

[17]Kramer (2002).

[18]Kramer (2002, p. 661).

[19]Ibid.

Meanwhile, the top-down *dorsolateral prefrontal cortex* is inactive while these REM signals are coursing through the brain, so no executive function is able to be applied to the dream content—no rationalising or logical assessment to shut down content creation before the dream is experienced. And of course there's a reduced capacity for working memory during this period, which is a stout explanation for why dreams aren't necessarily remembered. And although the resulting dream may *seem* more sophisticated than a reflexive or automatic neural event lacking executive control, the answer lies in the analogy of the sorting table I referred to earlier in the book, with different-sized holes that categorise different-sized rolling balls. Ascending electrochemical waves from the brainstem hit a higher neural network which unthinkingly compares these arriving packets of information with stockpiled templates of static images or memory fragments. Indeed for a very long time it was accepted that the jigsaws pieces of dream content are not novel creations, or necessarily from some past time, but very *recent* autobiographical inclusions that last only a few days as potential components of a dream sequence.[20]

In real time there is a "playback" of the best match of stored material to the incoming stimulation, resulting in an idiosyncratic dream with components of meaningful plot constantly changing to different plots, interspersed with seemingly random images. It's a strange patchwork quilt that, when remembered upon waking, may have additional meaning forced upon it once the executive centres have finally regained control of ordinary consciousness and throw their weight into what the dream *is really about.* In fact, a dream sequence might consist of nothing more than modules of short, simple content run in a random sequence and which the dreamer—*on waking* (if they remember anything about it)—might try and overlay a narrative after-the-fact. This notion sees the dream primarily as an ad hoc cognitive production, rather than a creative perceptual process.

The problem is, dreaming is only reported in 80% of subjects woken in sleep laboratories while displaying REM activity. The other 20% report

[20] Ibid.

dreams during *non*-REM slow wave sleep. This suggests REM is not a necessary requirement for dreams.[21] Determining what mental experiences take place in a non-REM state has been helped in recent years through improvement to investigative methods, with new styles of interview technique allowing dreamers to provide more nuanced commentary about their *oneiric*[22] experiences across every sleep stage.

The broader conclusion is that *any* random bottom-up brain signal can be pattern-matched to become a storyline, not solely REM-related activity like PGO waves. Yet the randomness of any ascending dream signals (whatever the source or whenever it arises) would mean no two dreams should be alike, so how does this bottom-up model account for *recurrent* dreams? Dreams also occasionally contain conversation and other content that appears too consistent to be arbitrarily spliced together. This suggests a top-down solution for dream origins, using the language of neurophysiology with flexible discussion of dynamic (dualistic) mental states; an oblique return to psychodynamics.

This all begins with the psychoanalytic purpose of dreams; they emerge as a combination two mental representations. The first is the *manifest* content, the remembered or accessible dream that is a compromise formation. That means the dream has been allowed to be "viewed" by the sleeping person, but only because the true connotations of the experience obscure. This actual meaning is found in the *latent* content—a wish or desire with the energy of a drive that is the mental derivative of the instinct.[23]

A self-sufficient unconscious becomes the producer of dreams, distinct from the ego and superego. Dream fabrication is termed *dream work* that expertly converts latent content to the more palatable manifest content. The processes available to the dream producer include *condensation* (where multiple latent elements are moulded in the manifest content), *displacement* (where trivial latent elements are promoted in the manifest content, and important latent elements are downplayed) and *dramatization* (where latent elements are converted into visual imagery).[24]

[21] Solomonova and Wei (2016).

[22] Meaning of, relating to, or suggestive of, dreams.

[23] Nye (1975).

[24] Nye (1975).

These methods are energised by a weak (hypothetical) force with enough power to hold the two components of latent and manifest content together temporarily (termed *perceptual identity*). There's no necessary reason why the manifest and latent content share this transitory bond—they don't need to have anything in common, which is why the distortion of the imagery by the force of the wish (conforming to its demands) often makes no obvious sense to the dreamer. And since there's no reason for any wish to team up with any image, the wish can attach itself to any image that might be available in the storehouse of the unconscious. *That* becomes the reason for rambling and incoherent dream sequence.[25]

To help fill the gap left by a dualistic producer, the second-path has adapted some of these psychoanalytic ideas *back* into neurophysiological language. One substrate is of particular interest—the ventral tegmental area (VTA). This is a reward- seeking circuit controlling the very purposeful notion of "wanting" and "wishing".[26] Activity of the VTA might even be labelled the equivalent of possessing "desires", although for social reasons these are often blocked from reaching *behavioural* fruition by the inhibitory input from anterior limbic structures. The outcome is a frustration of "volitional" behaviour, and the energy of VTA neural activity must go somewhere, so it is subsumed and provided an alternative form. This is as the dream, the arrangement of which is patterned by perceptual memories from the occipital association areas.

Dreams are therefore the equivalent of active perception in the *absence* of any perception, using the motivational energies from deep brain structures like the VTA and the same creative, imaginative process like *mind-wandering* found in waking consciousness.[27]

Despite the naming of neural substrates that (allegedly) select images to match wishes and desires, this *still* signals an inadvertent return to a (limited) producer and the seeming resurgence of a mind/body dichotomy, where a separate mind (holding wishes, and desires) examines the biological input from the various brain substrates and determines the outcome

[25]Clement (2007/2008).

[26]Stuber et al. (2012).

[27]Solomonova and Wei (2016).

(storyline and qualia). To rehash the longstanding argument about physical and mental states being entirely different, and the impossibility of getting from one state to the other;

> ...the functional anatomical explanations or the descriptions of the secretions from various cells of neurotransmitters or neuromodulators contribute little, if anything, to answer questions about the content, meaning, construction or function of the dream experience...there are no transduction rules to go from the discharge of neurons in the CNS or from the secretions of neurons to the concomitant mental states.[28]

What remains is the assumption that dreams (and by extension other phenomenological experiences that *seem* like they could be real, like Tom's ghost) are a form of spontaneous and ongoing creation, although the identity of the independent and cogent producer is never quite pinned down. There's only a hazy consideration that consciousness is something a bit more complicated than being an accident, but because it manifests holistically, dualism doesn't have to be explicitly mentioned;

> ...consciousness is not a passive computational process, nor is it an epiphenomenal feature of the brain activity. Rather, consciousness is a dynamic temporal process of sense-making ...that involves the whole embodied organism, its relationship with the world, intentionality (the aboutness of consciousness), and the way in which subjectivity is geared towards certain features/qualities in the world in a meaningful way.[29]

If pressed, the idea of a goal-directed want, wish or desire—even emerging from something as physical and scientifically respectable as the VTA, can be replaced using a modern cognitive science twist. It has been suggested that any emotions the dreamer has experienced during their waking day is held as a weakening trace (like the visual after-image) and retains enough energy to be the blind shaping force behind dream content that night during sleep.[30]

---

[28] Kramer (2002, p. 670).

[29] Solomonova and Wei (2016, p. 411).

[30] Clement (2007/2008).

First comes the emotional day residue, then a selected image in memory contextualizes the emotion to form the final dream image.[31]

Regarding the *structure* of the resulting dream;

...meaning in dreams is dictated by the unfolding of a linear narrative. However, the chain responsible for meaning is not composed of images but of units of lingering emotions or, emotional free radicals. Images present as the more recognizable and communicable aspects of the dream but emotions impart the movement and the tone, and as such serve as a narrative template to make a story. Only after the affective script has been formed does each emotional radical in the chain semi-randomly select an image so to match with it.[32]

And this explains the *nature* of dreams, which;

... unfold as if they were emotions; they emerge suddenly and unpredictably, they bear varying intensity and have no logic of their own. Moreover, dream images are plastic, they are unstable, follow no logical sequence, carry contradictions and incongruities; they condense together and can be displaced.[33]

Evidence for this "daily residue" theory of dream formation comes from recordings of brain activity during REM sleep. Heightened readings from baseline are found simultaneously in visual ("image mediating") extrastriate cortical areas, and emotional regions such as the amygdala. That's enough of a correlation to supersede the psychoanalytic focus on wish-fulfilment, even if the potential involvement of the VTA gives the latter a scientific veneer.

And unlike the psychoanalytic combination of unrelated image and wish or desire, there *is* a logical connection between dream imagery and persistent emotional activity in the limbic system—a similarity that facilitates binding and allows mental imagery to be selected as an affective

[31] Ibid.

[32] Clement (2007/2008, p. 250).

[33] Ibid., p. 252.

metaphor.[34] In terms of *who* does the selecting, if not a cogent producer then it must be something analogous to that lasting monist metaphor, the docile ball-sorter. While not definitive in this strong scientific model of dreaming, the expectation is that binding is a natural act, and simply *happens* by natural law, like magnetic attraction.

The result of this spontaneous selection process is that dream imagery no longer bears any trace of the emotion but only of the symbolic image representing it. So rather than feel the emotion, the dreamer witnesses the emotion as a projected storyline that doesn't necessarily make any "visual" sense, with no order or chronology and no beginning or end.[35]

For the second-path, the concept of dreaming, dream imagery and dream production is best represented by an experience that isn't completely random, but certainly not purposeful. And if a supporter of the second-path was to bother with the possibility of dream interpretation outside the framework of psychoanalysis, then this would involve an analysis of symbols found in the dream and guessing the potential emotions they represent—rather than trying to determine a repressed wish for some therapeutic (psychoanalytic) objective. A similar analysis can also be performed on a longer, spliced-together dream, picking apart the various emotions associated with each section. And when it comes to nightmares, there might just be one single intense negative emotion threaded through the terror-laden experience, with no storyline to speak of.[36] Finally, the dream model founded on *daily emotional residue* can be subpoenaed as an explanation for the content of NDEs and other sleep-related paranormal experiences, such as in this case of a diabetic patient who reported;

> ...a near-death experience during an episode of hypoglycaemia. During this episode, the patient was in a sleep-like state with rapid eye-movement (REM)—a common marker of dreaming and thought to underlie the consolidation of memories, a process that may explain life reviews during near-death experiences.[37]

[34]Ibid.

[35]Ibid.

[36]Clement (2007/2008).

[37]Mobbs and Watt (2011, p. 447).

Interestingly, these authors also state that the patient was not in danger of dying (implying the event could not have been "near death), yet at the same time admit that they were *resuscitated*—which is not uncommon for severe hypoglycaemia. This encourages suspicion that the patient might have been in a near death state after all!

References

Clement, P. (2007/2008). Dream Imagery as a Result of Emotions Matching with Images. *Imagination, Cognition and Personality, 27*(3), 245–257.

Edlow, B. L., Takahashi, E., Wu, O., Benner, T., Dai, G., Bu, L., et al. (2012). Neuroanatomic Connectivity of the Human Ascending Arousal System Critical to Consciousness and Its Disorders. *Journal of Neuropathology and Experimental Neurology, 71*(6), 531–546.

Goetzmann, L., Ruettner, B., & Siegel, A. (2018). Fantasy, Dream, Vision, and Hallucination: Approaches from a Parallactic Neuro-Psychoanalytic Perspective. *Neuropsychoanalysis, 20*(1), 15–31.

Hobson, J., Pace-Schott, E., & Stickgold, R. (2000). Dreaming and the Brain: Toward a Cognitive Neuroscience of Conscious States. *Behavioral and Brain Sciences, 23,* 793–842.

Kovalzon, V. M. (2016). Ascending Reticular Activating System of the Brain. *Brain Science Advances, 2*(4), 275–285.

Kraehenmann, R., Pokorny, D., Vollenweider, L., Preller, K., Pokorny, T., Seifritz, E., et al. (2017). Dreamlike effects of LSD on Waking Imagery in Humans Depend on Serotonin 2A Receptor Activation. *Psychopharmacology, 234*(13), 2031–2046.

Kramer, M. (2002). The Biology of Dream Formation: A Review and Critique. *Journal of the American Academy of Psychoanalysis, 30*(4), 657–671.

Mobbs, D., & Watt, C. (2011). There Is Nothing Paranormal About Near-Death Experiences: How Neuroscience Can Explain Seeing Bright Lights, Meeting the Dead, or Being Convinced You Are One of Them. *Trends in Cognitive Sciences, 15*(10), 447–449.

Nanay, B. (2016). Hallucination as Mental Imagery. *Journal of Consciousness Studies, 23*(7–8), 65–81.

Nye, R. D (1975/1981). *Three Psychologies* (2nd ed.). Belmont: Brooks/Cole.

Sayin, H. U. (2014). The Consumption of Psychoactive Plants During Religious Rituals: The Roots of Common Symbols and Figures in Religions and Myths. *Neuroquantology, 12*(2), 276–296.

Solomonova, E., & Wei, S. X. (2016). Exploring the Depth of Dream Experience. *Constructivist Foundations, 11*(2), 407–416.

Stuber, G. D., Britt, J. P., & Bonci, A. (2012). Optogenetic Modulation of Neural Circuits That Underlie Reward Seeking. *Biological Psychiatry, 71*(12), 1061–1067.

Stumbrys, T., & Erlacher, D. (2017). Inner Ghosts: Encounters with Threatening Dream Characters in Lucid Dreams. *Dreaming, 27*(1), 40–48.

Wickens, A. (2009). *Introduction to Biopsychology* (3rd ed.). Harlow, UK: Pearson Education Ltd.

27

The Producer Solved?

Do the arguments provided in the previous chapter mean the dualist producer of dreams (and by extension altered state hallucinations) can now be removed from discourse? Consider the generation of mental imagery and add the ingredient of creative thought, where you are conscious and *awake* and the producer of something *novel.* I've already alluded to the mental effort required to imagine a spoon sitting on a dining room table. You are effort-producing this portrait somewhere in your mental space. In terms of *how*, a sensible answer is that the rational, computational apparatus of your brain—your *neuromatrix*—consists of a vast neural network of nodes of "ideas" in countless different relationships with other nodes (represented by synaptic strengths). The apparatus receives and interprets a verbal, written or mental instruction from other deciphering networks and makes a probabilistic decision to activate the pattern representing (simplistically) two stored memories—a spoon and table. These are combined into an original picture—unless there's a readily available stored memory of a spoon on a table that can act as a substitute.

If the two concepts of spoon and table require merging, then the resulting process is messy, prone to distortion and far from stable. That accounts

T. Jinks, *Psychological Perspectives on Reality, Consciousness and Paranormal Experience*,
https://doi.org/10.1007/978-3-030-28902-7_27

for the ease with which the thinker switches to some other, unwanted thought process unless they are highly disciplined and able to persist with this new creation in their mental space.

And for the sake of argument, this creation may simply be the epiphenomenal experience of the underlying neural activity—the real outcome—that will allow the thinker to indulge in goal-seeking (or some other style) of behaviour.

Now allow the thinker's mental imagery apparatus to produce something much more creative using the following brief;

> Imagine a red car spinning three times then launching itself into the air and gradually turning into an elephant.

The cognitive response to these instructions involves accessing a hypothetical car template stored, or at least somehow extant, in the network. Perhaps it's not taxing for this template to be shaded red, since an "energetic" association between *car* and *red* might be already be in place. The brief also requests that this red car is to spin three times, but forcing motion onto a static image might be too challenging. Instead the cognitive *concept* of spinning—like a whirling circle—may be a shortcut that's overlayed on the image to provide an economical and less demanding response.

Creating the imagery of a red spinning car evolving into an elephant is an even more difficult transition, and it's likely the car will immediately transition to the formless lump of grey elephant on a blue-sky background. If more attention is paid to fashioning the graduation from car to elephant it might be possible to separately visualise single images in the process, such as an elephant's foot grow out of a front tyre. However any further effort will probably not be worth the exertion, unless the thinker has is skilled at calling on eidetic imagery.

Compare this mental struggle (for "normal" people") with the generation of dreams consisting of novel content. While themes from the day certainly play a role in shaping a dream experience, it's common for most of the material to be totally unfamiliar, like a dreamscape with never-seen-before skyline and structures. And irrespective of whether a wider set of images were grabbed by the day's emotions from unconscious storage, even places that have are interpreted within the dream as familiar;

like home and it's surrounds, or the local community or place of work or learning, possess only minimal shared features with what these places really look like. Moreover there is topographical coherence and the dreamscape can be navigated with precision. The wider geography can be grand and sophisticated, but specific features, like the tiles on a roof or a particular plant, can be attended to and examined in intricate detail.

Without resorting to a producer, it seems difficult to put this creativity down to the binding of recent imagery to recent emotions. The best bet is a *cryptic* brief, beneath awareness and stored in the network. This construct tasks the computational mechanisms normally dedicated to producing mental imagery to the creation of a new and wondrous realm—something perhaps only remembered because it was the last thing that was dreamed before waking.

To avoid reference to paranormality, the dreamscape *must* be a personal creation—but it takes no obvious mental effort. And it lasts long enough to *experience* (according to our earlier definition) and remember—unlike the spinning, flying and morphing image of a car which is almost impossible to retain in our mind-space. The same applies for hallucinations seen in normal waking consciousness, like a ghost, which take also no mental effort to achieve, persist for long enough to be recognised, and are cleverly grafted onto the background scene.

This consideration of dreams implies a consciousness/unconscious nexus more active than a mechanical neural network engaged in sourcing material and constructing a unique and innovative mental storyline. A distributed neural network might be an adequate solution when the stream of imagery is unintelligible (and often fleetingly remembered then forgotten), but with *complex*, *coherent* and *original* hallucinatory material like Ruth's content-specific delusion mentioned earlier, the drive to include or recognise a producer beyond ordinary consciousness is even more apparent.

The second-path provides a basic account for *how* dreamlike experiences are chaotically produced with ideas of bottom-up randomness or top-down automatic binding. However the notion of very ordered and focused content-specific delusions, and by association solid apparitions that remain in the visual field for a considerable period of time, is not so easy to explain in terms of the human brain operating as a non-sentient input/output network.

This argument leads to the charge of *premature parsimony*.[1] The principle of parsimony (also known as Occam's razor);

> ...requires that the inadequacy of materialistic explanations be demonstrated before other models can be proposed.[2]

Or more to the point, evaluating and making judgements about something—like a dream or hallucination—should not involve any *unnecessary assumptions.* Regarding the second-path's treatment of paranormal phenomena, it might be argued that the problem of the producer has been parsimoniously dealt with, insofar as knowledge is lacking about how complex arrays of neural activity specifically contribute to waking mental imagery or dream imagery.

However that conclusion might have been *prematurely* chosen as the simplest (or only) one available. This becomes an avenue for even simpler ("more parsimonious") first-path solutions to sneak in and be proffered. For instance, an alternative account for how apparitions and other hallucinations are "inserted" into visual space is the suggestion that the producer (*who*) is just someone else. Tyrrell's model of telepathic hallucinations provides a hint about *who* the producer is, using a theatre analogy. Like a stage play, your visual consciousness is created by a producer and stage carpenter, who are mid-level constituents of personality,[3] and these are enigmatically connected to other people's mid-level constituents. One of these people, the projector, provides *very basic* imagery to your brain—a simple "theme or motif"—without any storyline. There's no information transmission, as the mid-level constituents are somehow interconnected so multiple people can see the same apparition. When you receive this imagery your stage carpenter and producer get to work and mould this material (perhaps representation of the projector's appearance) into a more

[1]Tart (2008).
[2]Arnette (2012, p. 208).
[3]Price (1953, p. 12).

substantial vision, which is then cleverly forced to conform to the current environment.[4]

Alternately, the reason why dreams and a dreamscape might seem so foreign is that they are constructed from the memories, thoughts and feelings of *someone else's* mental world, slipping through a "leaky" Bergson-style filter.[5] These pieces are inserted into the puzzle board of the dreamer (or hallucinator's) waking consciousness. A favourite example of how this manifests in the real world is the case of *savancy* (originally from the Latin *sapare*; to comprehend, understand, or to know). When savant individuals display levels of unexpected knowledge or ability that they could not have personally achieved, this is traditionally attributed to a spirit. For instance, the medium Rosemary Brown allegedly communicated with the composer Liszt, allowing her to write music in his style (as well as that of Beethoven, Chopin, Schibert, Brahms and Debussy apparently, after the appropriate otherworldly introductions!).

> ...it seemed reasonably clear that Mrs Brown must be receiving the music from somewhere – it was not purely of her own invention....and one go the impression that she found some of it quite bewildering.[6]

Avoiding mediumship and the influence of *actual* spirit communication, unexplained savancy implies a change in the status of a filter that allows knowledge to enter consciousness as an additional element. This might be as a fully formed fact or statement. Regarding his mathematical ability, the savant Daniel Tammet stated the process of achieving a complicated numerical answer took only a few seconds and was spontaneous—meaning no cognitive effort on his behalf.[7]

Granted these abilities are legitimate (and there are always second-path sceptics who would debate the abilities of savants), the most popular view

[4]Ibid.

[5]Gauld (2012).

[6]Wilson (1971/1983, p. 652).

[7]Tammet (2006): The alternative view is as a complicated form of synesthesia.

is that the savant's brain is able to access a wider range of "computing networks" than might normally be found in other, ordinary people. Savancy is therefore a complicated extension of synaesthesia.[8]

However there's more to savancy than the rapid calculations of a neural supercomputer. Instead the answers the savant provides—mathematical or anything else—is said to be retrieved from an exogenous source not tied to any personal level of consciousness. For instance, in another case of music composition, and like Daniel Tammet's observations of his own abilities, it is suggested that the savant Jay Greenberg is simply "downloading" finished musical pieces directly into consciousness from somewhere else.[9] Support for this view of savancy is how quickly the knowledge arrives to the savant, the lack of concentrated effort required to access the knowledge (compared with my earlier example of creating the visual image of a rotating transforming red car), and the lack of an explanation as to how the knowledge was achieved. The savant doesn't agree that any of it came from a personal storage-house of memories, so to a first-path proponent it seems more *parsimonious* that they've tapped into some other information source.

Attempting to remain just a *little* bit grounded, the knowledge shown by savants, and by implication the unfamiliar experiences associated with dreams and hallucinatory visions, may indeed be existing chunks of replayed information that aren't constructed in the here-and-now of an individual's conscious space. An ingenious way to account for these astonishing plot devices is to suggest the snippets of information are stored in a deeper biological layer—locked into an individual's very own DNA.

> ...perhaps on some unknown, unconscious level the genetic encoder DNA provides a bridge to biological memories of all living things, and aura of unbounded awareness manifesting itself in the activated mind.[10]

This approach refers specifically to the imagery of psychedelic hallucinations, and is based (among other sources) on Terence McKenna's 1975

[8] Hughes et al. (2018).
[9] Brabant (2016).
[10] Hancock (2005, p. 459).

book *The Invisible Landscape* and Francis Crick's 1981 *Life Itself*, drops hint at intelligent seeding of DNA in universe. Tools such as psychedelics unlock sequences of the genetic code *not* associated with the production of component proteins, leading to a full component ASC where stored reels of information can be projected onto the reconfigured "screen" of consciousness. The logic for this starts with the linguist George Zipf, who in the first half of the twentieth Century noted that the frequency of the many different words in a language doesn't show uniform dispersion, or take the shape of a normal distribution, but is *inversely related to their rank*.[11] In other words, the product of rank order and frequency of elements in a code is a constant.[12] This is Zipf's law, and later this stable relationship was applied to another information system, chromosomes. The DNA they contain also displays a modified version of Zipf's law.[13] Taken to the extreme conclusion, the non-coded "junk" DNA that makes up most of chromosome is in the form of a hidden language—perhaps "engineered" by some form of ancestral, ancient astronaut—and that this can only be accessed when in an ASC![14]

This is taking the argument of premature parsimony to the extreme. Further criticism of the second-path focuses on the assumption that consciousness is made up of compartmentalised, neural sub-processes which have specific and limited function on their own (like the sense of vision, hearing or taste), but together build up the complete mind;

> The problem is why [this expression of] mind, if it is built up of these different modules, appears to us as undivided most of the time. After all, even though we might have different senses for hearing, tasting or seeing, it is the 'I' that hears, tastes and sees. We do not experience a division into modules. We therefore have to account for the link between all the different modules. It is at this stage that we run into problems with the modular model of the mind. We lack an explanation for the unity of consciousness. It is no problem dividing up the brain into different parts with each part fulfilling a certain role. However, we have to account for the subject of experience.

[11] Lestrade (2017).

[12] Bender and Gill (1986).

[13] Ibid.

[14] Hancock (2005, p. 484).

> This subject is the organising principle behind the different parts of the brain. If we take this subject to be the mind, and divide it, we create a new problem because we have to account for another subject as the organising principle behind the mind.[15]

Asking *who* or *what* is organising the organising principle is yet another expression of the problem of the producer.

Then there's the wider dissatisfaction with what some perceived to be the limitations of scientific reality—and how science is actually practiced. There's been a long-held complaint that the scientific process has been "distorted" from its objective standards. That is, scientific reality (and by extension the second-path) does not simply consist of the previously mentioned standards of reality, (monist) consciousness-as-brain-function, and the scientific process, but also a fourth aspect—*authoritative standards.* In short;

> … hypotheses are generally accepted more based on the authority of those who pronounce them, than in the scientific evidences available.[16]

These authoritative standards can affect an individual's interpretation of their own experience,[17] hence Tom might judge his ghost to be impossible based on his socio/cultural conditioning, and he will (unconsciously) feel obliged to re-interpret what he's seen as an error of perception rather than a real vision. This is really no different to forcing a paranormal explanation onto an ambiguous scene, so cognitive appraisal can be a two-edged sword.

Tom isn't influenced by authoritative pressure, and it helped that he verbally narrated his event soon after the sighting, assisting memory formation. However, as an IT engineer he understands the pressures of what is acceptable to scientific reality (a creed he endorses), so to this day is not completely convinced he saw an apparition, even though and is happy to accept it as something *he did see.* That's an example of cognitive dissonance that many paranormal witnesses are happy to live with.

---

15 Tuske (1999, p. 230).

16 Moreira-Almeida and Lotufo-Neto (2017, p. 284).

17 McLeodb et al. (1996).

Rather than a strict scientific paradigm of reality, we are now confronted with a broader *social* paradigm of reality where judgements of what is allowed to be legitimate (and what is not) are based on authoritative societal standards which include—but are not exclusively determined by—the scientific process.

In this context, it is taken for granted that scientific research is always impartial incorruptible, and there can be no context considered other than materialism.[18] The derogatory term for this is *scientism*[19]; "a faith in the ability of science to answer all questions, to solve all problems"[20] where commitment to scientific reality is maintained by the *politics* of science and its "protected assumptions" which are never themselves open to empirical testing.[21] Critics argue for the shifting of these assumptions, such as universal impartiality and reductionism, from dogma back to the lesser "the status of method".[22]

This reimagined approach to science demands broader parameters for paranormal evidence, and a more flexible theoretical framework to account for such evidence. An example is the assessment of personal reports that appear to contain highly improbable but correct information. As previously alluded to regarding Etzel Cardeña's work, parapsychologists have presented supportive results over many decades, despite the reproach that no such evidence exists (or at least is ignored by the second-path).

Here's one example from Charles Tart from his experimental work in the 1960s (and which he double checked for validity with the help of a trained magician). It involves an attempt by a participant, Miss Z, to "remote view" a 5-digit number while in a sleep-related ASC while her brain activity was recorded;

> On the first three laboratory nights Miss Z reported that in spite of occasionally being "out," she had not been able to control her experiences enough to be in position to see the target number (which was different each night).

[18] Stafford Betty (2004).

[19] Barušs (2010).

[20] Leahey (1987, p. 144).

[21] Tart (1998).

[22] Lampis (2004).

> On the fourth night, at 5:57 AM, there was a seven minute period of somewhat ambiguous EEG activity, sometimes looking like stage 1, sometimes like brief wakings. Then Miss Z awakened and called out over the intercom that the target number was 25132, which I wrote on the EEG recording. After she slept a few more minutes I woke her so she could go to work and she reported on the previous awakening that: "I woke up; it was stifling in the room. Awake for about five minutes. I kept waking up and drifting off, having floating feelings over and over. I needed to go higher because the number was lying down. Between 5:50 and 6:00 A.M. that did it. … I wanted to go read the number in the next room, but I couldn't leave the room, open the door, or float through the door. … I couldn't turn off the air conditioner! …. The number 25132 was indeed the correct target number. … The odds against guessing a 5-digit number by chance alone are 100,000 to 1, so this was a remarkable event! Note also that Miss Z had apparently expected me to have propped the target number up against the wall behind the shelf, but she correctly reported that it was lying flat.[23]

The second-path can easily fall back on the likelihood of an error in methodology, something akin to *sensory leakage* from the experimenters accidently providing Miss Z with a slightly above-chance idea of what the number might be; over time and replications, this total will account for any statistically-significant finding. There also isn't a consistent theoretical framework for the alleged evidence of her travelling. Presuming something anomalous did happen to Miss Z, was it because she left her physical body to seek out the number, or was the information transmitted to her using the channels of *General ESP*?

Nevertheless, the claim that anomalous events can be empirically documented give supportive researchers the justification to re-visit the ASC construct. To be clear, this is not necessarily as first-path perception of *real* energy through conventional, or even unconventional, senses. Paranormal experience is something "realer" than mere subjective hallucination. With enough doubt about how an altered state blindly constructs complex scenes that don't correspond to everyday reality, coupled the accusation of scientism tainting the outcomes or decisions of scientific practice, a *third path* to explain paranormal experiences is advocated. It's neither a right

[23]Tart (1998, p. 81).

nor left turn in the cul-de-sac of explanation, but heads off in an entirely different direction.

References

Arnette, J. K. (2012). Recollections of Death: A Medical Investigation. *Journal of Scientific Exploration, 26*(1), 207–211.

Baruš, I. (2010). Beyond Scientific Materialism: Toward a Transcendent Theory of Consciousness. *Journal of Consciousness Studies, 17*(7–8), 213–231.

Bender, M. L., & Gill, P. (1986). The Genetic Code and Zipf's Law. *Current Anthropology, 27*(3), 280–283.

Brabant, O. (2016). More Than Meets the Eye: Towards a Post-materialist Model of Consciousness. *Explore, 12*(5), 347–354.

Gauld, A. (2012). Living Consciousness: The Metaphysical Vision of Henri Bergson (Book Review). *Journal of the Society for Psychical Research, 76*(909), 222–227.

Hancock, G. (2005). *Supernatural: Meetings with the Ancient Teachers of Mankind.* London: Century.

Hughes, J. E. A., Gruffydd, E., Simner, J., & Ward, J. (2018). Synaesthetes Show Advantages in Savant Skill Acquisition: Training Calendar Calculation in Sequence-Space Synaesthesia. *Cortex, 113,* 67–82.

Lampis, R. (2004). The Theory of Reality: Or Why, If We Understand Reality, the Theory of Everything Is an Impossibility. Or How to Establish Basic Principles That Clarify Physical, Mental and Paraphysical Phenomena, and to Dispose of Concepts Such as Non-locality, Place. *International Journal of Humanities and Peace, 20*(1), 43–48.

Leahey, T. H (1987). *A History of Psychology* (2nd ed.). Englewood Cliffs, NJ: Prentice-Hall.

Lestrade, S. (2017). Unzipping Zipf's Law. *PLoS ONE, 12*(8), e0181987.

McLeod, C. C., Corbisier, B., & Mack, J. E. (1996). More Parsimonious Explanation for UFO Abduction. *Psychological Inquiry, 7*(2), 156–169.

Moreira-Almeida, A., & Lotufo-Neto, F. (2017). Methodological Guidelines to Investigate Altered States of Consciousness and Anomalous Experiences. *International Review of Psychiatry, 29*(3), 283–292.

Price, H. (1953/1963). Introduction. In G. N. M. Tyrell (Ed.), *Apparitions.* New York: Collier Books.

Stafford Betty, L. (2004). Mind, Paranormal Experience, and the Inadequacy of Materialism. *International Philosophical Quarterly, 44*(3), 373–392.
Tammet, D. (2006). *Born on a Blue Day; Inside the Extraordinary Mind of an Autistic Savant.* New York: Free Press.
Tart, C. T. (1998). Six Studies of Out-of-Body Experiences. *Journal of Near-Death Studies, 17*(2), 73–99.
Tart, C. T. (2008). Accessing State-Specific Transpersonal Knowledge: Inducing Altered States. *Journal of Transpersonal Psychology, 40*(2), 137–154.
Tuske, J. (1999). Being in Two Minds: The Divided Mind in the Nyayasutras. *Asian Philosophy, 9*(3), 229–239.
Wilson, C. (1971/1983). *The Occult.* St Albans: Grenada.

Part IV

The Third-Path

28

A Return to Dualism

The third-path is unabashedly dualistic. And there's no exertion spent trying to consolidate brain neurophysiology and qualia. As philosopher Frank Jackson writes;

> I think there are certain features of the bodily sensations especially, but also of certain perceptual experiences, which no amount of purely physical information includes. Tell me everything physical there is to tell about what is going on in a living brain, the kind of states, their functional role, their relation to what goes on at other times and in other brains, and so on and so forth, and be I as clever as can be in fitting it all together, you won't have told me about the hurtfulness of pains, the itchiness of itches, pangs of jealousy, or about the characteristic experience of tasting a lemon, smelling a rose, hearing a loud noise or seeing the sky.[1]

We've now returned to the question of *why* there is qualia, or the "feel" of conscious experience. And although the proposition that consciousness

[1] Jackson (1982, p. 127).

T. Jinks, *Psychological Perspectives on Reality, Consciousness and Paranormal Experience*,
https://doi.org/10.1007/978-3-030-28902-7_28

is epiphenomenal is discounted, epiphenomenalism is "intelligible"[2] as it implies some sort of distinction between brain and mind.

But what if there is an *actual* mind/brain separation? This is an entirely different agenda to scientific reality. By claiming that a non-physical consciousness is autonomous and akin to spirit, or soul[3] (or some other word detached from materialism) there's a lot more flexibility to provide a 5W1H solution to paranormal experiences—and of course a solution to the "problem of the producer". The trade-off is the inability to empirically verify the claims, although the third-path will point to the limitations of material evidence ever truly "explaining" paranormal experience.

The first argument in favour of a distinction between brain and mind (if not a complete separation) is found in the claims of the *placebo* and *nocebo*[4] effects, the most widely known being pain relief achieved from placebo analgesics like sugar pills.[5] In terms of an underlying cause;

> A real placebo effect… is a psychobiological phenomenon occurring in the patient's brain following the administration of an inert substance, or of a sham physical treatment such as sham surgery, along with verbal suggestions (or any other cue) of clinical benefit. Therefore, the effect that follows the administration of a placebo cannot be attributable to the inert substance alone, for saline solutions or sugar pills will never acquire therapeutic properties. Instead, the effect is due to the psychosocial context that surrounds the inert substance and the patient.[6]

The expressions "occurring in the patient's brain" and "psychosocial context" imply respectively (in a strict medical framework at least) that *conditioning* and *expectation* drive the effect. In particular, these are biological mechanisms associated with anxiety reduction, conditioned reinforcement, and social learning.

---

[2] Stafford Betty (2004).

[3] Ibid.

[4] Nocebo effects, which refer to reduced treatment success due to patients' negative expectations (Bartels et al. 2018).

[5] Haga et al. (2009).

[6] Carlino and Benedetti (2014, p. 37).

While anxiety reduction and conditioned reinforcement can be moulded into a neurological profile, reference to "social learning" might be taken as a commitment to mind having causal influence over the physical body, which is the antithesis of monism. In defence, a typical placebo effect does not necessarily concede any physical change; the effect is rather a *re-alignment of mental state* whereby the same biological signal is assigned a different subjective interpretation based on belief.[7] Put simply, the placebo is an example of *mind over mind*. Belief and the *feeling* of pain are both psychological; hence the former can negate the latter, but not the activity of those C-fibres (and other related pathways) that bring neural information from a damaged part of the body via the peripheral nervous system.

But what about *actual* physical changes to the body said to be caused by belief[8]—do these examples support dualism? Deftly adapting *mind over mind*, such effects might still be clarified by the second-path as representing *brain over matter*. Taking the position that mind is indistinguishable from brain activity, then there's no reason why activity in the prefrontal cortex (or any other cortical association area) cannot directly effect changes in heart rate or skin conductance following perception of social or dangerous environmental cues.

This is the insurmountable barrier between the second- and third-paths. They take fundamental and incompatible positions from the outset. For the third-path, there is an undisputable inner space of non-material mental reality and an external veridical physical world where the rules of scientific reality operate.[9] This is most importantly shown by the acceptance of mind-over-matter effects;

> …the hypothesis that the human mind should be considered causally ineffective or even entirely illusionary not only conflicts with our everyday experience, it is being refuted by a plethora of intention-based and belief-based phenomena, the most radical one being the mind's ability to cause the demise of the body.[10]

---

[7] Crane (2016).

[8] Kemper (2016).

[9] Meijer and Korf (2013).

[10] Brabant (2016, p. 348).

An even greater gulf concerns the position of consciousness. If consciousness is distinct (and does not originate) from the brain, then it doesn't even have to reside inside the skull at all! While the third-path may not depend on empiricism to support its position, there is some evidence that mental processes can be experienced independently of nervous system activity. When brain signals are measured using neuroimaging techniques, information is gained about the energy demands of neurons and their supporting cells—put simply, brain *metabolism*. Heavily active parts of the brain use a lot of energy to function, and for cortical regions it is assumed they are indulging in cognitive (and other) operations; mathematical calculation, social judgements, and a raft of other possibilities.

One example of an experiment measuring brain metabolism examines human observers reacting to rapid and subliminal light patterns with a very short duration (in the range of 50 milliseconds). The participants' energy consumption is very fast, and this is said to demonstrate information response times to match incoming stimuli happens in fewer than six successive action potentials.[11] The implication is that cognitions don't lag very far behind their alleged neurophysiological cause.

Furthermore, the perceived information need not be subliminal. Experimental participants can differentiate semantically valid (meaningful) words from nonsense words within the same short time period.[12] And with regards the processing of auditory information, there is an even faster processing speed. Measurements in individuals required to source the origin of noise shows time periods for pre-conscious decision-making in the order of 0.05 milliseconds. That's now argued to be *faster* than neurotransmission/action potential events. Taken together, this evidence is used as a foot in the door to announce;

> ... that complicated, culture-specific, information is recognized in our brain prior to reaching consciousness and also show that information processing by the brain is much faster than metabolism.[13]

[11] Meijer and Korf (2013, p. 6).
[12] Meijer and Korf (2013).
[13] Ibid., p. 6.

Is this apparent disconnect between brain and cognition supernatural? Not necessarily. A second-path response is that rapid mental decision-states are caused by events *preceding* action potentials but that are still physical; the extremely rapid, almost instantaneous changes occurring in the conformation of proteins (known as *conformational perturbations*) equivalent to opening and closing of ion channels. These are analogous to switches, or changes of processing state, in a computer system. These perturbations don't require much energy, and occur less than 0.1 milliseconds after a coupled receptor is activated, finishing within 2 milliseconds. Neuroimaging measures *gross* alterations in brain metabolism and is not compatible with detecting the earliest stages of the generation of consciousness. Metabolism represents the *restoration* of the brain's resting state. The initiation and conduction of information processing is argued to be well over before any measurable signals are initiated. This is usually in the order of half a second to two seconds for cerebral blood flow to increases enough to be noticed, before peaking at around four to six seconds.[14]

The third-path can wrest back control of the argument over the origins of consciousness by arguing that there are even more fundamental operations occurring in the mind/brain, much quicker and much smaller than proteins. These are very fast and very small quantum-like processes that allow intention (mental states) to;

> ... manifests physically [whereby]... mental influence can order the quantum fluctuations of a particle's momentum. And that such an influence can act on the brain, so that about 4,000 molecules would need to be ordered for sufficient sodium ion channels to open, thereby producing action potentials, resulting in a desired bodily action.[15]

The mind/brain axis is therefore physically extant, but "exists" at the most subtle quantum level, hence is not readily observable to conventional measuring instruments.

This unusual time-course of neural activity, behaviour and cognition is co-opted by the third-path to undermine the notion of an epiphenomenal

[14]Ibid.

[15]Barušs (2010, p. 224).

(helpless) mind. But that's not where the third-path rests. There's an even more extreme claim that there doesn't even *need* to be much (or any) brain activity to generate or accompany consciousness. Recorded neural activity corresponds to brain metabolism that might indeed occur *after* information processing has taken place (a notion that resembles time-symmetry and psi, where "what comes after, comes before"). Nevertheless, there's still a direct link between this metabolism and the preceding cognition. Typically there are increases in recorded activity (let's assume "restorative" activity), but these are offset by activity decreases in other brain regions. Conversely, areas of low activity observed in certain cerebral regions during cognitive tasks are compensated by activity increases in areas *remote* from these sites.[16]

And not surprisingly, psychedelics force interesting deviations to this routine. As we know, psilocin produces significant experiential effects (the hallucinogenic symptoms of an ASC), and yet when the brain is measured during a "trip" there is a pervasive decrease in general activity that's not accompanied by any significant increase, anywhere. Recall that psychedelics including psilocin *lower* interaction between the prefrontal cortex and cingulate cortex (the "connector hubs" of the brain) which amplifies phenomenological experience—that state of unconstrained cognition.[17] Rather than the second-path interpretation that these experiences are centred on illegitimate neurophysiological malfunction, the third-path uses this evidence to question the causal nature of neural activity in creating the pervasive visual narrative.

Naturally there is still *some* brain activity while psychedelics have their effects, termed the *neural correlates of consciousness*. This corresponds to the essential activity of a neural system required to generate basic conscious experience (with anything less inhibiting any sign of consciousness).[18] And yet this baseline brain activity is not equated with the sensory experiences initiated by psilocybin, or any other ASC-producing stimulant.

---

[16] Meijer and Korf (2013).

[17] Carhart-Harris et al. (2012).

[18] Polák and Marvan (2018).

Memories are also consolidated during the psychedelic state despite the minimal activity. Regarding NDEs, the counterargument that these experiences are no more than an REM-initiated dream, autoscopic hallucination or reaction to anaesthetics[19] is that individuals who vividly recollect their encounter upon awakening (or returning) are recalling memories that fashioned during a period of significant brain impairment[20]—a time where the neural correlates of consciousness might not even have been present.

> Both are times when, according to modern neuroscience, there should not be any awareness, cognitive processor memory formation whatsoever.[21]

That said, the second and third-paths are continually toing-and-froing with the interpretation of evidence, so it's not surprising that critics argue that an NDE (as a recalled "memory") is created *not* when the patient is dead, but at a time when the brain is partly functioning again, for instance at the moment of resuscitation. In that case, enough of the neural correlates of consciousness are present to assist visions and memory formation.[22]

References

Arnette, J. K. (2012). Recollections of Death: A Medical Investigation. *Journal of Scientific Exploration, 26*(1), 207–211.

Bartels, D., Van Laarhoven, A., Van De Karkhof, P., & Evers, A. (2018). Nocebo Effects and Scratching Behaviour on Itch. *Acta Dermato-Venereologica, 98*(10), 943–950.

Barušs, I. (2010). Beyond Scientific Materialism: Toward a Transcendent Theory of Consciousness. *Journal of Consciousness Studies, 17*(7–8), 213–231.

Brabant, O. (2016). More Than Meets the Eye: Towards a Post-materialist Model of Consciousness. *Explore, 12*(5), 347–354.

19 Arnette (2012).

20 Brabant (2016).

21 Ibid., p. 348.

22 Ibid.; Although this might be considered a moot point, if mind and brain are taken as independent (reiterating the disconnect between the second and third-paths).

Carhart-Harris, R. L., Erritzoe, D., Williams, T., Stone, J. M., Reed, L. J., Colasanti, A., et al. (2012). Neural Correlates of the Psychedelic State as Determined by fMRI Studies with Psilocybin. *Proceedings of the National Academy of Sciences of the United States of America, 109*(6), 2138–2143.

Carlino, E., & Benedetti, F. (2014). Placebo and Nocebo Effects. In D. I. Mostofsky (Ed.), *The Handbook of Behavioral Medicine* (pp. 36–57). Wiley-Blackwell: Hoboken, NJ.

Crane, G. S. (2016). Harnessing the Placebo Effect: A New Model for Mind-Body Healing Mechanisms. *International Journal of Transpersonal Studies, 35*(1), 39–51.

Haga, S. B., Warner, L. R., & O'Daniel, J. (2009). The Potential of a Placebo/Nocebo Effect in Pharmacogenetics. *Public Health Genomics, 12*(3), 158–162.

Jackson, F. (1982). Epiphenomenal Qualia. *Philosophical Quarterly, 32*(127), 127–133.

Kemper, K. J. (2016). Authentic Healing: Effects on Recipients and Healers. *Alternative and Complementary Therapies, 22*(3), 105–110.

Meijer, D. K. F., & Korf, J. (2013). Quantum Modeling of the Mental State: The Concept of a Cyclic Mental Workspace. *Syntropy, 1,* 1–41.

Polák, M., & Marvan, T. (2018). Neural Correlates of Consciousness Meet the Theory of Identity. *Frontiers in Psychology, 9,* 1–13.

Stafford Betty, L. (2004). Mind, Paranormal Experience, and the Inadequacy of Materialism. *International Philosophical Quarterly, 44*(3), 373–392.

29
The Prism

In the face of such effective retorts, how can the third-path distinguish itself from its competitors? This requires one very avant-garde concept; that brain activity *doesn't* cause consciousness. This corresponds to *active consciousness*, whereby authentic paranormal experience is not a product of passive sensory perception. Consciousness is not bound to the brain, experiencing the extraordinary through the breakdown of filters or the twisting of hypothetical tuners that drawn the curtain to reveal hidden information independent of the self. Rather, consciousness actively seeks information which the brain cannot do. This approach is well summarised by theology professor Lewis Stafford Betty;

> A brain cannot leave its body, but an immaterial self, which....follows very different laws from its body-may well be able to; nothing makes such an event impossible. A brain cannot know what another brain miles away, or even a person lacking a brain altogether, is communicating, but there is nothing we know about a self that makes either event impossible. A brain cannot work after it dies, but there is no law telling us that an immaterial

T. Jinks, *Psychological Perspectives on Reality, Consciousness and Paranormal Experience*,
https://doi.org/10.1007/978-3-030-28902-7_29

> structure-like a self or soul or mind-must cease to work just because the brain it formerly interacted with has died.[1]

This pure expression of mind/body dualism (or *ultra-dualism*[2]) has been around for a very long time. Plato's *Forms* are the true objects of knowledge in the *Realm of Being*, since people cannot garner eternal and unchangeable knowledge of a world of physical objects (the *Realm of Becoming*[3]) which is constantly in flux.[4] There are Forms for classes of objects and concepts (such as a dog, or the idea of "good"), and things perceived through the senses are imperfect reflections of these fundamental principles.[5]

Furthermore, only a non-physical mind can comprehend the Forms. The mind (or soul) also endeavours to break free of the (degraded or incomplete) physical body to enter the realm of Forms (Being), and this is where the production material of mental imagery derives.

A contemporary analogy applied to the separation of mind and body is the *prism*.[6] The prism corresponds to the CNS; a physical medium consisting of neurons and supporting cells (and everything else that makes up brain) that acts like an optical device. It has a different "refractive index" to the mind, and comparable to the way white light entering a glass medium will refract and split into colours of different wavelengths, an external mind will be refracted through the brain, split, and display an "internal theatre"[7] on the screen of personal consciousness. Mind is what is externally received, and consciousness is what is personally experienced. Taken this way, the prism analogy doesn't necessarily repudiate the orthodox understanding of perception. We certainly receive information through the conventional senses (eyes and ears and other organs), but that only corresponds to a narrow strain of submissive perceptual experience;

[1] Stafford Betty (2004, p. 387).

[2] Maher (1900).

[3] Capps (2015).

[4] Although there is debate over whether Plato actually *believed* in the realm of Forms as a literal, metaphysical "elsewhere" (which is the present context of the discussion), but used it for moral teaching about the "conception of one's place in the reality of the here and now" (Jonas 2016, p. 319).

[5] Leahey (1987, p. 43).

[6] Stafford Betty (2004).

[7] Hansen (2014); Charles Fernyhough makes this claim in his June 2006 essay for *The Psychologist.*

collecting local environmental information. Mind beamed through the prism meets this moulds this raw neural information into qualia that is observed in the auditorium of private consciousness.

The prism notion is also not a new idea (although certainly more recent than anything Plato wrote about). In his 1898 book *Human Immortality*, William James wrote the brain doesn't procreate the mind (in the way kettles produce steam), but transmits the mind. Therefore;

> … the function of the brain is not to produce consciousness but to transmit it, or permit its entry, from elsewhere.[8]

However this proposal is more complicated than the kettle analogy, since the kettle, the water, and the steam that's produced are homogeneous physical substances. Mind and brain are utterly different.[9]

And speaking of analogies, the filter/tuner mechanism is compatible with the prism, since Bergson himself did not view consciousness as a necessary product of brain function. The memories and other mental states strained by the filter do not necessarily originate from any part of the human brain.[10]

Indeed a modern take on the mind/brain interaction combines the filter/tuner and the prism, using a higher-order computer parallel. The brain is not just equivalent to a computer, it is equivalent to a *networked* computer.[11] "The internet" (and what it represents) is captured by the machinery and the vast range of content it covers is filtered to something personally manageable; conscious thought. A memory of what was viewed exists in the search history list and "cookies" (e.g., preferences), and this might give the impression that memories derive from the computer itself—somewhere deep in its processor. However these memories are simply a trace, or record, of more substantial *exogenous* material.[12]

[8]Gauld (2012, p. 223).

[9]Poller (2015).

[10]Gauld (2012) (Bergon gave this impression in *Matter and Memory*).

[11]Poller (2015).

[12]Ibid.

A particular point of interest—or complication—is how the mind (i.e., the content that is "beamed in" from elsewhere) and the brain interact if they are so different. This is troublesome to Plato (and Descarte's) dualism;

> How does this mind-body or body-mind interaction take place? How can a spiritual substance interact with a material substance, and vice-versa? Put differently, since all of the forms that bodies occur in involve only various extensional features, never mental ones; similarly, no form of thought involves extension. That means the realms of thought and extension is completely different. How then can mental events have anything to do with physical ones and vice-versa, since the one occurs in space and the other is nonextended thought with no physical properties whatsoever.[13]

This problem can be avoided if different "grades of matter" are conceived. Just as water and steam are different physical states, or glass and photons of light, so to can be mind and brain;

> I do not think that we know enough about the mysterious mind, or soul, to say what it can or cannot do if we conceive of it as material in some exquisitely subtle manner. It would not surprise me if it turned out that souls, though not reducible to a certain lump at some identifiable space in the brain, were nevertheless material in some subtle manner unknown to us mortals, and that their materiality allowed them to interact with the brain, not by some push-and-pull mechanism, but by something more akin to a force field between different grades of matter.[14]

And rather than a mind/body dichotomy;

> ... think of reality-all of it-as composed of a single kind of substance with innumerable grades of grossness and subtlety, ranging from the utterly dense stuff of what we call lifeless matter to the inconceivably subtle "matter" of the Divine Mind. The qualitative difference between these two extremes is impossible to exaggerate, yet both are expressions of only one substance....it is a lot easier to account for interaction between two things as different as

[13]Mohammed (2012, p. 100).

[14]Stafford Betty (2004, p. 390).

> mind and brain if they are both ultimately made of the same stuff, than it is if they are made of totally different stuff.[15]

Stafford Betty also points out this interaction is "unimaginably felicitous and intimate", but there's not much to add about how exactly mind and brain differ, other than they do;

> ...light is not identical to the prism, just as the self is not identical to the brain. Yet they work in the most intimate harmony; the prism makes it possible for refraction to occur, and the light gives the prism its *raison d'etre.*[16]

Charles Tart provides alternative suggestions to the interaction problem. Assuming the mind is now autonomous, that allows it to be imbued with particular abilities which facilitate interaction with a physical brain. However these abilities can only be described as psychic, therefore not effectively measurable. *Auto-clairvoyance* is the method by which the mind is able to determine the "physical state" of the brain (and body) that it is in relationship with.[17] In order to enact change in the world through behaviour, the mind influences the brain through *auto-psychokinesis*. These labels describe a volitional mind with complete control over a physical brain, which is the complete opposite to an epiphenomenal or even emergent mind-from-brain.

Regardless of whether mind/brain interaction is made possible by a continuum of grades of matter, or through psychic control, these ideas are completely at odds with the second-path conviction that humans reside *entirely* and *permanently* inside their bodies, and from the point of view of waking consciousness, inside their heads.[18] That's not to say it's a bad idea, because promotes survival through feeling "grounded" in reality, and gives the immaterial mind a convenient location in space.[19] However the third-path views this as an outmoded, inaccurate way to deliberate

---

[15]Stafford Betty (2004, p. 389).

[16]Ibid., p. 387.

[17]Tart (1998).

[18]Ibid.

[19]Stafford Betty (2004).

consciousness. A further technological analogy that provides a parallel to this (purported) mistake of believing "you" exist "inside" your head is the impression of participating in a computer-generated virtual reality world. When engrossed in a visual reality environment, a person;

> ..forgets where his or her physical body actually is, and becomes experientially located "in" the computer- generated world, it might be that our "souls" are actually located on some other planet, but we are so immersed in the biopsychological virtual reality our brains generate, and to which we are telepathically and psychokinetically connected.[20]

Of course, for the estranged mind to refract and become personal consciousness the CNS needs to be intact and functioning normally. Hence;

> As long as the prism retains its structure, so long does it refract light as intended - just as a healthy brain collaborates with the self in the production of conscious states.[21]

That means any adverse effects to the prism's structure—or damage to the brain, will affect the experience of consciousness. William James noted this even earlier than *Human Immortality*. In his 1890 work *The Principles of Psychology* he argued that even when the body is intact and functioning normally, personal consciousness is *altered* with a "few ounces of alcohol or grains of opium or hashish, or a whiff of chloroform or nitrous oxide gas", and can *completely abolished* following a head blow, loss of blood or "the pressure of an apoplectic haemorrhage".[22] He reasoned;

> ..if ... a man loses consciousness so soon as his brain is injured, it is clearly as good an explanation to say the injury to the brain destroyed the mechanism by which the manifestation of consciousness was rendered possible, as to say that it destroyed the seat of consciousness.[23]

---

[20]Tart (1998, p. 94).

[21]Stafford Betty (2004, p. 387).

[22]Presti (2011, p. 22).

[23]Gauld (2012, p. 223).

With the prism temporarily distorted, the "light" of the mind will not refract correctly and an ASC will transpire because ordinary consciousness cannot be effectively produced and maintained. With the prism permanently chipped, chronic problems with the ability to perceive everyday reality will result. As revealed in Charles Bonnet Syndrome, what is produced in consciousness from the damaged CNS locations is not a blank emptiness, but alternative visual content. In the second-path this is due to release of stored memories "filling in" perceptual gaps. In the third-path, damage to the brain-prism (which perhaps loosens a filter/tuner mechanism in the process), will cause the mind-content "beamed" from these regions to be entirely different to what is expected from conventional reality.

And when the brain is completely destroyed, the "light" of the mind will not be able refract at all as personal consciousness. But that doesn't mean the light/mind doesn't cease to exist. Rather "it simply shines elsewhere."[24]

The implications of this statement are quite profound. It suggests;

> The light exists in its own right and does not in any way depend on the prism for its existence. In the same way, the self, the analogue of white light, is independent of the brain for its existence. Indeed there is no reason to assume that the self cannot pre-exist the body, just as the light pre-exists the prism.[25]

Physicist Oliver Lodge made similar claims with regards telepathic abilities in an early SPR article, although not with a prism in mind;

> Just as the energy of an electric charge, though apparently on the conductor, is not on the conductor, but in all the space round it; just as the energy of an electric current, though apparently in the copper wire, is certainly not all in the copper wire, and possibly not any of it; so it may be that the sensory consciousness of a person, though apparently located in the brain, may be conceived of as also existing like a faint echo in space, or in other brains, though these are ordinarily too busy and pre-occupied to notice it.[26]

[24] Stafford Betty (2004, p. 387).
[25] Ibid.
[26] Aspre (2011, p. 146).

Mind (of whatever substance) divorced of brain also allows for information acquisition beyond of the sensory organs. This is the canny third-path explanation for OBEs, NDEs and other forms of mystical travel.[27] Some OBEs (as well as experiences occurring under hypnosis) are truly hallucinatory. Consciousness behaves as though it is located in the skull and receives information during the ASC that appears novel, but is simply a rearrangement of the jigsaw components of memory that are entirely endogenous. Alternatively, some of the components might derive from outside sources and arrive at the brain to be passively perceived, but give the *impression* of disembodied journeying. But in some special cases, such as that of Miss Z, the information retrieval best fits a circumstance where personal consciousness is resumed into broader mind, and is able to "go" to other places in the physical world, and presumable places in a non-physical realm typified by the astral journey. The problem is;

> This is a messy situation in some ways, especially because all three of these types of [OBE-like experiences] may seem experientially identical to the person having them, at least at rough levels of description.[28]

A further repercussion of the prism analogy is that there's no theoretical reason a single mind must enter a single brain-prism. There could conceivably be an infinite number of "minds", or offshoots of mind, that make use of the machinery of a solitary brain. William James recognised this possibility by suggesting in much-quoted statement from *The Varieties of Religious Experience* (1902);

> ...our normal waking consciousness, rational consciousness as we call it, is but one special type of consciousness, while all about it, parted from it by the filmiest of screens, there lie potential forms of consciousness entirely different.[29]

If these offshoots are the personal consciousness of other people, or the deceased, hitching a ride on the broader mind before sliding down the

[27] Brabant (2016).
[28] Tart (1998, p. 92).
[29] James (1902, pp. 307–308).

chute to *your* personal mind, then it would account for all sorts of anomalous phenomena, like savancy;

> Hyslop (1909) reports the case of a man who found his body suddenly invaded by a foreign intelligence, suspected to be a recently deceased landscape artist. With little artistic training, this man produced several paintings very closely resembling unfinished sketches of the relevant artist never shown publicly.[30]

In this context savancy, alongside reincarnation memories and déjà vu, is simply an expression of a foreign consciousness entering a brain-prism as a *drop-in* or *walk-in*. To use yet another analogy, the *film reel* has been replaced in the brain *projector*. In fact, this new experience is not strictly an ASC (by current definitions) because there hasn't been as *alteration* of ordinary consciousness through distortion of the global state, or a rearrangement of the component pieces. Instead there's been a complete replacement of ordinary consciousness with someone else's (or some*thing* else's!).

This entails someone else's consciousness has left *their* brain and refracts through *your* one, such that their memories and perceptions are now experienced by you. Or, if given free reign, the consciousness is from a deceased individual (or anything else with a similar style of non-physical mental energy) that beams into your brain and provide recollections of their life, objects and places, from the past present or future. That's because in this context of free mind, there is no passage of time so past, present and future are "melded into eternity".[31]

This does present a problem. Apart from a fugue state, where there is no recollection of the experience at all (if such a situation can even be *considered* an authentic experience, if there is no self-awareness),[32] how is it possible for you to enter another consciousness when your own consciousness (or part of it) must still be attached to the machinery of your brain-prism? Presumably you are still to *some* extent "yourself", and can form memories about your impression of this alien consciousness (or

30 Hanes (2012, p. 27).

31 Poller (2015, p. 259).

32 An important suggestion from an anonymous reviewer of the manuscript.

many). That suggests there's competition for brain-prism resources. As a quick fix, the jigsaw puzzle analogy might be recycled and grafted to the brain-as-prism conceptualisation. When another consciousness "drops-in" to your mental space, you still retain the global state (the rectangular puzzle frame) but you perceive a complete arrangement of the elements of another's consciousness without *their* self-awareness. That's a very easy addition to make—there are no rules or regulations to this third-path journey of paranormal experience; extraordinary experiences can be caused by your consciousness travelling free of the confines of a physical brain in this (or other) worlds, or entering someone else's brain to "see" what they see. Or even someone (or something) entering *your* brain to provide you with surreptitious information from literally anywhere or any time!

References

Asprem, E. (2011). Pondering Imponderables: Occultism in the Mirror of Late Classical Physics. *Aries, 11*(2), 129–165.

Brabant, O. (2016). More Than Meets the Eye: Towards a Post-materialist Model of Consciousness. *Explore, 12*(5), 347–354.

Capps, M. (2015). Body and Soul in Plato: How Our Bodies Can Serve Our Philosophical Pursuits. *Dialogue, 58*(1), 19–29.

Gauld, A. (2012). Living Consciousness: The Metaphysical Vision of Henri Bergson (Book Review). *Journal of the Society for Psychical Research, 76*(909), 222–227.

Hanes, K. (2012). Unusual Phenomena Associated with a Transcendent Human Experience: A Case Study. *Journal of Transpersonal Psychology, 44*(1), 26–47.

Hansen, D. (2014). Heavenly and Hellish—Writers on Hallucinogens. *The Psychologist, 27*(9), 680–682.

James, W. (1902/1985). *The Varieties of Religious Experience.* Cambridge: Harvard University Press.

Jonas, M. E. (2016). Three Misunderstandings of Plato's Theory of Moral Education. *Educational Theory, 66*(3), 301–322.

Leahey, T. H (1987). *A History of Psychology* (2nd ed.). Englewood Cliffs, NJ: Prentice-Hall.

Maher, M. (1900). *Psychology: Empirical and Rational* (4th ed.). London: Longmans, Green and Co.

Mohammed, A. A. (2012). A Critique of Descartes' Mind-Body Dualism. *Kritike: An Online Journal of Philosophy, 6*(1), 95–112.
Poller, J. (2015). Beyond the Subliminal Mind: Psychical Research in the Work of Aldous Huxley. *Aries, 15*(2), 247–266.
Presti, D. E. (2011). Neurochemistry and Altered Consciousness. In E. Cardeña & M. Winkelman (Eds.), *Altering Consciousness: Multidisciplinary Perspectives: History, Culture, and the Humanities; Biological and Psychological Perspectives* (Vol. 2, pp. 21–41). Santa Barbara: Praeger.
Stafford Betty, L. (2004). Mind, Paranormal Experience, and the Inadequacy of Materialism. *International Philosophical Quarterly, 44*(3), 373–392.
Tart, C. T. (1998). Six Studies of Out-of-Body Experiences. *Journal of Near-Death Studies, 17*(2), 73–99.

30
The Otherreal

Let's play around with the implications of this new approach. According to the third-path, paranormal events from alien abductions to zooforms are legitimate because they *conform to the definition of experience*, but many cases are not tangibly real in because they have not been perceived through the senses. Instead what is encountered is another kind of real, the *otherreal*.

Take for instance Grahame Hancock's impression of a DMT-induced ASC. He found himself in a room where;

> …the technology of the room seemed to be about displaying some sort of recording that had been set in playback mode by *the arrival of my consciousness.* (emphasis added)[1]

His consciousness had *arrived* in that room. Similarly following ingestion of ayahuasca;

[1] Hancock (2005, p. 429).

T. Jinks, *Psychological Perspectives on Reality, Consciousness and Paranormal Experience*,
https://doi.org/10.1007/978-3-030-28902-7_30

> ...our bodies remained in the material world while the 'plant spirits' temporarily liberated some aspect of our consciousness to travel in other, non-physical realms to interact with intelligences there.[2]

Ostensibly this would require an untold number of receptive prisms allowing our consciousness to be focussed elsewhere, not just on "everyday" earth and through your own brain. These prisms don't even have to be constructed from a neural network, and indeed might be better at refracting free-floating consciousness than brain matter.

A feature the third-path uses to justify the otherreal is the "realer than real"[3] phenomenological nature of experiences. Paranormal events possess are said to possess sense of extreme lucidity more substantial than life impressions, let alone a conventional hallucination;

> For those who have experienced these alternate realities there is a clarity of perception and a vividness that is absent in our normal everyday world.[4]

For instance, "survivors" of an NDE are said to "scoff" at the second-path claim they are imaginary because they *know* they are real.[5] They also assert a clearness of mental capacity that's at least equal to if not superior ordinary consciousness.[6] This sense of hyper-reality is also applied to the lucid dream;

> When materialists encounter the challenge that physical manifestation does not actually have the solidity with which it presents itself to our sensory perceptions, they have sometimes responded by invoking Samuel Johnson's refutation of Bishop Berkley's idealism, which consisted of kicking a large stone. And with that they think that they have dispelled the challenge. Of course that is a logically flawed 'argument' but I sometimes hear it used anyway. During my lucid dream [in a highrise building], I remembered this 'argument' and decided to take the materialists at their word. So I walked over to one of the walls and pounded on it with my fist, hard, a

[2]Ibid., p. 435.

[3]von Lucadou and Wald (2014).

[4]Hume (1995, p. 25).

[5]Stafford Betty (2004).

[6]Brabant (2016, p. 220).

> number of times. The wall was solid. There was no question about it. The environment in which I found myself was indistinguishable from physical manifestation. Clearly, finding the dream wall to be solid does not prove that my dreamscape is made out of billiard balls. Nor, of course, does such action while awake prove that the landscape of my ordinary waking state is made out of billiard balls.[7]

Alongside clarity of vision, the intensity of emotion during the otherreal encounter is taken as evidence that the experience is a *special kind* of authentic. That's found in John Mack's treatment of the alien abduction experience, which is construed as happening *to* consciousness in an otherreal space;

> …I want to stress what I believe is fundamental in distinguishing this phenomenon from fantasy or delusions, namely, the intense distressing affect that is associated with the recall of these experiences. Abductees literally shake with fear when they reach the memories of how they were paralyzed or saw the little aliens with the big black eyes. Their whole bodies tremble, and they may scream with terror as they recall the intrusive procedures that occurred on the ships. Any effort to simulate these experiences, like the works of Alan Lawson, fails to recreate what an abduction encounter is really like. People can talk about what an abduction would be like, but it is the affective power that accompanies the memories that gives authenticity to them.[8]

This is the same otherreal destination arrived at by the psychedelic drug user;

> …the beings that ayahuasca and ibogaine brought me literally face-to-face with might just be illusions of disturbed brain chemistry, as Western scientists claim, but they certainly hadn't seemed that way to me. On the contrary, very thing that these two plant based hallucinogens had put me through had felt very strongly and convincingly like the opening of doors on to other and entirely different levels of reality. The experiences had been

[7] Barušs (2010).

[8] Mack (1993, p. 28).

> profound, deeply moving, through provoking in a lasting way an also, in one case, weirdly menacing and terrifying.[9]

It's one thing to use the brain-as-a-prism analogy to describe how an otherreal might be experienced in waking consciousness. It's another to conceive how this non-reality *really* exists. A basic scaffold may be found with the concept of *potentialities*, as opposed to the substantial objects of scientific reality.[10] Think back to Heisenberg's superpositional state, whereby "quantum entities" do not reside in a specific conformation like everyday objects. Indeed, quantum entities in superposition aren't *real* like everyday objects—when they are observed or measured they have the potential to *become* something real. This corresponds to reality being of dual nature, and founded on quantum entities transitioning from the "possible" [or potential] to the "actual" and back again.[11] For example, a hydrogen atom consists of many potential states, which all hypothetically "exist", but not in a "factual-material" form. And yet;

> ...they exist in the sense that their logical or mathematical order is part of the constitution of the system, completely determined and fixed by the conditions of the system, and can be calculated a priori; that is, the order of virtual states is preestablished and entirely predictable before it is actual.[12]

This order is not able to be seen, or experienced or measured in any way so it is not conventionally real, but transcendent. For this reason, it is termed a *transcendent order*, and due to its latent ability to manifest, it is also an *immanent order*. When "actualised", this order becomes everyday reality which is the stuff of scientific reality (the second-path) and even the paranormal things people "see" in the first-path as ghosts and other strange things. But being transcendent or immanent is not to be non-real in the sense that it possesses no information; in this framework non-material entities possess a strong claim to reality.[13]

---

[9]Hancock (2005, p. 289).
[10]Poynton (2012).
[11]Schäfer (2006).
[12]Ibid., p. 576.
[13]Ibid.

This immanent order corresponds to physicist David Bohm's *implicate order*, where the actualisation is in the shape of the *explicate* that has emerge from the implicate order of potentiality.[14] Bohm's model is practically conceived through holographic (or "holonomic") analogy[15]—a two dimensional plane can store the potential for a full three dimensional representation, which is expressed (actualised) when the appropriate medium is passed through. In the holographic analogy this medium is light. In terms of what is happening to actualise reality, this is a deeper state, or grounding, which is difficult to comprehend using any available language.

For the third-path, the holonomic analogy is the prism, so learning about what might be residing or existing in the immanent state is achieved by the actions of non-material consciousness. Specifically, consciousness (or mind) manifests immanent or implicate information from a clandestine storage "plate".

And what *is* this storage plate? Exploring the idea of the universe at large being conceptualised a "flowing hologram" (or "holomovement") in more detail, this involves information-as-energy cycling between the implicate and explicate order.[16] That's where the tiny protein microtubules in living cells come in once again. These hollow crystalline cylinders are only 50 micrometers in length and in their various conformational arrangements can hold "bits" of information, hence form the basis of a quantum computer.[17] This is excitedly referred to by some[18] as a doorway to a multiverse of otherreal possibilities, using a range of terms and theories we're already familiar with;

> ...a wormhole that uses the principle of an Einstein-Rosen Bridge to project consciousness into a parallel reality... [containing] information....within the zero-point field (Akashic Field).[19]

[14]Poynton (2012).
[15]Meijer and Korf (2013).
[16]Joye (2017); another expression is physicist Karl Pribram's Holoflux.
[17]Das (2016).
[18]Poynton (2012).
[19]Ibid., pp. 116–117.

For the third-path, active consciousness triggers the contents of this aetheric, zero-point or Akashic Field, rather than the field manifesting itself to be perceived passively by the human observer (a first-path treatment of consciousness). Not surprisingly, these ideas have been criticised as overwrought, receiving cold consideration because they give the impression of being the answer to everything—from telepathy, precognition and remote viewing to reincarnation memories and ghosts,[20] and irrevocably leads to the notion that consciousness *creates matter*—an idea that goes back in Western physics at least as far as Max Planck in 1931;

> I regard consciousness as fundamental. I regard matter as derivative from consciousness.[21]

However that just means it's even more popular in contemporary third-path explanation for the paranormal. After all, the same idea goes back to the Iron-age Indian Vedas, so there's an ancient and mystical pedigree that surely deserves respect?;

> .... the whole precedes the parts, and the parts emerge as limitations of the whole. Instead of the reductive physicalist worldview that consciousness and mind are epiphenomenal and fundamentally nonexistent, they are the *basis* of matter and direct its expressions all along. Non-physical mind underlies physical parts and intentionally guides unitary physical behavior at every stage.[22]

Therefore, consciousness is not only responsible for the inner universe of personal mental states, but makes use of a prism or hologram (choose your analogy) accessed through quantum processes occurring at the very edge of physical reality to outwardly project its "energy". That's how ordinary macro-reality is established. In this tongue-twisting flip, the otherreal is

---

[20] Poynton (2012): There's also the "serious question as to whether any model developed within the context of physics is capable of extension into non-physical experience".

[21] Brooks (2012, p. 42).

[22] Boyer (2014, p. 115); John Poynton does suggest these in many modern attempts to create a novel theory of reality, classic Vedic ideas have merely been "....dressed in terms of quantum physics and cell biology" to make them... "more kosher to the contemporary mind" (Poynton 2012, p. 118).

actually the *real* real, and the everyday physical reality we take for granted is the *fabricated* real—a true illusion.

References

Baruš̌s, I. (2010). Beyond Scientific Materialism: Toward a Transcendent Theory of Consciousness. *Journal of Consciousness Studies, 17*(7–8), 213–231.
Boyer, R. W. (2014). Unless We Are Robots, Classical and Quantum Theories Are Fundamentally Inadequate. *NeuroQuantology, 12*(1), 102–125.
Brabant, O. (2016). More Than Meets the Eye: Towards a Post-materialist Model of Consciousness. *Explore, 12*(5), 347–354.
Brooks, M. (2012). Does Consciousness Create Reality? *New Scientist, 215*(2884), 42–43.
Das, T. (2016). Origin of Consciousness and Zero-Point Field. *Neuroquantology, 14*(1), 45–48.
Hancock, G. (2005). *Supernatural: Meetings with the Ancient Teachers of Mankind.* London: Century.
Hume, L. (1995). By Knot of One the Spell's Begun, By Knot of Two It Cometh True… Magic, Altered States, and Multiple Realities. *Social Alternatives, 14*(4), 23–27.
Joye, S. R. (2017). Tuning the Mind in the Frequency Domain: Karl Pribham's Holonomic Brain Theory and David Bohm's Implicate Order. *Cosmos and History, 13*(2), 166–184.
Mack, J. E. (1993). The UFO Abduction Phenomenon: What Might It Mean for the Evolution of Human Consciousness? *ReVision, 16*(1), 26–32.
Meijer, D. K. F., & Korf, J. (2013). Quantum Modeling of the Mental State: The Concept of a Cyclic Mental Workspace. *Syntropy, 1,* 1–41.
Poynton, J. (2012). The Out-of-Body Experience: The History and Science of Astral Travel (Book Review). *Journal of the Society for Psychical Research, 76*(907), 116–120.
Schäfer, L. (2006). A Response to Ervin Laszlo: Quantum and Consciousness. *Zygon: Journal of Religion and Science, 41*(3), 573–582.
Stafford Betty, L. (2004). Mind, Paranormal Experience, and the Inadequacy of Materialism. *International Philosophical Quarterly, 44*(3), 373–392.
von Lucadou, W., & Wald, F. (2014). Extraordinary Experiences in Its Cultural and Theoretical Context. *International Review of Psychiatry, 26*(3), 324–334.

31
Consciousness and Dimensionality

If nothing more can be offered regarding prism and holographic analogies, a more grounded third-path solution to consciousness and reality may be offered—one that involves paranormal experiences *actively* perceived in greater dimensional space. To begin, this requires a rejection of the conventional (second-path) four dimensional space-time model (e.g., Einstein's theories of relativity). These are no longer considered the immutable boundaries of all that can be considered. Rather, a *hierarchy of reality* is proposed which spreads into further mathematical dimensions.[1] The first step can be a five dimensional structure going back to the early 1920s when physicists Theodore Kaluza and Oskar Klein extended relativity through unification of electromagnetic and gravitational fields. This fifth dimension is said to consist of that dark energy/dark matter we're familiar with from the first-path, and it's worth exploring the concept in a little more detail.

Dark matter is a "rich form of matter" corresponding to the ratio of a substance's current (decaying) level of energy relative to its initial energy

[1] Neppe and Close (2015).

T. Jinks, *Psychological Perspectives on Reality, Consciousness and Paranormal Experience*,
https://doi.org/10.1007/978-3-030-28902-7_31

value. As the ratio decreases the fifth dimension is no longer empirically observable, so it retreats to become mathematical construct.[2] Nevertheless dark energy and associated dark matter still "exists" in some imperceptible manner to make up approximately 70% and 25%, respectively, of the universe's total energy density (the 5% left over is the observable energy we can measure and perceive).[3] Loosely speaking, these cosmic "sectors" could be affiliated with the old-time concept of an invisible aether.

By the 1980s and 1990s ideas of dark energy and dark matter were reformed (or "rehabilitated") by supergravity and superstring theorists.[4] This lead to speculation of hypothetical dimensions beyond four or five that suit whatever mathematical model you fancy. For instance, one research team has recently developed a tongue-twisting theory involving six dimensions. The title of their work is *supersymmetrization of supergravity grounded in superstring theory surveyed in six dimensions.*[5] And more levels can be added, such as a reality of eight dimensions, one that proposes a ten dimensional reality based on dual four dimensional *branes* (a mathematical but also physical membrane permeating space and assigned dimensions) held together by a thinner two dimensional brane (four plus four plus two make up the ten). Or if you'd prefer, the ten dimensions might consist of a double five dimensional structure! Then there's an eleven dimensional model, which consists of a five dimensional *brane* within in a six dimensional bulk[6] (the bulk is the "container" for the brane, so to speak; brane dynamics is a part of the interestingly-named field of *braneworld cosmology*).[7]

Once you've opened the door to multi-dimensional speculation, and conveniently ignore the mathematical nature of the dimensions, you can ponder what might be "contained" in these additional planes. There could be those other-dimensional denizens mentioned earlier that are popular with the first-path. If we can only passively perceive three spatial and a

[2] Zakirov (2017).
[3] Pasqua et al. (2018).
[4] Beichler (2004).
[5] Novak et al. (2017); I've taken the liberty to turn their difficult statement into a tongue-twister.
[6] Beichler (2004).
[7] La Camera (2008).

single time dimension through the ordinary senses, then what is out there has to do all the work to get noticed. These entities;

> ...must appear or come to us in embodied forms. For the language and domain of the physical is all we can perceive. Another intelligence can change our consciousness only by manifesting in a physical form. It may be for this reason that the UFOs and the aliens enter our world in the seemingly familiar high-tech style or form that they do.[8]

Furthermore;

> Paradoxically these non-physical entities seem to have mastered a 'technology' and I use the word advisedly that enables them to enter our material world and manifest in it in a quasi-physical shape-shifting sometimes therianthropic sometimes large headed small bodied humanoid form but always evanescent ephemeral somehow luminous always but not quite transparent.[9]

That seems very much a first-path interpretation. The third-path considers multi-dimensional space as a realm where embodied self is *expanded*, allowing putative higher dimensional space—the otherreal habitat—to be encountered in a dynamic manner. Physicist Edward Close and Psychiatrist Vernon Neppe are supporters of a nine dimensional solution to expanded reality,[10] and to envisage how an other-dimensional experience might be revealed to an observer they present a thought experiment; imagine being inside a box and observing the enclosing walls around you. This represents your immersion in ordinary consciousness and by looking around with your eyes (and picking up other information from additional sensory organs) you accept three spatial dimensions (and the passing of time) as totality. This is equivalent to *mind embodied*; that is, mind as limited by the physical body and information perceived through he senses.[11]

---

[8]Mack (1993).

[9]Hancock (2005, p. 403).

[10]Close and Neppe (2015).

[11]Tart (1998).

Anything outside of this frame of reference appears non-existent, but actually *does* exist although it is *nonlocal* (since spatial locality only refers to the three spatial dimensions of length, breadth and width). It is a "covert reality" enmeshed with four dimensional "overt reality". With this in mind, you can now imagine being outside the box looking at the 3 spatial dimensions it's made from, as well as everything else surrounding the box—a broader view of higher dimensional space which is more fundamental *reality* and corresponds to a deeper conscious experience. You are now in expanded consciousness. The drawback is that if you're in this frame of reference, outside the box, it's not inevitable that you see everything inside it. Expanded consciousness does not automatically grant you omniscient powers—something that's clear with individuals who report global and full component ASCs. They don't achieve greater insight into everyday reality. Their focus is much broader.

Turning to an example, the NDE or OBE separation experience is analogous to you—the traveller's—consciousness extended beyond the three dimensional space box, with time thrown in as the fourth dimension. From your relative position, what you experience still seems "local", but it would appear non-local to someone still inside the box of conventional reality. Of course, which or how many "extra" dimensions are part of the experience is not comprehensible, but in the example of an OBE the three spatial and single time dimension *might occasionally* be incorporated into the impression (in other words, the ability to look back back into the box), so there would doubtless be familiar impressions of your personal reality available to you. Where there is no longer any insight into the space-time box, the disconnect leads to a completely NDE or astral journey.

If *expanded* or *extended* are ill-fitting words to account for extraordinary experiences, then the prism analogy can be reconsidered. In this schema, consciousness reveals a more substantial reality through a *wider lens*. That's how hidden knowledge becomes available, and those intelligent denizens are potentially met, although what is recounted by the traveller will be quite different to a standard sensory experience. There's the ongoing problem of justifying how sight and sound can be perceived in this otherreal location without physical eyes or ears. However if a vision or a sound is presumed to be an elementary experience—something fundamental—then eyes and ears might merely provide these sensations from the four-dimensional world. The five senses aren't mandatory in producing equivalent impressions in higher-dimensional space.

The third-path literature[12] is enamoured with the similarity between a shamanic travelling, fairy and alien encounters,[13] and the intelligent things reported during psychedelic drug studies—anything from an elf, mythological dwarf, imp or gremlin to more modern conceptions of ghosts, reptilian aliens and even clowns.[14] There's the physical similarity of spatial distortion, will the experience of being transported through solid walls (e.g., into spacecraft, like the fellows on the Allagash waterway mentioned previously) and once inside finding the location bigger on the inside than the outside. And in terms of *knowledge*, when physician and mathematician Jerome Cardan's father Facius Cardan came face-to-face with sylphs in 1491 (presumably in an ASC), he was introduced to the philosophical theory of *continuing creation*. This was uncanonical, unscientific and incomprehensible to Cardan. However;

> …what seems extraordinary in retrospect is that the concept of continuous creation should have been mentioned at all in an account of an experience that most modern psychologists would consider hallucinatory. If the beings who conversed with Facius Cardan were only figments of his imagination, how could they have expressed ideas that must have been quite foreign to him?..If they were no figments of their imagination, what were they?[15]

But why even assume these entities are autonomous and live separate "lives" (if the word "alive" applies to them) in higher dimensional, or some other aetheric or vibrational space? Returning to the notion of expanded consciousness, the traveller might be encountering *themselves* "spread" over a multidimensional framework. That's an idea bearing similarity to ideas expressed by the idealist Buddhist school of Vijn̄ānavāda, where;

[12]Better to call it "fringe", but it does correspond to the third-path.

[13]Sturma (2002).

[14]Luke et al. (2012).

[15]Holroyd (1980, p. 53).

> …individual being represents an erroneous limited projection of the universal consciousness, of the storehouse consciousness.[16]

The mind (manas) represents the universal consciousness "appropriating" a personal identity, but this individual self (ātman) is an ontological illusion and shouldn't be considered as real. If it *is* mistakenly considered real the individual becomes entrapped, hence;

> The anxiety and the suffering that characterize human existence are due to this erroneous identification of the human nature with the illusory identity appropriated by the mind.[17]

On a more positive note;

> … when a person "realizes whole mind" and recognizes themselves in greater dimensional space, they might interpret the being(s) they encounter as independent entities - as aliens or strange beasts. However, the alien or strange beast is another, non-physical aspect of their own selves! The goal of life to "realize self", to realize who the alien or beast is. It is about unifying the fragmented components of this dispersed, quasi-mystical personality.[18]

The multidimensional "spread" of consciousness is not limited to meeting strange entities in other realms. Rather than Neppe's box, a fabric analogy is proposed by physicist James Beichler. Beichler favours a five dimensional model, with the fifth dimension a mathematical (not practically imaginable) sphere-like sheet which consist of elementary particles indulging in "exchanges of energy".[19] The result is an energetic field called the *biofield*. Living within this fifth dimensional sheet, the observer cannot actually see all curves and bends, wrinkles and folds caused by the energetic particles (like an actual, scrunched-up bed-sheet might possess). However it's alleged these distortions are subtly present, and can be sensed as gravity,

[16] Nedu (2013, p. 327).

[17] Ibid.

[18] Jinks (2012, p. 233).

[19] Beichler (2004).

electric and magnetic fields, and of course solid matter.[20] It's too simplistic to say the observer *lives in* the field, for they are really a part of it. Any particular "woven pattern" of a localised portion of the fabric represents the observer's personal impact on the elementary particles at this location. The observer's energetic sensory perceptions are imprinted as memories on the sheet, which implies that memories are stored in a more nebulous way than fixed to the neural network of a physical human brain.

When there is awareness of this imprinted "mind", the observer has achieved sentient consciousness; an awareness of not simply the physical body in three spatial dimensions, but the non-locality of mental components as cognitions and stored memories. And in the fashion of Neppe's box analogy it may be possible to step beyond the confines of four dimensional space-time as a consequence of an acute or chronic CNS dysfunction, the altered state experience. Expanded (or extended) consciousness permits recognition of the equilibrium patterns in the *fifth* dimension, and all that this dimension contains.

This entry into expanded consciousness allows more than the temporary separation experience of an OBE, NDE or astral journeying, for it is an account of LAD. Using quantum psychics terminology, final death results in the decoherence of the patterns in the dimensional fabric due to degradation of the physico-chemical processes that maintain life within the five dimensional conformation. Initially this process is chaotic, but soon smooths out in the four dimensional space-time continuum. And pressed into the sphere-like sheet of the fifth dimension as a density pattern there remains the deceased individual's life-imprint. That is, the weaves and fibres of the fabric maintain the person's unique pattern even if the bumps and wrinkles that once represented their physical presence have dispelled. That's a new analogy for the disconnect between physical nature of a material body and the more persistent *life force* contained in the biofield. These retain the traces of the substance of mind, and the higher level of departed consciousness—like autobiographical memories. This gives some basis to free consciousness as a remaining portion of the deceased who has "settled" into a subtle form in the aether of five dimensional

[20] Beichler (2004).

space, ostentatiously named a *self-cohering field density variation pattern.*[21] This allows the deceased *ghost* potentially interacting with the still-living, and doesn't rely on the former just being seen by the later; the weaves and patterns of the remaining ghostly consciousness is in intimate relationship with the weaves and patterns of living consciousness, and that is where the experience takes place—in a non-local, non-physical but above all *unspeakable* higher dimensional "place".

References

Beichler, J. E. (2004). "I See Dead People" Don't You? *Proceedings of the Academy of Religion & Psychical Research, 2004,* 43–55.

Close, E. R., & Neppe, V. M. (2015). Translating Fifteen Mysteries of the Universe by Applying a Nine Dimensional Spinning Model of Finite Reality: A Perspective, the Standard Model and Triadic Dimensional-Distinction Vortical Paradigm. *Part I. NeuroQuantology, 13*(2), 205–217.

Hancock, G. (2005). *Supernatural: Meetings with the Ancient Teachers of Mankind.* London: Century.

Holroyd, S. (1980). *Alien Intelligence.* London: Sphere Books (Original work published 1979).

Jinks, A. (2012). *An Introduction to the Psychology of Paranormal Belief and Experience.* Jefferson, NC: McFarland.

La Camera, M. (2008). Bran Cosmology with (4)R Term in the Bulk. *Acta Physica Polonica B, 39*(8), 1795–1809.

Luke, D., Zychowicz, K., Richterova, O., Tjurina, I., & Polonnikova, J. (2012). A Sideways Look at the Neurobiology of Psi: Precognition and Circadian Rhythms. *Neuroquantology, 10*(3), 580–590.

Mack, J. E. (1993). The UFO Abduction Phenomenon: What Might It Mean for the Evolution of Human Consciousness? *ReVision, 16*(1), 26–32.

Nedu, O. C. (2013). The Mind (Manas) and the Illusory projection of the Afflicted (Klista) Individual Self (Atman) in Vijnanavada Buddhism. *Danubius, 31,* 327–373.

[21] Beichler (2004).

Neppe, V. M., & Close, E. R. (2015). The Concept of Relative Non-locality: Theoretical Implications in Consciousness Research. *Explore: The Journal of Science and Healing, 11*(2), 102–108.
Novak, J., Ozkan, M., Pang, Y., & Tartaglino-Mazzucchelli, G. (2017). Gauss-Bonnet Supergravity in Six Dimensions. *Physical Review Letters, 119,* 111602.
Pasqua, A., Chattopadhyay, S., & Myrzakulov, R. (2018). Consequences of Three Modified Forms of Holographic Dark Energy Models in Bulk-Brane Interaction. *Canadian Journal of Physics, 96*(1), 112–125.
Sturma, M. (2002). Aliens and Indians: A Comparison of Abduction and Captivity Narratives. *Journal of Popular Culture, 36*(2), 318–334.
Tart, C. T. (1998). Six Studies of Out-of-Body Experiences. *Journal of Near-Death Studies, 17*(2), 73–99.
Zakirov, U. N. (2017). Model of Decaying Dark Energy Based on the Kaluza-Klein Theory. *Russian Physics Journal, 60*(4), 678–684.

32
A Transpersonal Alternative

Did Tom's ghost (metaphorically) rub shoulders with Tom in some higher dimensional density field? If that seems short on purpose or substance, then a more mature third-path perspective of expanded consciousness and paranormal experience can be found in the discipline of transpersonal psychology. This is grounded in the recognition of states of consciousness beyond the personal ego—that is, "ordinary mind".[1] In simple terms, these deeper structure(s) of the psyche are the progenitor of both personal ego and the material world (the use of singular or plural to characterise this origin is futile).

The transpersonal state is therefore neither physical or mental, but rather the realm of expanded consciousness—in this context famous expression *collective unconscious* can be applied. In contemporary (non-transpersonal) writings the collective unconscious is frequently made synonymous with the pre-physical substrate of the implicate order,[2] or the invisible aether

[1] Hartelius et al. (2007).

[2] Barušs (2010).

T. Jinks, *Psychological Perspectives on Reality, Consciousness and Paranormal Experience*,
https://doi.org/10.1007/978-3-030-28902-7_32

corresponding to the Dirac sea[3] where consciousness is an inherent property of all matter (and energy). This is *mind at large*,[4] Huxley's term for the expanded field of information located beyond personal consciousness.[5] This indicates these apparently very distinct concepts with their own terminology in fact share a conceptual common ground.

The attractiveness of viewing personal mind and environmental matter as emerging from a more fundamental substrate (rather than conscious as a product of brain activity) is that it allows the interaction between mind and body to take place in this unobservable and non-local realm which operates with a set of laws very different to those present in actualised reality.[6] This post-materialist approach[7] seems a neat (if not overly hasty) resolution to the mind-body problem.

The transpersonal is the where mystical experience derive, but it is not strictly spiritual, or personal, or even paranormal.[8] Mystical refers to the divine, and that's not what these other terms refer to. Even the word *spiritual* can imply rather prosaic personal insights, such as an uplifting feeling when thinking happy thoughts. The transpersonal, on the other hand, is far advanced of these common mental processes. The product involves;

> … intimate insights into…absolute reality"…. Transpersonal experiences have unique tendencies to pull the practitioner out of the world, as it were, by virtue of being in such stark contrast to the ordinary day-to-day life experiences: the sense of complete freedom, deep sense of wholeness, pure love, peace, joy and happiness, Oneness, Omnipresence, utter lack of any kind of reference - unlike anything experienced in the daily affairs.[9]

Indeed, this experience;

---

[3]Song (2018).
[4]Brabant (2016).
[5]Poller (2015).
[6]Brabant (2016).
[7]Ibid.
[8]Cigale (2009).
[9]Ibid., p. 11.

.... cuts so deeply that it undermines the epistemological value of the ordinary waking state as much, if not more, than that of some transcendent states, so that the waking state *itself* can come to be regarded as a hallucination and delusion.[10]

Then again, it's equally accepted that paranormal experiences *do* emerge from a non-local, grounded or collective domain that's also given the name transpersonal. Hence the scuffle is over the words used to delineate what different researchers mean by an information storehouse residing beyond personal consciousness. This might be attributed to a degree of social construction of the experience where;

...an unmediated, pure consciousness event....may ...be interpreted or constructed in accordance with particular conceptual, linguistic or cultural frameworks.[11]

Perhaps the best-known manifestation of the collective unconscious is *synchronicity*, which is paranormal to the extent it the experience is *beyond normal* and involves a mental impression that has an indisputable, meaningful link to a physical effect in everyday reality.

There are various ways synchronicity can be articulated. To Wolfgang Pauli, mental and physical states were "two sides of the same coin" governed by common ordering principles, and most accurately understood as "complementary aspects of the same reality".[12] Carl Jung shared this view, but also argued that scientific reality (our second-path) consisted of the three straightforward principles of space, time and causality. However there's a fourth principle which is synchronicity, or acausal connection.[13]

During World War I several women of an extended family joined by individuals from next door were sitting together knitting, sewing, and mending. Suddenly one of the women who had been deaf for several years said she had heard a loud bang. And from that moment on she could hear again.

---

[10] Barušs (2010, p. 222).

[11] Hanes (2012, p. 27).

[12] Meijer and Korf (2013).

[13] Main (2014).

> The next day she received a telegram informing her that her son had been shot dead in battle around the time she had heard that bang.[14]

In this story there has been an inner phenomenological event (a mental noise) and an outer physical event (the gunshot), and these had no causal connection in *conventional* reality. The nature of the connection was what Jung was most interested in, rather than the "meaningfulness" of any perceived set of mental and physical events. And yet the link remains incomprehensible because it is an expression of a more fundamental, unspeakable grounded state.

In terms of how *acausality* might "work", it might be seen as a cause-and-effect process after all, but nothing like the process we are familiar with in the natural world. Instead, an individual's personal consciousness manifests volitional intentions (the *mind* component of synchronicity) but rather than instigating a subsequent logical behaviour (and resulting physiological action) in the here-and-now, these intentions work *backwards* to the pre-physical substrate to initiate unexpected physical changes elsewhere that *resemble* the original intention (or for many people, the event is meaningfully linked to the original intention, even if the individual is oblivious to what that intention might be).

This is a concept found in multi-dimensional models of reality, where events separated by space, time and configuration in a lower dimensional framework (like ideas, as compared with objects) are adjacent in higher dimensions, hence;

> ...events that seem impossible because they are discontinuous and apparently disconnected may be connected when observed from higher dimensions ("top-down"). We could say from the lower dimensions that there is a disconnection in space (e.g., as in "remote viewing"), time (e.g. as in "precognition" or "retrocognition"), or both (e.g., precognitive remote viewing)....but from a greater level of dimensions there is no disconnection.[15]

[14]Giegerich (2012, p. 504).
[15]Neppe and Close (2015, p. 107).

This is the definitive mind-over-matter explanation, and has contemporary expression in Rupert Sheldrake's theory of *morphic fields*. This idea Sheldrake developed from previous suggestions by biologists Hans Spemann and Paul Weiss in the late 1930s[16] and Jung's collective unconscious,[17] as well as his own early professional work in plant biology. *Morphogenesis* describes how the features of the plant (such as leaves) take shape. Since cells all contain the same genes (for the coding and synthesis of proteins), the problem Sheldrake pondered was how different cells use these genes in different ways for different outcomes of *form*. This applies to humans—Sheldrake uses the distasteful example of grinding up hands and feet separately, and the product being indistinguishable. For this reason the *morphogenetic field* was proposed in the 1920s (replacing the archaic term "vital force") to describe the *self-organising regions of influence* that are framework for biological development (but the *how* and *why* are unknown).[18] This doesn't solely apply to the formation of plant or animal body parts, but also mental and behavioural function, and even social function.[19] In this way, Sheldrake sees the morphic field (a shorthand of morphogenetic) as compatible with both the implicate order and Laszlo's Akashic field.

Sheldrake's next stride beyond *standard* morphic field theory is more ambitious. Morphic field structure (wherever it resides) is *also* influenced by past events. That "past" could consist of anything at all, which is stored in the field like a memory. The more the event is repeated, the more strongly it is stored, so morphic fields are said to be *habitual*.

This is best understood when applied to a simple chemical process;

> …when a new chemical compound is synthesized by research chemists and crystallized, it may take a long time for the crystal to form for the first time. There is no pre-existing morphic field for the lattice structure. But when the first crystals form, they will make it easier for similar crystals to appear

[16] Mishkind (1993).
[17] Bache (1996).
[18] Wilber (1984).
[19] Sheldrake (2006, p. 33).

> anywhere in the world. The more often the compound is crystallized, the easier it should be to crystallize.[20]

Applied to the human sphere;

> ...every individual experience leaves a trace on the subliminal level, where it remains forever stored.[21]

For this reason, morphic fields account for rapid learning of the structure of language in children, or simultaneous IQ advances amongst different countries. In terms of the latter, that's because IQ tests are performed, and the experiences "shared", in the morphic field. The more people complete them, the easier they are to do well on, rather than people's intelligences increasing. That's an observation known as the *Flynn Effect*.[22]

Regarding the paranormal implications of the morphic field, since the field is contained in the pre-physical substrate and acts as the continuously learning blueprint of potentialities "it" has the ability to initiate a physical display from a mental drive. This is a model for synchronicity. Intention allows a substitute configuration of matter to be selected and actualised in place of what might be "typically" expected to emerge from the morphic field. When locally focussed, a similar mental effect on physical systems can generate gross psychokinesis manifesting as spoon-bending, fire-walking, bloodless skin-piercing and energy healing.[23]

"Perturbations" in the morphic field also give rise to various psychical events; telepathy, clairvoyance, clairaudience[24] and lucid dreaming with a noetic quality or a psi facilitating outcome.[25] All these outcomes allegedly involve the bidirectional movement of information amongst people in

---

[20] Ibid.

[21] Brabant (2016, p. 352): "Such a view is actually not new, but has already been expressed in the mid-nineteenth century by Gustav Fechner, pioneer in experimental psychology and father of psychophysics. He postulated the existence of "a great reservoir in which the memories of earth's inhabitants are pooled and preserved, and from which, when the threshold lowers or the valve opens, information ordinarily shut out leaks into the mind of exceptional individuals among us."

[22] Sheldrake (2006).

[23] Cigale (2009).

[24] Ibid.

[25] Krippner (1981).

close social and emotional connection.[26] This is the same process as synchronicity, only the intention of one person generates a mental state in one or more other people (a thought or emotion), rather than a physical one.

Presumably this individual's intention would have to be very strong to influence the blueprint and create a telepathic or psychokinetic event, since the field is highly stable due to the repetitive (and reinforcing) actions of many, many individuals.[27] For that reason mass intention is proposed as a more realistic way to alter the strength of potentialities in the field, as in the so-called "power of prayer".[28]

The confrontation with a subliminal "informational blueprint" (or morphic field) by the *active* "informational collector" (supraliminal self) allows the personal mind to have access to information beyond the normal range of the senses but which is intrinsically a part of the psyche. Indeed, Sheldrake is quite happy to conceptualise personal consciousness as another type of field, a *perceptual field* that radiates from the brain the way a magnetic field radiates from magnetised iron.[29] Where this perceptual field interacts with the world is the location where seeing and hearing things occurs. To Sheldrake, reading this book is happening at the page itself, not somewhere inside the brain. There is a connective projection from the observer to the object or event which is the sensing of reality. This infers people have the potential to affect what is seen external to them, analogous to the role of the observer determining quantum states. The sense of being stared at is Sheldrake's favourite example of this effect. Stare at someone who isn't looking at you, and they'll become uncomfortable very quickly because of perceptual field interaction.

In the manner of synchronicity, savancy also fits the expanded consciousness model. Inaccessible knowledge isn't gained through the reception of perceptual information from a broader storehouse as previously implied (removing the conundrum of perception without the actions of

[26] Sheldrake (2006).

[27] Brabant (2016).

[28] Levin (1996).

[29] There's no indication whether the perceptual field does derived from brain activity entirely, like the analogy of a magnetic field emanating from iron, or uses the brain prismatically in order to manifest.

perceptive organs), but is rather an extremely well-ordered, immediate knowledge of "everything" that's stored in the grounded state (whether termed a morphic or Akashic field, a higher dimensional space or any other expression). That's because individuals are simply "seeing" more of what they already have access to—like taking off a set of restrictive blinkers.

> This is precisely what makes geniuses and prodigies so exceptional: they have a natural and easy access to a certain type of subliminal information (individual and/or collective while most people can only access that information some- times, either randomly or after having purposefully entered an ASC. The latter would be the case of shamans, who use specific consciousness-altering rituals in order to initiate a vision quest or a shamanic journey.[30]

These blinkers are clearly a type of filter/tuner. And yet to reiterate, the mechanism's release does not let more information *in*to be experiences "in the head" by personal consciousness (the first-path), but allows consciousness *out*, or better still, to be what consciousness should be; limitless (as held by the third-path). Practically speaking, the outcomes are identical, although the third-path allows for some interesting, extra options to reside in the expanded conscious space, and that's where we'll take our final discussion.

References

Bache, C. M. (1996). Expanding Grof's Concept of the Perinatal: Deepening the Inquiry into Frightening Near-Death Experiences. *Journal of Near-Death Studies, 15*(2), 115–139.

Barušs, I. (2010). Beyond Scientific Materialism: Toward a Transcendent Theory of Consciousness. *Journal of Consciousness Studies, 17*(7–8), 213–231.

Brabant, O. (2016). More Than Meets the Eye: Towards a Post-materialist Model of Consciousness. *Explore, 12*(5), 347–354.

Cigale, E. (2009, February). The Value of the Transpersonal in Psychotherapy and in Everyday Life. *Europe's Journal of Psychologym, 5*(Special section), 1–16.

---

[30] Brabant (2016, p. 353).

Giegerich, W. (2012). A Serious Misunderstanding: Synchronicity and the Generation of Meaning. *Journal of Analytical Psychology, 57*(4), 500–511.
Hanes, K. (2012). Unusual Phenomena Associated with a Transcendent Human Experience: A Case Study. *Journal of Transpersonal Psychology, 44*(1), 26–47.
Hartelius, G., Caplan, M., & Rardin, M. A. (2007). Transpersonal Psychology: Defining the Past, Divining the Future. *The Humanistic Psychologist, 35*(2), 135–160.
Krippner, S. C. (1981). Dreams and Other Altered Conscious States. *Parapsychological Journal of South Africa, 2*(2), 18–34.
Levin, J. S. (1996). How Prayer Heals: A Theoretical Model. *Alternative Therapies in Health and Medicine, 2*(1), 66–73.
Main, R. (2014). The Cultural Significance of Synchronicity for Jung and Pauli. *Journal of Analytical Psychology, 59*(2), 174–180.
Meijer, D. K. F., & Korf, J. (2013). Quantum Modeling of the Mental State: The Concept of a Cyclic Mental Workspace. *Syntropy, 1,* 1–41.
Mishkind, M. (1993). A Test for Morphic Resonance in Behavioural Responses to Multiple Choice Stimuli. *Journal of Analytical Psychology, 38*(3), 257–271.
Neppe, V. M., & Close, E. R. (2015). The Concept of Relative Non-locality: Theoretical Implications in Consciousness Research. *Explore: The Journal of Science and Healing, 11*(2), 102–108.
Poller, J. (2015). Beyond the Subliminal Mind: Psychical Research in the Work of Aldous Huxley. *Aries, 15*(2), 247–266.
Sheldrake, R. (2006). Morphic Fields. *World Futures: the Journal of General Evolution, 62*(1/2), 31–41.
Song, D. (2018). Visualization of Information. *Neuroquantology, 16*(1), 32–36.
Wilber, K. (1984). Sheldrake's Theory of Morphogenesis. *Journal of Humanistic Psychology, 24*(2), 107–116.

33

Witnessed Experience and the Third-Path

In terms of *seeing* paranormal things like ghosts, locked within the expanded realm is the *potentiality* of unrealised states which includes both the here-and-now awareness of mental, physical and perceptual elements, and the vast plain of unrecognised material. Some of this material comprises shared information that has the potential to enter ordinary consciousness,[1] but which is not derived from everyday experience. These are *archetypal* forms—concepts that people don't see or experience passively through first-path perception, but are dynamically *apprehended*. For Jung, archetypes represent a shared interpretation and patterns of behaviour fundamental to human experience that reside in the psyche[2] which seem to be present "always and everywhere".[3] These universal psychic predispositions are not physical or event fixed mental elements, but fluid residents of the psyche (that part which is the collective unconscious).[4] Archetypes are best

[1] Brabant (2016).

[2] A term for an expanded or extended consciousness.

[3] Jung (1959/1990).

[4] Weitz (1976): Very roughly, there might be some reference to archetypes as patterned in a morphic field, but that might upset stringent depth psychologists.

T. Jinks, *Psychological Perspectives on Reality, Consciousness and Paranormal Experience*,
https://doi.org/10.1007/978-3-030-28902-7_33

represented as sinks or vessels that draw fundamental themes or representations that manifest in an individual's personal consciousness, and these are fathomed through their *symbolic nature.* Archetypal qualities such as self-sacrifice, wisdom and love have a long-standing connection to physical symbology; culturally-determined figures like mythological beings, gods and saints, and products in the form of folktales, drawings or paintings and images.[5]

Are archetypes paranormal or supernatural? Not according to the view that archetypal disclosure is merely a product of culture—archetypes have no biological basis at all, but rather apprehended or "grown into" as a consequence of shared beliefs and behaviours.[6] They're just a form of social behaviour.

On the other hand, archetypes may be biologically inherited, although their legacy is again based on shared environmental stimuli. Take the symbol of Ourobóros, the serpent devouring its own tale.[7] Rather than innately emerging from an ephemeral collective unconscious, Ourobóros simply represents;

> …a highly visible and conspicuous cause in the sky, one impressive enough to survive for millennia as a pervasive theme. As traditional images of astronomical content typically portray the entire object as it appears to the eye, one would expect a celestial prototype of the ourobóros to have looked like a complete ring—unlike the rainbow or the ecliptic band, which require a sufficient level of astronomical sophistication to be extrapolated to a circle. The aurorae, also known as the northern and southern lights, are promising candidates for an interpretation along these lines.[8]

There's be a place for a genetic component to the archetypal symbol,[9] although Jung never mentioned archetypes as necessarily inherited.[10] Nor is it a tacit acceptance of an archetypal language in the genes conforming

[5]Kaminker (2016).

[6]Stevens (1998).

[7]van der Sluijs and Peratt (2009); also represented as a rounded object, four pillars, axis mundi, a rainbow, and lightning.

[8]Ibid., p. 21.

[9]Nunn (1998) and Huston (1999).

[10]Martin-Vallas (2013).

to Zipf's law and embedded in junk DNA (rather than the spliced-in language of reptilian overlords!). If an inherited nature of archetypes is can be conceived, it is as *apprehension* through *emergence.* The physical organization of the brain attracts (or is a receptacle for) archetypal expression which becomes an emergent structure, with a completely different standard of complexity to anything perceived through the senses.[11] This might be attributed to a brain component such as the limbic system with proximity to the inferior temporal lobes. Here neurons respond to specific "neural imagery" (like geometric patterns) and the resulting elementary (symbolic) images intersect with emotional energy to procreate the mystical archetype.[12]

Archetypal symbols are encountered in art and fairy tales, and of course dreams

> …an unceasing stream, or perhaps an ocean of images and figures which drift into consciousness in our dreams or in abnormal states of mind.[13]

The dream is identified as an ASC where the individual is more open to receiving information from the expanded psyche, so in this context the dream *producer* is grounded in the transpersonal—That solves the producer problem!

Archetypal dreams are also not to be mistaken with ordinary dreams.[14] They are;

> …identifiable by their allusions to mythological, religious, alchemical, and other archaic symbolism … and further typically display irrational exposition and development of plot, remoteness from events of everyday life, and intensity of associated affect … These characteristics may be used to differentiate them from 'personal dreams' … which manifest a preponderance of ontogenetically based images pertaining more to personal and subjective demands of the dreamer.[15]

---

[11]Ibid.

[12]Joseph (2001); Of course, people share everything else in the brain, but that's not the point!

[13]Kaminker (2016, p. 93).

[14]Zhu (2013).

[15]Welman and Faber (1992, p. 65).

Concepts expressed in the archetypal dream includes the *self* (the totality of personality),[16] the *Hero* (corresponding to the goal-driven ego), the *Shadow* (a negative, unfamiliar component of personality), the *Anima* (man's feminine constituent), the *Animus* (woman's masculine constituent), the *Wise Old Man* (referring to wisdom) and *Magna Mater* (the Great Mother).[17]

There can also be *waking* apprehension of archetypes and their symbols through *projection.* Projective tests used in humanistic psychology are presumed to force symbolic interpretation of ambiguous pictorial stimuli. For instance, if you recognise an animal in a messy ink-blot, that's an indicator of your fixation with mother and father figures. Furthermore, interpreting the ambiguous image as a bear is a reflection of a "benign and sympathetic father figure", whereas seeing a spider is argued to symbolize a "wicked mother"![18]

Applied to paranormal experience, a *witnessed* archetypal projection is far more profound. Most famously, Jung wrote a short book about flying saucers and how these objects could be viewed as a representation of the *self. Flying Saucers; a Modern Myth of Things Seen in the Skies* gives the impression that flying saucers might be legitimate extraterrestrial craft, or illusions stemming from environmental stimuli. It really doesn't matter what they are physically, or where they come from. The important point is that there's an essential need for the affected individual to project onto their real or hallucinatory shape the archetypal message as a *subjective psychic prefiguration.*[19] The witness's *self* is fractured (a very messy concept), and presumably needs some degree of personal recognition if there's any hope of restructure or healing.

Complicating matters is the boundless flexibility regarding the identity of the symbol. The archetype of *self* could constellate around the flying saucer disc form, but a similar transformation might accompany an egg, a diamond shape, a child, jewel, flower or chalice.[20] It could also be expressed

[16] Urban (2005).
[17] Bradshaw and Storm (2013).
[18] Lash and Polyson (1987).
[19] Jung (1959/1991).
[20] Rosen et al. (1991).

through divided shapes like a square or circle with a cross inside it.[21] There are a *whole lot* of possibilities, which make the diagnosis of an archetypal experience very subjective, meaning the method is sidelined as a mainstream psychological therapy.

But we're not talking about therapy, but paranormal experience. Sticking with the animal motif, Jung identified these as emblematic of "man's primitive and instinctual nature".[22] And while self-aware appreciation of its meaning is not necessary, the witness might still feel there was a hidden purpose to the sighting. At its most mundane, this might be profound significance radiating from a spider on the bathroom ceiling or a pigeon in the park.

However it is the encounter with a *strange* or *numinous* creature that's particularly relevant to the expression of the archetype.[23] These are *para-psychic* entities; a tulpa or thought-form. Wolfgang Pauli believed that failure to face a personal, psychological problem may allow archetypal manifestation in parapsychic form.[24] The "energy" of the disturbance is an attractor to an appropriate (allied) archetype, and the symbol is the perceived disclosure. A lovely example relates to a fixation with modernity and technology—computers and gaming consoles. That's ultimately troublesome to the psyche, and under the right circumstances may result in the expression of the *opposite* somewhere unexpected; like a wild, hairy hominid known as a Bigfoot seen in a carpark. The archetype has been divulged through a para-animal form which;

> ...contains hints of tremendous power, of total freedom from rules and restraint.[25]

While most "Bigfoot"-like stories are low-information sightings of seemingly flesh-and-bone hominids—mostly in wilderness areas but sometimes in the centre of a large city! (which is *very* archetypal) And every so often

---

21 Fordham (1953).

22 Lash and Polyson (1987).

23 Jung's proposition about the paranormal, made in his 1959 writings.

24 van Erkelens (1991).

25 Clarke and Coleman (1975, p. 238).

a report conforms to an archetypal mould. Here's a story I was once told;

> … of a man, living rough in mountainous bushland on the outskirts of Sydney, Australia, who encountered a Yowie (Australia's equivalent to a Bigfoot). This huge, smelly and hairy hominid came out of the scrub and right up to the witness's shack, yet for three days following the event the witness forgot that it had ever happened. Only later did he remember what he'd seen, and told a neighbour with some bemusement.[26]

And there's more to the story. It turns out the man had moved into the shack (in reality a glorified tent) to remove himself from stressful everyday living and "commune with nature". As you might expect, he had been meditating frequently, and probably using quite a few different types of drugs, in order to achieve an ASC (not his particular use of word, but that was his intentions). He remembers the morning he confronted the hominid as it emerged from the bush, and says he wasn't scared or surprised. The beast looked at him and disappeared back into the scrub, and its appearance was immediately and temporarily forgotten for about three days until it came back to him as a rushed stream of images. It doesn't matter whether the thing he saw was an actual cryptozoological relic, or a tree truck made illusionary, or a pure hallucination, this story ticks all the boxes of an archetypal encounter (or evidence of meaningless second-path psychosis, take your pick).

Other styles of entity have been ascribed to the archetypes, rather than being a mere psychosocial artefact. "Contactee" aliens in the heyday of the 1950s might be viewed as waking materialisations of Jung's technological angels,[27] and post-Jung the grey aliens encountered in a modern abduction[28] are technological demons.[29] The possibilities are endless. For instance, if someone believes they have seen in real-life (or in a particularly vivid dream) a vampire or some other blood or spirit-sucking creature, then this;

[26]Jinks (2012, p. 19).
[27]Kottmeyer (2006).
[28]Williams (2005).
[29]Kottmeyer (2006).

> ...might be the personification of the shadow archetype, the sworn enemy of mankind, the tempter or the principle of evil incarnate (in other words a variation of the devil): One might speak of the vampire as the radical otherness of the human being, as the shadow of man, a form of existence both inhuman and devilish. In Jungian terms the shadow is a personification of the principle of evil, i.e. the devil.[30]

In a concluding, flowery statement, encountering the transpersonal psyche is realising the expanded self in the "home" of the archetypes. Meeting the archetype will be barely comprehensible, even though it might be a direct result of mundane personal concerns there won't be a guide book to assist interpretation. This can lead to a severe disturbance due to the threat posed to the dominant ego, nevertheless but consequence can be ultimately empowering (or destructive). That's seen in the initiation process of the shaman, or abductee, or Joseph Campbell's mythological Hero journey[31] which is a th*reshold experience* on the "road of trials". That is where;

> ...gods and monsters, daimons and demons who are collectively otherworldly in exercising their power over us.[32]

References

Bolea, S. (2015). The Shadow Archetype of the Vampire in Three Horror Movies. *Caietele Echinox, 28,* 281–289.

Brabant, O. (2016). More Than Meets the Eye: Towards a Post-materialist Model of Consciousness. *Explore, 12*(5), 347–354.

Bradshaw, S., & Storm, L. (2013). Archetypes, Symbols and the Apprehension of Meaning. *International Journal of Jungian Studies, 5*(2), 154–176.

Clarke, J., & Coleman, L. (1975). *The Unidentified: Notes Towards Solving the UFO Mystery.* New York: Warner Paperback Library.

[30]Bolea (2015, p. 281).

[31]Williams (2005).

[32]Ibid., p. 134.

Fordham, F. (1953/1959). *An Introduction to Jung's Psychology*. Harmondsworth: Penguin Books.
Huston, H. (1999). The Evolutionary Significance of Archetypal Dreams. In D. H. Rosen & M. C. Luebbert (Eds.), *Evolution of the Psyche* (pp. 188–195). Westport, CT: Greenwood Publishing Group.
Jinks, A. (2012). *An Introduction to the Psychology of Paranormal Belief and Experience*. Jefferson, NC: McFarland.
Joseph, R. (2001). The Limbic System and the Soul: Evolution and the Neuroanatomy of Religious Experience. *Zygon, 36*(1), 105–136.
Jung, C. G. (1990). The Archetypes and the Collective Unconscious (10th ed.; R. F. C. Hull, Trans.). Princeton: Princeton University Press (Original work published 1959).
Jung, C. G. (1991). *Flying Saucers: A Modern Myth of Things Seen in the Skies* (R. F. C. Hull, Trans.). Princeton: Princeton University Press (Original work published 1959).
Kaminker, J. (2016). Images, Figures and Qualities: Clarifying the Relationship Between Individual and Archetype. *International Journal of Transpersonal Studies, 35*(2), 93–102.
Kottmeyer, M. (2006, February 25). Letters to Ye Olde Editor. *Saucer Smear, 53*(2).
Lash, S. J., & Polyson, J. A. (1987). The Gender Relevance of Projected Animal Content. *Journal of Clinical Psychology, 43*(1), 145–150.
Martin-Vallas, F. (2013). Are Archetypes Transmitted or Emergent? A Response to Christian Roesler. *Journal of Analytical Psychology, 58*(2), 278–285.
Nunn, C. M. H. (1998). Archetypes and Memes: Their Structure, Relationships and Behaviour. *Journal of Consciousness Studies, 5*(3), 344–354.
Rosen, D. H., Smith, S. M., & Huston, H. L. (1991). Empirical Study of Associations Between Symbols and Their Meanings: Evidence of Collective Unconscious (Archetypal) Memory. *Journal of Analytical Psychology, 36*(2), 211–228.
Stevens, A. (1998). Response to P. Pietikainen. *Journal of Analytical Psychology, 43*, 345–355.
Urban, E. (2005). Fordham, Jung and the Self: A Re-examination of Fordham's Contribution to Jung's Conceptualization of the Self. *Journal of Analytical Psychology, 50*(5), 571–594.
van der Sluijs, M. A., & Peratt, A. L. (2009). The Ourobóros as an Auroral Phenomenon. *Journal of Folklore Research, 46*(1), 3–41.
van Erkelens, H. (1991). Wolfgang Pauli's Dialogue with the Spirit of Matter. *Psychological Perspectives: a Semi-Annual Journal of Jungian Thought, 24*(1), 34–53.

Weitz, L. J. (1976). Jung's and Freud's Contributions to Dream Interpretation: A Comparison. *American Journal of Psychotherapy, 30*(2), 289–293.
Welman, M., & Faber, P. A. (1992). The Dream in Terminal Illness: A Jungian Formulation. *Journal of Analytical Psychology, 37*(1), 61–81.
Williams, R. J. (2005). *The Symbol of the Hybrid Human/Alien Child in the Abduction Phenomenon: Rebirthing Within the Psyche and the Psychosomatic Imagination.* Proceedings of the Eighth Australian and International Religion, Literature and the Arts Conference 2004. http://hdl.handle.net/2123/1263.
Zhu, C. (2013). Jung on the Nature and Interpretation of Dreams: A Developmental Delineation with Cognitive Neuroscientific Responses. *Behavioral Sciences, 3*(4), 662–675.

34

The Three Paths and Tom's Ghost

So what did Tom actually experience? Did he *see* something *real* or not? Applied to his seeing an apparition, there are strengths and weaknesses to each of the three paths.

Regarding the first-path, Tom sees what's out there; detectable physical reality. The *what* and *how* of the 5W1H method is well expressed, although there are many possibilities that are not amenable to systematic empirical investigation and Tom's ghost, like the vast majority (maybe all) of sightings, is spontaneous. That amplifies the uncertainty surrounding the *who* and the *why*. Who was the apparition and why did "he" appear, or was there even a reason to appear? Are witnessed events caused by a purposeful spirit or a meaningless video-style replay? Putting aside the unfeasible notion of witnessing something real that is simultaneously unknown and unmeasurable, at least the first-path has no trouble identifying *who* the producer is; it doesn't get tangled-up justifying no producer, or an infinite regress of producers. Whether the ghost *meant* to appear to Tom, or not, it was nonetheless an external process causing his visual system to respond like any other life event.

T. Jinks, *Psychological Perspectives on Reality, Consciousness and Paranormal Experience*,
https://doi.org/10.1007/978-3-030-28902-7_34

The second-path is more powerful in restricting the range of possibilities for what causes a ghost sighting, and the *what* and *how* are by far the most empirically established. Mistaken perception is enforced by an *altered state of consciousness* that's grounded in barely-understood disruptions of CNS function. While the nuances might be uncertain, there's enough practical evidence that ASCs can be induced through biological dysfunction and acute chemical disruption (the actions of psychedelics is powerful evidence) leading to extraordinary delusions presented as both global and component changes in perceptual experience.

With the first-path, the unprecedented nature of the sighting can be justified—the ghost was a *fixed* spiritual resident of the corridor, and Tom only ever visited there one weekend in his life so he was only going to see the ghost once! That's where the second-path runs into some trouble. It's a long stretch to argue a healthy, drug-free individual confronts a "non-existent" ghost as a singular event in their lives because of a "one-off" ASC with an unknown neurological origin. And yet it's not unreasonable to suggest that due to some unknown but minor disturbance of diet, fatigue or work related stress (or a combination of all of these), a brief, partial component experience of a man's figure is accidently overlayed somewhere between his retina and his visual cortex as a bottom-up error, or as a top-down fault derived from complex interplay amongst higher cortical association areas. The monist imperative of the second-path does create a headache for the content of sophisticated hallucinations, which includes the superimposition of the moving figure of a man onto a stabilised backdrop of a well-lit hotel corridor. That is difficult to attribute to mere accidental playback of stored memories onto the visual field, which is why the second-path tends to skirt the *who* and *why* of the sighting—most forcefully, by claiming these two requirements are irrelevant to explaining apparitions. And whatever the mechanism of fabrication, the hallucination's producer can't have had much practice in refining its abilities if Tom had never seen a ghost before—unless he regularly sees these figures in streets and parks and on trains but thinks nothing of them, and the event in the corridor was an inadvertent realisation that what he perceives are not real (but wouldn't he also see them at home, alone?)

Another pertinent issue, not mentioned until now, is that Tom remains adamant that during the experience his mental state was not perceptually or affectively different to what he understands is his *normal* state of consciousness. Granted his brief partial component hallucination (if that's what it was) might be expected to feel experientially weaker than an altered state leading to an NDE, OBE or astral travel, a spacecraft abduction or anything else so extreme. But even daydreaming or drunkenness are mild varieties of ASC, and he understands these sensations well. Can he really be said to have entered and ASC, if he doesn't believe he did? How can an altered state of consciousness then be said to differ, or even be distinguished from an ordinary state of consciousness?

This is also troublesome for the third-path. A ghost sighting might not be a function of passive perception, but there's no firm grasp on *what* is happening to allow Tom to see his ghost, let alone *how* it happens, *why* it took the form it did, or *who* the producer is. That's because the workings of the higher dimensional realm, or fundamental collective unconscious, or whatever name is preferred, remains unspeakable. Was Tom's ghost the cameo actualisation of a shapeless potentiality? Did Tom's disconnected consciousness accidently and momentarily stream through another person's prism from a time past or present in the hotel corridor? Or, based on current idiosyncrasies of this hallway, an interaction was allowed between the living individual—Tom—resonating with the remnant "complex" of a deceased participant resonating in higher dimensional space but also accessible "to his 5D mind. The product was not passively observed but experienced as a pattern match at the level of the higher cortex, the way the objects can be matched by the 5D mind when viewed in novel perspectives.[1] Alternately, did Tom's expanded psyche create a thought-form corresponding to the archetype of the "shadow" or "old man" for his personal psyche to apprehend? He didn't report Bigfoot, so we can probably rule out his having concerns about his IT job! And what Tom described seeing, and what hs *personal appearance* looks like, are quite similar. Was it "himself" that he saw, as a representation of the *self* (rather than an egg, or a UFO)? With all these miscellaneous options, the third-path provides

---

[1] Beichler (2004).

an unrestrained freedom to choose from a hardware store of build-your-own solutions without constrictive rules. And that's also its fundamental problem.

But is there any necessity to travel down only one of the three paths? Favourite elements could be selected from each and reassembled in a hybrid model of paranormal experience. Tom stubbornly believes what he perceived was real. From his frame of reference, that's all a visual impression can be, and seeing something is unquestionably linked to activation of a biological visual system. Path One. However an altered state of consciousness is required to facilitate visual perception, and that comes about through changes in the ordinary working state of the central nervous system. So what he sees is a special kind of "seen". It's not completely real in the sense of being of the same composition as the carpet on the floor or the painted wood panelling on the walls, so by definition the apparition is hallucinatory in nature. Path Two. However the substance of the vision comes from a non-physical and non-mental grounded state that is actualised from the quantum soup of potentialities through the enterprise of active observation, and intention, by a dynamic and expansive consciousness. Path Three.

However there's another aspect to Tom's story that I haven't so far mentioned; he was profoundly affected in the days following his brush with the paranormal. At home on the Thursday evening he reports having a severe attack of (in his words) "the heebie jeebies". To paraphrase what he told me;

> I couldn't relax at all. I had a great few days and we had a lot of fun at dinner talking about ghosts. But I guess it took time to sink in, and when I got home it really hit me and I was a bit of a wreck. I didn't sleep at all that night.

That doesn't require Tom's ghost to be best explained with a meaningful *why*. Tom also reports that nothing in his life noticeably changed in the short or long term. He'd recovered from his panic by Saturday, and went back to work as usual the following Monday finding the new week passed like any other. His personal or professional life weren't interrupted in a positive or negative way. He's still in the same job, doing the same things, with the same group of friends, same interests and same life perspectives. For this reason no solution offered by the three paths, or even the hybrid creation, is better than any other in assisting Tom with really knowing the *how, what, who and why* of his brief but extraordinary life moment. They are three paths travelling in different directions offering manifestly different explanations for paranormal experience. But in reality they are roads leading to the same, unresolved destination.

Reference

Beichler, J. E. (2004). "I See Dead People" Don't You? *Proceedings of the Academy of Religion & Psychical Research, 2004,* 43–55.

References

Abbott, E. J., Connor, G. B., Artes, P. H., & Abadi, R. V. (2007). Visual Loss and Visual Hallucinations in Patients with Age-Related Macular Degeneration (Charles Bonnet Syndrome). *Investigative Ophthalmology and Visual Science, 48*(3), 1416–1423.

Agafonov, A. I. (2018). Massless Composite Bosons Formed by the Coupled Electron-Positron Pairs and Two-Photon Angular Correlations in the Colliding Beam Reaction $e^{-} e^{+} \rightarrow B\gamma\gamma$ with Emission of the Massless Boson. *Advances in High Energy Physics,* 1–18 (Article ID: 6137380). https://doi.org/10.1155/2018/6137380.

Akinci, E., Oncu, F., & Topcular, B. (2016). Tactile Hallucination and Delusion Following Acute Stroke: A Case Report. *Journal of Psychiatry & Neurological Sciences, 29*(1), 79–84.

Alcock, J. (2004, December 21). Electronic Voice Phenomena: Voices of the Dead? *Skeptical Inquirer.* Retrieved August 16, 2018, from https://www.csicop.org/specialarticles/show/electronic_voice_phenomena_voices_of_the_dead?/specialarticles/evp.html.

Altschul, B. (2017). Why Cerenkov Radiation May Not Occur, Even When It Is Allowed By Lorentz-Violating Kinematics. *Symmetry, 9* (11), 250–260.

Alvarado, C. S. (2014a). The Term 'Out-of-Body Experience'. *Journal of the Society for Psychical Research, 78*(914), 61.

T. Jinks, *Psychological Perspectives on Reality, Consciousness and Paranormal Experience,*
https://doi.org/10.1007/978-3-030-28902-7

Alvarado, C. S. (2014b). On W. J. Crawford's Studies of Physical Mediumship. *Journal of Scientific Exploration, 28*(2), 351–357.

Alvarado, C. S. (2016). Out-of-Body-Experiences During Physical Activity: Report of Four New Cases. *Journal of the Society for Psychical Research, 80*(922), 1–12.

Amaral, D. G. (2003). The Amygdala, Social Behavior, and Danger Detection. *Annals of the New York Academy of Sciences, 1000*(1), 337–347.

Apelle, S. (1995). Point Counterpoint. *Mercury, 24*(1), 9.

Appel, K., Pipa, G., & Dresler, M. (2018). Investigating Consciousness in the Sleep Laboratory—An Interdisciplinary Perspective on Lucid Dreaming. *Interdisciplinary Science Reviews, 43*(2), 192–207.

Arnette, J. K. (2012). Recollections of Death: A Medical Investigation. *Journal of Scientific Exploration, 26*(1), 207–211.

Arnulf, I. (2013). Dream Imagery, Rapid Eye Movement Sleep Behavior Disorder, and Hallucinations. *Sleep and Biological Rhythms, 11*(Suppl. 1), 15–20.

Asprem, E. (2011). Pondering Imponderables: Occultism in the Mirror of Late Classical Physics. *Aries, 11*(2), 129–165.

Austad, C. S., & Gendron, M. S. (2018). Biofeedback: Using the Power of the Mind–Body Connection, Technology, and Business in Psychotherapies of the Future. *Professional Psychology: Research and Practice, 49*(4), 264–273.

Aynsworth, C., Nemat, N., Collerton, D., Smailes, D., & Dudley, R. (2017). Reality Monitoring Performance and the Role of Visual Imagery in Visual Hallucinations. *Behaviour Research and Therapy, 97*, 115–122.

Bache, C. M. (1996). Expanding Grof's Concept of the Perinatal: Deepening the Inquiry into Frightening Near-Death Experiences. *Journal of Near-Death Studies, 15*(2), 115–139.

Bancaud, J., Brunet-Bourgin, F., Cahauvel, P., & Halgren, E. (1994). Anatomical Origin of Déjà Vu and Vivid 'Memories' in Human Temporal Lobe Epilepsy. *Brain, 117*(1), 71–90.

Barker, S. A., Borjigin, J., Lomnicka, I., & Strassman, R. (2013). LC/MS/MS Analysis of the Endogenous Dimethyltryptamine Hallucinogens, Their Precursors, and Major Metabolites in Rat Pineal Gland Microdialysate. *Biomedical Chromatography, 27*(12), 1690–1700.

Barker, S. A., McIlhenny, E. H., & Strassman, R. (2012). A Critical Review of Reports of Endogenous Psychedelic N, N-Dimethyltryptamines in Humans: 1955–2010. *Drug Test Analysis, 4*, 617–635.

Bartels, D., Van Laarhoven, A., Van De Karkhof, P., & Evers, A. (2018). Nocebo Effects and Scratching Behaviour on Itch. *Acta Dermato-Venereologica, 98*(10), 943–950.

Bartholomew, R. E., Basterfield, K., & Howard, G. S. (1991). UFO Abductees and Contactees: Psychopathology or Fantasy Proneness? *Professional Psychology: Research and Practice, 22*(3), 215–222.

Bartocci, G. (2004). Transcendence Techniques and Psychobiological Mechanisms Underlying Religious Experience. *Mental Health, Religion & Culture, 7*(2), 171–181.

Barušs, I. (2010). Beyond Scientific Materialism: Toward a Transcendent Theory of Consciousness. *Journal of Consciousness Studies, 17*(7–8), 213–231.

Beichler, J. E. (2004). "I See Dead People" Don't You? *Proceedings of the Academy of Religion & Psychical Research, 2004,* 43–55.

Belanger, J. (2007). *The Ghost Files: Paranormal Encounters, Discussion, and Research.* Franklin Lakes, NJ: Career Press.

Bender, M. L., & Gill, P. (1986). The Genetic Code and Zipf's Law. *Current Anthropology, 27*(3), 280–283.

Benford, M. S. (2001). Can a Theory Derived from Recent Experimental Data Explain Precognition and Other Mysterious Spiritual Phenomena? *Journal of Religion and Psychical Research, 24*(3), 132–142.

Bensley, D. A. (2003). Can Minds Leave Bodies? *Skeptical Inquirer, 27*(4), 34–40.

Bernstein, E. M., & Putnam, F. W. (1986). Development, Reliability, and Validity of a Dissociation Scale. *Journal of Nervous and Mental Disease, 174*(12), 727–735.

Besterman, T., & Gatty, O. (1934). Investigation of Paraphysical Phenomena. *Nature, 133*(3363), 569–570.

Bian, J., & Wang, J. (2016). A Deduction from Dirac Sea Model. *Physics Essays, 29*(2), 218–221.

Bidney, D. (1949). Towards a Psychocultural Definition of the Concept of Personality. In S. S. Sargent & M. W. Smith (Eds.), *Culture and Personality* (pp. 31–55). New York: Viking Fund.

Bidoli, V., Casolino, M., De Pascale, M. P., Furano, G., Morselli, A., Narici, L., et al. (2000). Study of Cosmic Rays and Light Flashes on Board Space Station Mir: The Sileye Experiment. *Advances in Space Research, 25*(10), 2075–2079.

Biederman, I. (1987). Recognition-By-Components: A Theory of Human Image Understanding. *Psychological Review, 94*(2), 115–147.

Bierman, D. (2010). Consciousness-Induced Restoration of Time Symmetry (CIRTS). A Psychophysical Theoretical Perspective. *Journal of Parapsychology, 24,* 273–300.

Billock, V. A., & Tsou, B. H. (2012). Elementary Visual Hallucinations and Their Relationships to Neural Pattern-Forming Mechanisms. *Psychological Bulletin, 138*(4), 744–774.

Blackmore, S. J. (1986). Where Am I? Perspectives in Imagery, Memory and the OBE. In D. H. Weiner & D. I. Radin (Eds.), *Research in Parapsychology* (pp. 108–111). Metuchen, NJ: Scarecrow Press.

Blackmore, S. J. (1992). *Beyond the Body: An Investigation of Out-of-the-Body Experiences.* Chicago: Academy Chicago Publishers.

Blainey, M. (2015). Forbidden Therapies: Santo Daime, Ayahuasca, and the Prohibition of Entheogens in Western Society. *Journal of Religion and Health, 54*(1), 287–302.

Blanke, O., & Metzinger, T. (2008). Full-Body Illusions and Minimal Phenomenal Selfhood. *Trends in Cognitive Sciences, 13*(1), 7–13.

Blanke, O., Ortigue, S., Landis, T., & Seeck, M. (2002). Stimulating Illusory Own-Body Perceptions. *Nature, 419,* 269–270.

Blavatsky, H. P. (1889). *Dialogue on the Mysteries of the After Life: On the Constitution of the Inner Man and Its Division.* Theosophical Articles, Vol. II, Retrieved September 5, 2018, from https://theosophy.org/Blavatsky/Articles/DialogueOnTheMysteriesOfTheAfterLife.htm.

Blavatsky, H. P. (1966). *Collected Writings* (Vol. 6). Wheaton, IL: Quest Books (Original work published 1954).

Blavatsky, H. P. (2012). *Isis Unveiled: A Master-Key to the Mysteries of Ancient and Modern Science and Theology. Volume 1: Science.* New York: Cambridge University Press (Original work published 1877).

Blum, H. (2013). Dissociation and Its Disorders. *Psychoanalytic Inquiry, 33,* 427–438.

Blum, H., De Boer, J. A., Bosker, F. J., & Slaap, B. R. (2000). Serotonergic Drugs in the Treatment of Depressive and Anxiety Disorders. *Human Psychopharmacology: Clinical and Experimental, 15*(5), 315–336.

Bolea, S. (2015). The Shadow Archetype of the Vampire in Three Horror Movies. *Caietele Echinox, 28,* 281–289.

Booth, J. N., Koren, S. A., & Persinger, M. A. (2008). Increased Theta Activity in Quantitative Electroencephalographic (QEEG) Measurements During Exposure to Complex Weak Magnetic Fields. *Electromagnetic Biology and Medicine, 27*(4), 426–436.

Borutta, R. (2009). Paranormal Phenomena, Physics and Dirac Energy. *Journal of Spirituality & Paranormal Studies, 32*(4), 192–200.

Boyer, R. W. (2014). Unless We Are Robots, Classical and Quantum Theories Are Fundamentally Inadequate. *NeuroQuantology, 12*(1), 102–125.

Brabant, O. (2016). More Than Meets the Eye: Towards a Post-materialist Model of Consciousness. *Explore, 12*(5), 347–354.

Bradley, A., Zhang, H., Applegate, R. A., Thibos, L. N., & Elsner, A. E. (1998). Entoptic Image Quality of the Retinal Vasculature. *Vision Research, 38*(17), 2685–2696.

Bradshaw, S., & Storm, L. (2013). Archetypes, Symbols and the Apprehension of Meaning. *International Journal of Jungian Studies, 5*(2), 154–176.

Braithwaite, J. J., Watson, D. G., & Dewe, H. (2017). Predisposition to Out-of-Body Experience (OBE) Is Associated with Aberrations in Multisensory Integration: Psychophysiological Support from a "Rubber Hand Illusion" Study. *Journal of Experimental Psychology: Human Perception and Performance, 43*(6), 1125–1143.

Braud, W. G. (1975). Psi-Conducive States. *Journal of Communication, 25*(1), 142–152.

Bridge, H., Harrold, S., Holmes, E., Stokes, M., & Kennard, C. (2012). Vivid Visual Mental Imagery in the Absence of the Primary Visual Cortex. *Journal of Neurology, 259*(6), 1062–1070.

Briehl, W. (1995). Wilhelm Reich. *Psychoanalytic Pioneers, 1*(1), 430–439.

Brooks, M. (2012). Does Consciousness Create Reality? *New Scientist, 215*(2884), 42–43.

Brooks, M. (2015). The Secret Life of Reality. *New Scientist, 225*(3002), 26–29.

Brown, A. (2003, October 26). Thoughts That Keep Going Bump in the Night. *The Sunday Times*, 3.

Brown, J. R. (1985). von Neumann and the Anti-realists. *Erkenntnis, 23*(2), 149–159.

Budinger, T. F., Lyman, J. T., & Tobias, C. A. (1972). Visual Perception of Accelerated Nitrogen Nuclei Interacting with the Human Retina. *Nature, 239*(5369), 209–211.

Bühler, K.-E., & Heim, G. (2011). Etiology, Pathogenesis, and Therapy According to Pierre Janet Concerning Conversion Disorders and Dissociative Disorders. *American Journal of Psychotherapy, 65*(4), 281–309.

Campbell, N. R. (1921). Metaphysics and Materialism. *Nature, 108*(2717), 399–400.

Cancelli, I., Marcon, G., & Balestrieri, M. (2004). Factors Associated with Complex Visual Hallucinations During Antidepressant Treatment. *Human Psychopharmacology: Clinical & Experimental, 19*(8), 577–584.

Capps, M. (2015). Body and Soul in Plato: How Our Bodies Can Serve Our Philosophical Pursuits. *Dialogue, 58*(1), 19–29.

Cardeña, E. (2018). The Experimental Evidence for Parapsychological Phenomena: A Review. *American Psychologist, 73*(5), 663–677.

Cardeña, E., & Krippner, S. (2010). The Cultural Context of Hypnosis. In S. J. Lynn, J. W. Rhue, & I. Kirsch (Eds.), *Handbook of Clinical Hypnosis* (2nd ed., pp. 743–771). Washington, DC: American Psychological Association.

Cardoso, A. (2012). A Two-Year Investigation of the Allegedly Anomalous Voices or EVP. *Neuroquantology, 10*(3), 492–514.

Carhart-Harris, R. L. (2013). Psychedelic Drugs, Magical Thinking and Psychosis. *Journal of Neurology, Neurosurgery, and Psychiatry, 84*(9, e1).

Carhart-Harris, R. L., Erritzoe, D., Williams, T., Stone, J. M., Reed, L. J., Colasanti, A., et al. (2012). Neural Correlates of the Psychedelic State as Determined by fMRI Studies with Psilocybin. *Proceedings of the National Academy of Sciences of the United States of America, 109*(6), 2138–2143.

Carlino, E., & Benedetti, F. (2014). Placebo and Nocebo Effects. In D. I. Mostofsky (Ed.), *The Handbook of Behavioral Medicine* (pp. 36–57). Wiley-Blackwell: Hoboken, NJ.

Carlson, N. R., & Birkett, M. A. (2017). *Physiology of Behaviour* (12th ed.). Harlow: Pearson.

Carr, B. (2008). Can Psychical Research Bridge the Gulf Between Matter and Mind? *Proceedings of the Society for Psychical Research, 59*(221), 1–96.

Carr, H. W. (1921). The Metaphysical Aspects of Relativity. *Nature, 106*(2677), 809–811.

Casolino, M., Bidoli, V., Morselli, A., Narici, L., De Pascale, M. P., Picozza, P., et al. (2003). Space Travel: Dual Origins of Light Flashes Seen in Space. *Nature, 422*(6933), 680.

Cassol, H., Pétré, B., Degrange, S., Martial, C., Charland-Verville, V., Lallier, F., et al. (2018). Qualitative Thematic Analysis of the Phenomenology of Near-Death Experiences. *PLoS ONE, 13*(2), 1–14.

Castillo, R. J. (2003). Trance, Functional Psychosis, and Culture. *Psychiatry, 66*(1), 9–21.

Chase, H. W. (1912). Consciousness and the Unconscious. *Psychological Bulletin, 9*(1), 20–25.

Cheyne, J. A., Newby-Clarke, I. R., & Rueffer, S. D. (1999). Relations Among Hypnagogic and Hypnopompic Experiences Associated with Sleep Paralysis. *Journal of Sleep Research, 8,* 313–317.

Choong, C., Hunter, M. D., & Woodruff, P. W. (2007). Auditory Hallucinations in Those Populations That Do Not Suffer from Schizophrenia. *Current Psychiatry Reports, 9*(3), 206–212.

Cigale, E. (2009, February). The Value of the Transpersonal in Psychotherapy and in Everyday Life. *Europe's Journal of Psychologym, 5*(Special section), 1–16.

Clarke, J., & Coleman, L. (1975). *The Unidentified: Notes Towards Solving the UFO Mystery*. New York: Warner Paperback Library.
Cleary, A. M., & Claxton, A. B. (2018). Déjà Vu: An Illusion of Prediction. *Psychological Science, 29*(4), 635.
Clement, P. (2007/2008). Dream Imagery as a Result of Emotions Matching with Images. *Imagination, Cognition and Personality, 27*(3), 245–257.
Close, E. R., & Neppe, V. M. (2015). Translating Fifteen Mysteries of the Universe by Applying a Nine Dimensional Spinning Model of Finite Reality: A Perspective, the Standard Model and Triadic Dimensional-Distinction Vortical Paradigm. *Part I. NeuroQuantology, 13*(2), 205–217.
Coleman, L. (2013). Are Zooforms Cryptids? *Cryptozoonews*. Retrieved July 17, 2018, from http://www.cryptozoonews.com/zooforms-cryptids/.
Collerton, D., Perry, E., & McKeith, I. (2005). Why People See Things That Are Not There: A Novel Perception and Attention Deficit Model for Recurrent Complex Visual Hallucinations. *Behavioral and Brain Sciences, 28*(6), 737–757.
Collerton, D., Taylor, J.-P., Tsuda, I., Fujii, H., Nara, S., Aihara, K., et al. (2016). How Can We See Things That Are Not There? Current Insights into Complex Visual Hallucinations. *Journal of Consciousness Studies, 23*(7–8), 195–227.
Comparelli, A., De Carolis, A., Kotzalidis, G. D., Masillo, A., Ferracuti, S., & Conesa, J. (2000). Geomagnetic, Cross-Cultural and Occupational Faces of Sleep Paralysis: An Ecological Perspective. *Sleep and Hypnosis, 2*(3), 105–111.
Cossins, D. (2018). Total Wipeout: The First Gravitational Waves Validated Einstein's Theory Of Gravity—The Ones to Come Could Destroy It. *New Scientist, 238*(3183), 32–35.
Crane, G. S. (2016). Harnessing the Placebo Effect: A New Model for Mind-Body Healing Mechanisms. *International Journal of Transpersonal Studies, 35*(1), 39–51.
Culicover, P. W., & Jackendoff, R. (2010). Quantitative Methods Alone Are Not Enough: Response to Gibson and Fedorenko. *Trends in Cognitive Sciences, 14*(6), 234–235.
Dacher, E. S. (2014). A Brief History of Mind-Body Medicine. *International Journal of Transpersonal Studies, 33*(1), 148–157.
Dale, I. (2015). *Responses to Self Harm: An Historical Analysis of Medical, Religious, Military and Psychological Perspectives*. Jefferson, NC: McFarland.
Dane, L. (1975–1976). Astral Travel: A Psychological Overview. *Journal of Altered States of Consciousness, 2*(3), 249–258.
Darby, G. (2010). Quantum Mechanics and Metaphysical Indeterminacy. *Australasian Journal of Philosophy, 88*(2), 227–245.

Das, T. (2016). Origin of Consciousness and Zero-Point Field. *Neuroquantology, 14*(1), 45–48.

De Gennaro, L., Ferrara, M., Urbani, L., Bertini, M., De Gennaro, L., Ferrara, M., et al. (2000). Complementary Relationship Between Wake and REM Sleep in the Auditory System: A Pre-sleep Increase of Middle-Ear Muscle Activity (MEMA) Causes a Decrease of MEMA During Sleep. *Experimental Brain Research, 130*(1), 105–112.

Descartes, R. (1985). *The Philosophical Writings of Descartes* (Vols. 1–2; J. Cottingham, R. Stoothoff, & D. Murdoch, Trans.). Cambridge: Cambridge University Press.

Deveraux, P. (2001). *Haunted Land: Investigations into Ancient Mysteries and Modern Day Phenomena.* London: Piatkus Ltd.

Dowben, J. S., Kowalski, P. C., & Keltner, N. L. (2017). Formication, Tactile Hallucinations, Delusional Parasitosis, and Morgellons: Enough to Make Your Skin Crawl. *Perspectives on Psychiatric Care, 53,* 220–221.

Drinkwater, K., Dagnall, N., & Bate, L. (2013). Into the Unknown: Using Interpretative Phenomenological Analysis to Explore Personal Accounts of Paranormal Experiences. *Journal of Parapsychology, 77,* 281–294.

Drinkwater, K., Dagnall, N., Grogan, S., & Riley, V. (2017). Understanding the Unknown: A Thematic Analysis of Subjective Paranormal Experiences. *Australian Journal of Parapsychology, 17*(1), 23–46.

Dudley, R., Aynsworth, C., Cheetham, R., McCarthy-Jones, S., & Collerton, D. (2018). Prevalence and Characteristics of Multi-modal Hallucinations in People with Psychosis Who Experience Visual Hallucinations. *Psychiatry Research, 269,* 25–30.

Durham, R. B. (2015). *Charles Webster Leadbeater: The Astral Plane, the Devachanic Plane, Clairvoyance, the Inner Life* (Vol. 1 & 2). Morrisville, NC: Lulu Enterprises.

Dyck, E., & Farrel, P. (2018). Psychedelics and Psychotherapy in Canada: Humphry Osmond and Aldous Huxley. *History of Psychology, 21*(3), 240–253.

Ebon, M. (1978). History of Parapsychology. In M. Ebon (Ed.), *The Signet Handbook of Parapsychology* (pp. 17–40). New York: Lombard Associates.

Edlow, B. L., Takahashi, E., Wu, O., Benner, T., Dai, G., Bu, L., et al. (2012). Neuroanatomic Connectivity of the Human Ascending Arousal System Critical to Consciousness and Its Disorders. *Journal of Neuropathology and Experimental Neurology, 71*(6), 531–546.

Eliade, M. (1972). *Shamanism: Archaic Techniques of Ecstasy.* Princeton, NJ: Princeton University Press.

Elliott, C. (2018, April). Meaning in Randomness. *Psychologist,* 22–25.

Evans, H. (1984). *Visions, Apparitions, Alien Visitors.* London: Book Club Associates.

Fallik, B., & Eliot, J. (1986). An Examination of Possible Age Differences in Interrelationships Between Intuition, Cognitive Style, and Hemispheric Preference Variables. *Perceptual and Motor Skills, 63*(3), 1251–1257.

Feiman, G. (2017). Eidetic Memory and School Age. *Journal of Russian & East European Psychology, 54*(2), 130–191.

Field, G. D., & Sampath, A. (2016). Behavioural and Physiological Limits to Vision in Mammals. *Philosophical Transactions of the Royal Society B, 372,* 20160072. https://doi.org/10.1098/rstb.2016.0072.

Finster, F. (2011). A Formulation of Quantum Field Theory Realizing a Sea of Interacting Dirac Particles. *Letters in Mathematical Physics, 97,* 165–183.

Fish, W. (2005). Disjunctivism and Non-disjunctivism: Making Sense of the Debate. *Proceedings of the Aristotelian Society, 105*(1), 119–127.

Fodor, N. (1933). *Encyclopedia of Psychic Science.* London: Arthurs Press.

Fordham, F. (1953/1959). *An Introduction to Jung's Psychology.* Harmondsworth: Penguin Books.

Fosse, R., Stickgold, R., & Hobson, A. (2004). Thinking and Hallucinating: Reciprocal Changes in Sleep. *Psychophysiology, 41,* 298–305.

Fowler, R. E. (2001). Allagash Abductions. In R. D. Story (Ed.), *The Mammoth Encyclopedia of Extraterrestrial Encounters* (pp. 54–59). London: Constable and Robinson.

Foy, K. (2011). Leprechauns and Lewy Body Disease. *Transcultural Psychiatry, 48*(3), 362–364.

Frank, P. (1957). Philosophical Uses of Science. *Bulletin of the Atomic Scientists, 13*(4), 125–130.

Furuya, H., Ikezoe, K., Ohyagi, Y., Miyoshi, T., & Fujii, N. (2006). A Case of Progressive Posterior Cortical Atrophy (PCA) with Vivid Hallucination: Are Some Ghost Tales Vivid Hallucinations in Normal People? *Journal of Neurology, Neurosurgery and Psychiatry, 77*(3), 424–425.

Fyfe, S., Williams, C., Mason, O. J., & Pickup, G. J. (2008). Apophenia, Theory of Mind and Schizotypy: Perceiving Meaning and Intentionality in Randomness. *Cortex, 44*(10), 1316–1325.

Gao, X., Zhang, W., Yao, L., Xiao, Y., Liu, L., Liu, J., et al. (2018). Association Between Structural and Functional Brain Alterations in Drug-Free Patients with Schizophrenia: A Multimodal Meta-Analysis. *Journal of Psychiatry and Neuroscience, 43*(2), 131–142.

Gauld, A. (2012). Living Consciousness: The Metaphysical Vision of Henri Bergson (Book Review). *Journal of the Society for Psychical Research, 76*(909), 222–227.

Geesink, H. J. H., & Meijer, D. K. F. (2016). Quantum Wave Information of Life Revealed: An Algorithm for Electromagnetic Frequencies That Create Stability of Biological Order, with Implications for Brain Function and Consciousness. *Neuroquantology, 14*(1), 106–125.

Gengerelli, J. A. (1976). Eidetic Imagery in Two Subjects After 46 Years. *Journal of General Psychology, 95*(2), 219–225.

Gentili, P. L., Rightler, A. L., Heron, B. M., & Gabbutt, C. D. (2016). Extending Human Perception of Electromagnetic Radiation to the UV Region Through Biologically Inspired Photochromic Fuzzy Logic (BIPFUL) Systems. *Chemical Communications, 52*(7), 1474–1477.

Geyer, M. A., & Vollenweider, F. X. (2008). Serotonin Research: Contributions to Understanding Psychoses. *Trends in Pharmacological Sciences, 29,* 445–453.

Gibson, J. J. (2002). A Theory of Direct Visual Perception. In N. Alva & E. Thompson (Eds.), *Vision and Mind: Selected Readings in the Philosophy of Perception* (pp. 77–91). Cambridge: MIT Press (Original work published 1972).

Giegerich, W. (2012). A Serious Misunderstanding: Synchronicity and the Generation of Meaning. *Journal of Analytical Psychology, 57*(4), 500–511.

Girard, T. A., & Cheyne, J. A. (2004). Individual Differences in Lateralization of Hallucinations Associated with Sleep Paralysis. *Brain and Cognition, 9*(1), 93–111.

Glicksohn, J. (1990). Belief in the Paranormal and Subjective Paranormal Experience. *Personality and Individual Differences, 11*(7), 675–683.

Goetz, C. G., Vaughan, C. L., Goldman, J. G., & Stebbins, G. T. (2014). I Finally See What You See: Parkinson's Disease Visual Hallucinations Captured with Functional Neuroimaging. *Movement Disorders: Official Journal of the Movement Disorder Society, 29*(1), 115–117.

Goetzmann, L., Ruettner, B., & Siegel, A. (2018). Fantasy, Dream, Vision, and Hallucination: Approaches from a Parallactic Neuro-Psychoanalytic Perspective. *Neuropsychoanalysis, 20*(1), 15–31.

Gonzalez, S., Moreno-Delgado, D., Moreno, E., Perez-Capote, K., Franco, R., et al. (2012). Circadian-Related Heteromerization of Adrenergic and Dopamine D4 Receptors Modulates Melatonin Synthesis and Release in the Pineal Gland. *PLoS Biol, 10*(6), e1001347. https://doi.org/10.1371/journal.pbio.1001347.

González-Casillas, A., Parra, L., Martin, L., Avila-Contreras, C., Ramirez-Pedraza, R., Vargas, N., et al. (2018). Towards a Model of Visual Recognition

Based on Neurosciences. *Biologically Inspired Cognitive Architectures, 25*(1), 119–129.

Gordon, R. S. (1992/2000). *The Paranormal: An Illustrated Encyclopedia.* London: Caxton.

Grammenos, D., & Barker, S. (2015). On the Transmethylation Hypothesis: Stress, N, N-Dimethyltryptamine, and Positive Symptoms of Psychosis. *Journal of Neural Transmission, 122*(6), 733–739.

Greaney, J., & MacRae, A. W. (1992, May). The Order of Visual Processing: Top-Down, bottom-Up, middle-Out, or None of These? *Bulletin of the Psychonomic Society, 30*(3), 255–257.

Green, A. (1973/1976). *Ghost Hunting: A Practical Guide.* St Albans: Mayflower books.

Greene, F. G. (1999). A Projection Geometry for Separation Experiences. *Journal of Near-Death Studies, 17*(3), 151–191.

Gregory, R. (1970). *The Intelligent Eye.* London: Weidenfeld and Nicolson.

Gu, Y. (2018). Vestibular Signals in Primate Cortex for Self-Motion Perception. *Current Opinion in Neurobiology, 52,* 10–17.

Guedj, E., Aubert, S., McGonigal, A., Mundler, O., & Bartolomei, F. (2010). Déjà-vu in Temporal Lobe Epilepsy: Metabolic Pattern of Cortical Involvement in Patients with Normal Brain MRI. *Neuropsychologia, 48*(7), 2174–2181.

Haddock, A. (2011). The Disjunctive Conception of Perceiving. *Philosophical Explorations, 14*(1), 23–42.

Haga, S. B., Warner, L. R., & O'Daniel, J. (2009). The Potential of a Placebo/Nocebo Effect in Pharmacogenetics. *Public Health Genomics, 12*(3), 158–162.

Halberstadt, A. L., Nichols, D. E., & Geyer, M. A. (2012). Behavioral Effects of α, α, β, β Tetradeutero-5-MeO-DMT in Rats: Comparison with-MeO-DMT Administered in Combination with a Monoamine Oxidase Inhibitor. *Psychopharmacology, 221*(4), 709–718.

Hall, C. S. (1959). *A Primer of Freudian Psychology.* New York: Mentor Books.

Hancock, G. (2005). *Supernatural: Meetings with the Ancient Teachers of Mankind.* London: Century.

Hanes, K. (2012). Unusual Phenomena Associated with a Transcendent Human Experience: A Case Study. *Journal of Transpersonal Psychology, 44*(1), 26–47.

Hannan, B. (1993). Don't Stop Believing: The Case Against Eliminative Materialism. *Mind and Language, 8*(2), 165–179.

Hansen, D. (2014). Heavenly and Hellish—Writers on Hallucinogens. *The Psychologist, 27*(9), 680–682.

Hartelius, G., Caplan, M., & Rardin, M. A. (2007). Transpersonal Psychology: Defining the Past, Divining the Future. *The Humanistic Psychologist, 35*(2), 135–160.

Harvey, J. (2004). The Photographic Medium: Representation, Reconstitution, Consciousness, and Collaboration in Early-Twentieth-Century Spiritualism. *Technoetic Arts: An International Journal of Speculative Research, 2*(2), 109–124.

Heller, N. H., & Davidenko, N. (2018). Dissociating Higher and Lower Order Visual Motion Systems by Priming Illusory Apparent Motion. *Perception, 47*(1), 30–43.

Helmes, E., Brown, J.-M., & Elliott, L. (2015). A Case of Dissociative Fugue and General Amnesia with an 11-Year Follow-Up. *Journal of Trauma and Dissociation, 16*(1), 100–113.

Hennig-Fast, K., Meister, F., Frodl, T., Beraldi, A., Padberg, F., Engel, R. R., et al. (2008). A Case of Persistent Retrograde Amnesia Following a Dissociative Fugue: Neuropsychological and Neurofunctional Underpinnings of Loss of Autobiographical Memory and Self-Awareness. *Neuropsychologia, 46*(12), 2993–3005.

Hirakawa, H., Terao, T., Kanehisa, M., Ninomiya, T., & Ishii, N. (2016). Coexistence of Delusional Parasitosis and Complex Visual Hallucinations with Micropsia. *The Journal of Neuropsychiatry and Clinical Neurosciences, 28*(1), e10–e12.

Hobson, J., Pace-Schott, E., & Stickgold, R. (2000). Dreaming and the Brain: Toward a Cognitive Neuroscience of Conscious States. *Behavioral and Brain Sciences, 23,* 793–842.

Holroyd, S. (1980). *Alien Intelligence.* London: Sphere Books (Original work published 1979).

Hoffman, A. (1979/1983). *LSD: My Problem Child: Reflections on Sacred Drugs, Mysticism, and Science.* Saline, MI: McNaughton and Gunn.

Honorton, C. (1977). Psi and Attentional Internal States. In B. B. Wolman (Ed.), *Handbook of Parapsychology* (pp. 435–472). Jefferson, NC: McFarland.

Houran, J., Ashe, D. D., & Thalbourne, M. A. (2003). Encounter Experiences in the Context of Mental Boundaries and Bilaterality. *Journal of the Society for Psychical Research, 67*(873), 260–280.

Houran, J., Kumar, V. K., Thalbourne, M. A., & Lavertue, N. E. (2002). Haunted by Somatic Tendencies: Spirit Infestation as Psychogenic Illness. *Mental Health, Religion and Culture, 5*(2), 119–133.

Howard, R. G. (2011). If Paranormal Phenomena Are Information Received by the Conscious Psyche, Then What Is the Conscious Psyche? *Journal of Spirituality and Paranormal Studies, 34*(2), 62–72.

Howes, O. D., & Kapur, S. (2009). The Dopamine Hypothesis of Schizophrenia: Version III—The Final Common Pathway. *Schizophrenia Bulletin, 35*(3), 549–562.
Hubert, E. (1977). Gazetteer of Mysterious Sites Around the World. In *The World's Last Mysteries* (pp. 297–318). London and New York: Reader's Digest Association.
Hughes, J. E. A., Gruffydd, E., Simner, J., & Ward, J. (2018). Synaesthetes Show Advantages in Savant Skill Acquisition: Training Calendar Calculation in Sequence-Space Synaesthesia. *Cortex, 113,* 67–82.
Hume, L. (1995). By Knot of One the Spell's Begun, By Knot of Two It Cometh True… Magic, Altered States, and Multiple Realities. *Social Alternatives, 14*(4), 23–27.
Huston, H. (1999). The Evolutionary Significance of Archetypal Dreams. In D. H. Rosen & M. C. Luebbert (Eds.), *Evolution of the Psyche* (pp. 188–195). Westport, CT: Greenwood Publishing Group.
Hut, P., & Shepard, R. N. (1996). Turning 'the Hard Problem' Upside Down and Sideways. *Journal of Consciousness Studies, 3*(4), 313–329.
Huxley, A. (2011). *Doors of Perception and Heaven and Hell.* London: Thinking Ink (Original work published 1952/1954).
Inglis, B. (1985/1986). *The Paranormal: An Encyclopedia of Psychic Phenomena.* London: Paladin.
Ingram, R., Kang, H., Lightman, S., Jane, D., Bortolotto, Z. A., Collingridge, G. L., et al. (2018). Some Distorted Thoughts About Ketamine as a Psychedelic and a Novel Hypothesis Based on NMDA Receptor-Mediated Synaptic Plasticity. *Neuropharmacology, 142,* 30–34.
Irwin, H. J. (2007). Science, Nonscience and Rejected Knowledge: The Case of Parapsychology. *Australian Journal of Parapsychology, 7*(1), 8–32.
Irwin, H. J. (2018). Stress Sensitivity and Minimal-Self Dysfunction as Predictors of Anomalous Experiences and Paranormal Attribution. *Journal of the Society for Psychical Research, 82*(1), 1–14.
Irwin, H. J., Dagnall, N., & Drinkwater, K. (2013). Parapsychological Experience as Anomalous Experience Plus Paranormal Attribution: A Questionnaire Based on a New Approach to Measurement. *Journal of Parapsychology, 77*(1), 39–53.
Irwin, H. J., & Watt, C. A. (2007). *An Introduction to Parapsychology* (5th ed.). Jefferson, NC: McFarland.
Irwin, K. D. (2006). Seeing with Superconductors. *Scientific American, 295*(5), 86–92.

Jackson, F. (1982). Epiphenomenal Qualia. *Philosophical Quarterly, 32*(127), 127–133.

Jacobs, G. H. (2010). Recent Progress in Understanding Mammalian Color Vision. *Ophthalmic and Physiological Optics, 30*(5), 422–434.

James, W. (1902/1985). *The Varieties of Religious Experience.* Cambridge: Harvard University Press.

Jinks, A. (2012). *An Introduction to the Psychology of Paranormal Belief and Experience.* Jefferson, NC: McFarland.

Jinks, A. (2016). *The Disappearing Object Phenomenon: An Investigation.* Jefferson, NC: McFarland.

Johnson, T. (2003). Shamans, Medicine Men, and Other Healers. *Native American Medicine, 2003,* 20–32.

Jonas, M. E. (2016). Three Misunderstandings of Plato's Theory of Moral Education. *Educational Theory, 66*(3), 301–322.

Jones, H. A. (2017). Reincarnation. *Journal for Spiritual and Consciousness Studies, 40*(2), 140–149.

Jonsson, K., Grim, K., & Kjellgren, A. (2014). Do Highly Sensitive Persons Experience More Nonordinary States of Consciousness During Sensory Isolation? *Social Behavior and Personality, 42*(9), 1495–1506.

Joseph, R. (2001). The Limbic System and the Soul: Evolution and the Neuroanatomy of Religious Experience. *Zygon, 36*(1), 105–136.

Joye, S. R. (2017). Tuning the Mind in the Frequency Domain: Karl Pribham's Holonomic Brain Theory and David Bohm's Implicate Order. *Cosmos and History, 13*(2), 166–184.

Jung, C. G. (1990). The Archetypes and the Collective Unconscious (10th ed.; R. F. C. Hull, Trans.). Princeton: Princeton University Press (Original work published 1959).

Jung, C. G. (1991). *Flying Saucers: A Modern Myth of Things Seen in the Skies* (R. F. C. Hull, Trans.). Princeton: Princeton University Press (Original work published 1959).

Kaminker, J. (2016). Images, Figures and Qualities: Clarifying the Relationship Between Individual and Archetype. *International Journal of Transpersonal Studies, 35*(2), 93–102.

Keel, J. (1975/1976). *Strange Creatures from Time and Space.* London: Sphere Books.

Keel, J. A. (1973). *Operation Trojan Horse.* London: Sphere Books.

Kelly, M. T. (1978). Sensory Deprivation and Research on Telepathy. *Parapsychology Review, 9*(4), 24–26.

Kemper, K. J. (2016). Authentic Healing: Effects on Recipients and Healers. *Alternative and Complementary Therapies, 22*(3), 105–110.
Krippner, S. C. (1981). Dreams and Other Altered Conscious States. *Parapsychological Journal of South Africa, 2*(2), 18–34.
Krippner, S. C. (2002). Conflicting Perspectives on Shamans and Shamanism: Points and Counterpoints. *American Psychologist, 57*(11), 3–66.
Kocsis, S., Braverman, B., Ravets, S., Stevens, M. J., Mirin, R. P., Shalm, L. K., et al. (2011). Observing the Average Trajectories of Single Photons in a Two-Slit Interferometer. *Science, 332*(6034), 1170–1173.
Koethe, D., Hoyer, C., & Leweke, F. M. (2009). The Endocannabinoid System as a Target for Modelling Psychosis. *Psychopharmacology, 206*(4), 551–561.
Kokoszka, A. (1999–2000). Altered States of Consciousness: A Comparison of Profoundly and Superficially Altered States. *Imagination, Cognition and Personality, 19*(2), 165–184.
Kosman, K. A., & Silbersweig, D. A. (2018). Pseudo-Charles Bonnet Syndrome with a Frontal Tumor: Visual Hallucinations, the Brain, and the Two-Hit Hypothesis. *The Journal of Neuropsychiatry and Clinical Neurosciences, 30*(1), 84–86.
Kottmeyer, M. (2006, February 25). Letters to Ye Olde Editor. *Saucer Smear, 53*(2).
Kottmeyer, M. S. (2001). Grays. In R. D. Story (Ed.), *The Mammoth Encyclopedia of Extraterrestrial Encounters* (pp. 273–281). London: Constable and Robinson Ltd.
Kovalzon, V. M. (2016). Ascending Reticular Activating System of the Brain. *Brain Science Advances, 2*(4), 275–285.
Kraehenmann, R., Pokorny, D., Vollenweider, L., Preller, K., Pokorny, T., Seifritz, E., et al. (2017). Dreamlike effects of LSD on Waking Imagery in Humans Depend on Serotonin 2A Receptor Activation. *Psychopharmacology, 234*(13), 2031–2046.
Kramer, M. (2002). The Biology of Dream Formation: A Review and Critique. *Journal of the American Academy of Psychoanalysis, 30*(4), 657–671.
Labib, B. (2017). Reflecting on Entoptic Phenomena: Vivid Images of the Retinal Vasculature Often Appear to Our Patients During Exam. Are They a Cause for Concern? *Review of Optometry, 154*(3), 24–25.
La Camera, M. (2008). Bran Cosmology with (4)R Term in the Bulk. *Acta Physica Polonica B., 39*(8), 1795–1809.
Lampis, R. (2004). The Theory of Reality: Or Why, If We Understand Reality, the Theory of Everything Is an Impossibility. Or How to Establish Basic Principles That Clarify Physical, Mental and Paraphysical Phenomena, and

to Dispose of Concepts Such as Non-locality, Place. *International Journal of Humanities and Peace, 20*(1), 43–48.

Lanciano, T., & Curci, A. (2011). Memory for Emotional Events: The Accuracy of Central and Peripheral Details. *Europe's Journal of Psychology, 7*(2), 323–336.

Lange, R., & Houran, J. (1998). Delusions of the Paranormal: A Haunting Question of Perception. *Journal of Nervous and Mental Disease, 186*(10), 637–645.

Lash, S. J., & Polyson, J. A. (1987). The Gender Relevance of Projected Animal Content. *Journal of Clinical Psychology, 43*(1), 145–150.

Laszlo, E. (2004). Matter and Mind: The New Holism and the Greater Humanity. In D. Loye (Ed.), *The Great Adventure: Toward a Fully Human Theory of Evolution* (pp. 39–52). New York: State University of New York Press.

Leahey, T. H (1987). *A History of Psychology* (2nd ed.). Englewood Cliffs, NJ: Prentice-Hall.

Lehmann, D., Faber, P. L., Achermann, P., Jeanmonod, D., Gianotti, L. R., & Pizzagalli, D. (2001). Brain Sources of EEG Gamma Frequency During Volitionally Meditation-Induced, Altered States of Consciousness, and Experience of the Self. *Psychiatry Research, 108*(2), 111–121.

Lestrade, S. (2017). Unzipping Zipf's Law. *PLoS ONE, 12*(8), e0181987.

Leventhall, G. (2007). What Is Infrasound? *Progress in Biophysics and Molecular Biology, 93*(1–3), 130–137.

Levin, J. S. (1996). How Prayer Heals: A Theoretical Model. *Alternative Therapies in Health and Medicine, 2*(1), 66–73.

Lewis, T., & Kahn, R. (2005). The Reptoid Hypothesis: Utopian and Dystopian Representational Motifs in David Icke's Alien Conspiracy Theory. *Utopian Studies, 16*(1), 45–74.

Libet, B. (2006). Reflections on the Interaction of the Mind and Brain. *Progress in Neurobiology, 78*(3–5), 322–326.

Liechti, M. E., Dolder, P. C., & Schmid, Y. (2017). Alterations of Consciousness and Mystical-Type Experiences After Acute LSD in Humans. *Psychopharmacology, 234*(9–10), 1499–1510.

Ligeza, T. S., Tymorek, A. D., & Wyczesany, M. (2017). Top-Down and Bottom-Up Competition in Visual Stimuli Processing. *Acta Neurobiologiae Experimentalis, 77*(4), 305–316.

Limoli, C. L. (2017). Deep-Space Deal Breaker. *Scientific American, 316*(2), 54–59.

Lin, W., Xie, Y.-C., Cheng, P.-Y., Dong, L.-Y., Hung, G.-U., & Chiu, P.-Y. (2018). Association of Visual Hallucinations with Very Mild Degenerative Dementia Due to Dementia with Lewy Bodies. *PLoS ONE, 13*(10), 1–8.

Lindeman, M., & Svedholm, A. M. (2012). What's in a Term? Paranormal, Superstitious, Magical and Supernatural Beliefs by Any Other Name Would Mean the Same. *Review of General Psychology, 16*(3), 241–255.

López Rodrigo, A. M., Paíno Piñeiro, M. M., Martínez Suárez, P. C., Caro, M. I., & Lemos Giráldez, S. (1997). Hallucinations in a Normal Population: Imagery and Personality Influences. *Psychology in Spain, 1*(1), 10–16.

Ludwig, A. M. (1966, March 4–6). Possession States in Primitive People, from the Symposium Sponsored by R. M. Bucke Society, Montreal. *Archives of General Psychiatry, 15.*

Luke, D. (2008). Inner Paths to Outer Space: Journeys to Alien Worlds Through Psychedelics and Other Spiritual Technologies (Book Review). *Journal of Scientific Exploration, 22*(4), 564–569.

Luke, D. (2011a). Anomalous Phenomena, Psi, and Altered Consciousness. In E. Cardeña & M. Winkelman (Eds.), *Altering Consciousness: Multidisciplinary Perspectives: History, Culture, and the Humanities; Biological and Psychological Perspectives* (Vol. 2, pp. 355–374). Santa Barbara: Praeger.

Luke, D. (2011b). Discarnate Entities and Domethyltryptamine (DMT): Psychopharmacology, Phenomenology and Ontology. *Journal of the Society for Psychical Research, 75*(902), 26–42.

Luke, D. (2012). Psychoactive Substances and Paranormal Phenomena: A Comprehensive Review. *International Journal of Transpersonal Studies, 31*(1), 97–156.

Luke, D., Zychowicz, K., Richterova, O., Tjurina, I., & Polonnikova, J. (2012). A Sideways Look at the Neurobiology of Psi: Precognition and Circadian Rhythms. *Neuroquantology, 10*(3), 580–590.

Lynn, S. J., & Evans, J. (2017). Hypnotic Suggestion Produces Mystical-Type Experiences in the Laboratory: A Demonstration Proof. *Psychology of Consciousness: Theory, Research, and Practice, 4*(1), 23–37.

Lynn, S. J., Milano, M., & Weekes, J. R. (1991). Hypnosis and Pseudomemories: The Effects of Prehypnotic Expectancies. *Journal of Personality and Social Psychology, 60*(2), 318–326.

Mack, J. E. (1993). The UFO Abduction Phenomenon: What Might It Mean for the Evolution of Human Consciousness? *ReVision, 16*(1), 26–32.

Maher, M. (1900). *Psychology: Empirical and Rational* (4th ed.). London: Longmans, Green and Co.

Maher, M. (1999). Riding the Waves in Search of the Particles: A Modern Study of Ghosts and Apparitions. *Journal of Parapsychology, 63*(1), 47–80.

Mahowald, M. W., Woods, S. R., & Schenk, C. H. (1998). Sleeping Dreams, Waking Hallucinations, and the Central Nervous System. *Dreaming, 8*(2), 89–102.

Maillard, L., Vignal, J.-P., Gavaret, M., Guye, M., Biraben, A., McGonigal, A., et al. (2004). Semiologic and Electrophysiologic Correlations in Temporal Lobe Seizure Subtypes. *Epilepsia, 45*(12), 1590–1599.

Main, R. (2014). The Cultural Significance of Synchronicity for Jung and Pauli. *Journal of Analytical Psychology, 59*(2), 174–180.

Mannu, G. S. (2014). Retinal Phototransduction. *Neurosciences, 19*(4), 275–280.

Martin-Vallas, F. (2013). Are Archetypes Transmitted or Emergent? A Response to Christian Roesler. *Journal of Analytical Psychology, 58*(2), 278–285.

Mazzoni, G., & Vannucci, M. (2007). Hindsight Bias, the Misinformation Effect, and False Autobiographical Memories. *Social Cognition, 25*(1), 203–220.

McLeod, C. C., Corbisier, B., & Mack, J. E. (1996). More Parsimonious Explanation for UFO Abduction. *Psychological Inquiry, 7*(2), 156–169.

McNally, R. J., & Clancy, S. A. (2005). Sleep Paralysis, Sexual Abuse and Space Alien Abduction. *Transcultural Psychiatry, 42*(1), 113–122.

Meijer, D. K. F., & Korf, J. (2013). Quantum Modeling of the Mental State: The Concept of a Cyclic Mental Workspace. *Syntropy, 1,* 1–41.

Merali, Z. (2006). Ether Returns in Bid to Oust Dark Matter. *New Scientist, 191*(2566), 8–9.

Metzinger, T. (2009). Why Are Out-of-Body Experiences Interesting for Philosophers? The Theoretical Relevance of OBE Research. *Cortex, 45*(2), 256–258.

Mikles, N. L., & Laycock, J. P. (2015). Tracking the Tulpa: Exploring the 'Tibetan' Origins of a Contemporary Paranormal Idea. *Nova Religio: The Journal of Alternative and Emergent Religions, 19*(1), 87–97.

Midena, E., & Vujosevic, S. (2015). Metamorphopsia: An Overlooked Visual Symptom. *Ophthalmic Research, 55*(1), 26–36.

Mills, J. A. (2010). Hallucinogens as Hard Science: The Adrenochrome Hypothesis for the Biogenesis of Schizophrenia. *History of Psychology, 13*(2), 178–195.

Mishkind, M. (1993). A Test for Morphic Resonance in Behavioural Responses to Multiple Choice Stimuli. *Journal of Analytical Psychology, 38*(3), 257–271.

Mishor, Z., McKenna, D. J., & Callaway, J. C. (2011). DMT and Human Consciousness. In E. Cardeña & M. Winkelman (Eds.), *Altering Consciousness: Multidisciplinary Perspectives: History, Culture, and the Humanities; Biological and Psychological Perspectives* (Vol. 2, pp. 85–119). Santa Barbara: Praeger.

Mobbs, D., & Watt, C. (2011). There Is Nothing Paranormal About Near-Death Experiences: How Neuroscience Can Explain Seeing Bright Lights, Meeting the Dead, or Being Convinced You Are One of Them. *Trends in Cognitive Sciences, 15*(10), 447–449.

Modestino, E. J. (2016). Neurophenomenology of an Altered State of Consciousness: An fMRI Case Study. *Explore: The Journal of Science & Healing, 12*(2), 128–135.

Mohammed, A. A. (2012). A Critique of Descartes' Mind-Body Dualism. *Kritike: An Online Journal of Philosophy, 6*(1), 95–112.

Møller, H., & Pedersen, C. S. (2004). Hearing at Low and Infrasonic Frequencies. *Noise and Health, 6*(23), 37–57.

Moody, R. A. (1975/2001). *Life After Life: The Investigation of a Phenomenon—Survival of Bodily Death.* New York: HarperOne (Original work published 1975).

Moreira-Almeida, A., & Lotufo-Neto, F. (2017). Methodological Guidelines to Investigate Altered States of Consciousness and Anomalous Experiences. *International Review of Psychiatry, 29*(3), 283–292.

Morganti, M. (2015). Dependence, Justification and Explanation: Must Reality Be Well-Founded? *Erkenntnis, 80*(3), 555–572.

Moulin, C. J. A. (2018). *The Cognitive Neuropsychology of Déjà Vu.* London: Routledge.

Mussacchio, J. M. (2005). The Ineffability of Qualia and the Word-Anchoring Problem. *Language Sciences, 27*(4), 403–435.

Myers, F. (1903). Phantasms of the Dead. In M. Ebon (Ed.) (1978), *The Signet Handbook of Parapsychology* (pp. 273–285). New York: Lombard Associates.

Naish, P. (2005). Detecting Hypnotically Altered States of Consciousness. *Contemporary Hypnosis, 22*(1), 24–30.

Nanay, B. (2016). Hallucination as Mental Imagery. *Journal of Consciousness Studies, 23*(7–8), 65–81.

Nedu, O. C. (2013). The Mind (Manas) and the Illusory projection of the Afflicted (Klista) Individual Self (Atman) in Vijnanavada Buddhism. *Danubius, 31*, 327–373.

Neppe, V. M., & Close, E. R. (2015). The Concept of Relative Non-locality: Theoretical Implications in Consciousness Research. *Explore: The Journal of Science and Healing, 11*(2), 102–108.

Ness, R. C. (1978). The Old Hag Phenomenon as Sleep Paralysis: A Biocultural Interpretation. *Culture, Medicine and Psychiatry, 2*(1), 15–39.

Neubauer, D. N. (2009). Fearful Sleeping and Waking. *Primary Psychiatry, 16*(11), 16–18.

Nishino, S., Riehl, J., Hong, J., Kwan, M., Reid, M., & Mignot, E. (2000). Is Narcolepsy a REM Sleep Disorder? Analysis of Sleep Abnormalities in Narcoleptic Dobermans. *Neuroscience Research, 38,* 437–446.

Noakes, R. (2008). The 'World of the Infinitely Little': Connecting Physical and Psychical Realities Circa 1900. *Studies in History and Philosophy of Science Part A, 39*(3), 323–334.

Novak, J., Ozkan, M., Pang, Y., & Tartaglino-Mazzucchelli, G. (2017). Gauss-Bonnet Supergravity in Six Dimensions. *Physical Review Letters, 119,* 111602.

Nunn, C. M. H. (1998). Archetypes and Memes: Their Structure, Relationships and Behaviour. *Journal of Consciousness Studies, 5*(3), 344–354.

Nye, R. D (1975/1981). *Three Psychologies* (2nd ed.). Belmont: Brooks/Cole.

O'Meara, S. J. (1996). Space-Time and the 'Ether' Bunny. *Odyssey, 5*(9), 16–21.

Palczewska, G., Vinberg, F., Stremplewski, P., Bircher, M. P., Salom, D., Komar, K., et al. (2014). Human Infrared Vision Is Triggered by Two-Photon Chromophore Isomerization. *Proceedings of the National Academy of Sciences of the United States of America, 111*(50), E5445–E5454.

Parker, A., Frederiksen, A., & Johansson, H. (1997). Towards Specifying the Recipe for Success with the Ganzfeld: Replication of the Ganzfeld Findings Using a Manual Ganzfeld with Subjects Reporting Prior Paranormal Experiences. *European Journal of Parapsychology, 13,* 15–27.

Parra, A. (2006). 'Seeing and Feeling Ghosts': Absorption, Fantasy Proneness, and Healthy Schizotypy as Predictors of Crisis Apparition Experiences. *Journal of Parapsychology, 70*(2), 357–372.

Parra, A. (2015). Seeing Rare Things with the Mind's Eye: Visual Imagery Vividness and Paranormal/Anomalous Experiences. *Australian Journal of Parapsychology, 15*(1), 37–51.

Parra, A. (2018). Perceptual-Personality Variables Associated with Entity Encounter Experiences. *Australian Journal of Parapsychology, 18*(1), 23–48.

Parsons, S. T. (2012). Infrasound and the Paranormal. *Journal of the Society for Psychical Research, 76*(908), 150–174.

Partridge, C. (2004). Alien Demonology: The Christian Roots of the Malevolent Extraterrestrial in UFO Religions and Abduction Spiritualities. *Religion, 34,* 163–189.

Pasqua, A., Chattopadhyay, S., & Myrzakulov, R. (2018). Consequences of Three Modified Forms of Holographic Dark Energy Models in Bulk-Brane Interaction. *Canadian Journal of Physics, 96*(1), 112–125.

Penman, L. T. I. (2015). The History of the Word Paranormal. *Notes and Queries, 62*(260), 31–34.

Persinger, M. A. (1979). Possible Infrequent Geophysical Sources of Close UFO Encounters: Expected Physical and Behavioral-Biological Effects. In R. F. Haines (Ed.), *UFO Phenomena and the Behavioral Scientist* (pp. 396–433). Metuchen, NJ: Scarecrow Press.

Persinger, M. A., Bureau, Y. R. J., Peredery, O. P., & Richards, P. M. (1994). The Sensed Presence as Right Hemispheric Intrusions into the Left Hemispheric Awareness of Self: An Illustrative Case Study. *Perceptual and Motor Skills, 78*(3), 999–1009.

Persinger, M. A., & Cameron, R. A. (1986). Are Earth Faults at Fault in Some Poltergeist-Like Episodes? *Journal of the American Society for Psychical Research, 80,* 49–73.

Picard, F. (2013). State of Belief, Subjective Certainty and Bliss as a Product of Cortical Dysfunction. *Cortex, 49*(9), 2494–2500.

Pinel, J. P. J., & Barnes, S. (2017). *Biopsychology* (10th ed.). Boston: Pearson.

Pizzi, R., Wang, R., & Rossetti, D. (2016). Human Visual System as a Double-Slit Single Photon Interference Sensor: A Comparison between Modellistic and Biophysical Tests. *PLoS ONE, 11*(1), 1–24.

Pockett, S. (2004). Does Consciousness Cause Behaviour? *Journal of Consciousness Studies, 11*(2), 23–40.

Pokazanyeva, A. (2016). Mind Within Matter: Science, the Occult, and the (Meta)Physics of Ether and Akasha. *Zygon: Journal of Religion and Science, 51*(2), 318–346.

Polák, M., & Marvan, T. (2018). Neural Correlates of Consciousness Meet the Theory of Identity. *Frontiers in Psychology, 9,* 1–13.

Polito, V., Langdon, R., & Brown, J. (2010). The Experience of Altered States of Consciousness in Shamanic Ritual: The Role of Pre-existing Beliefs and Affective Factors. *Consciousness and Cognition: An International Journal, 19*(4), 918–925.

Poller, J. (2015). Beyond the Subliminal Mind: Psychical Research in the Work of Aldous Huxley. *Aries, 15*(2), 247–266.

Poynton, J. (2012). The Out-of-Body Experience: The History and Science of Astral Travel (Book Review). *Journal of the Society for Psychical Research, 76*(907), 116–120.

Preller, K. H., & Vollenweider, F. X. (2018). Phenomenology, Structure, and Dynamic of Psychedelic States. *Current Topics in Behavioral Neurosciences, 36,* 221–256.

Presti, D. E. (2011). Neurochemistry and Altered Consciousness. In E. Cardeña & M. Winkelman (Eds.), *Altering Consciousness: Multidisciplinary Perspectives:*

History, Culture, and the Humanities; Biological and Psychological Perspectives (Vol. 2, pp. 21–41). Santa Barbara: Praeger.

Price, B. (2016). Hallucinations: Insights and Supportive First Care. *Nursing Standard, 30*(21), 49–58.

Price, H. (1953/1963). Introduction. In G. N. M. Tyrell (Ed.), *Apparitions*. New York: Collier Books.

Puhle, A., & Parker, A. (2017). An Exploratory Study of Lucid Dreams Concerning Deceased Persons. *Journal of the Society for Psychical Research, 81*(3), 145–160.

Purves, D., Augustine, G. J., Fitzpatrick, D., Katz, L. C., LaMantia, A.-S., McNamara, J. O., et al. (Eds.). (2001). *Neuroscience*. Sunderland: Sinauer Associates.

Rabeyron, T., & Caussie, S. (2016). Clinical Aspects of Out-of-Body Experiences: Trauma, Reflexivity and Symbolisation. *L'Évolution Psychiatrique, 81*(4), e53–e71.

Ramachandran, V. S., & Blakeslee, S. (1998). *Phantoms in the Brain*. London: Fourth Estate.

Read, S. J., & Rosson, M. B. (1982). Rewriting History: The Biasing Effects of Attitudes on Memory. *Social Cognition, 1*(3), 240–255.

Reinhuber, E. (2017). Phasmagraphy: A Potential Future for Artistic Imaging. *Technoetic Arts: An International Journal of Speculative Research, 15*(3), 261–273.

Reissig, C., Carter, L., Johnson, M., Mintzer, M., Klinedinst, M., & Griffiths, R. (2012). High Doses of Dextromethorphan, an NMDA Antagonist, Produce Effects Similar to Classic Hallucinogens. *Psychopharmacology, 223*(1), 1–15.

Reznik, D., Gertner-Saad, L., Even-Furst, H., Henik, A., Ben Mair, E., Shechter-Amir, D., et al. (2018). Oneiric Synesthesia: Preliminary Evidence for the Occurrence of Synesthetic-Like Experiences During Sleep-Inertia. *Psychology of Consciousness: Theory, Research, and Practice, 5*(4), 374–383.

Rich, A. N., & Mattingly, J. B. (2002). Anomalous Perception in Synaethesia: A Cognitive Neuroscience Perspective. *Nature Reviews—Neuroscience, 3*(1), 43–52.

Rickard, B., & Michell, J. (2007). *The Rough Guide to Unexplained Phenomena* (2nd ed.). London: Rough Guides Ltd.

Riggs, P. J. (2013). Why a Spaceship Cannot Reach the Speed of Light from the Perspective of the Spaceship's Rest Frame. *Latin-American Journal of Physics Education, 7*(4), 604–608.

Ring, K. (1989). Near-Death and UFO Encounters as Shamanic Initiations. *ReVision, 11*(3), 1–22.

Rioux, D. (1996). Shamanic Healing Techniques: Toward Holistic Addiction Counseling. *Alcoholism Treatment Quarterly, 14*(1), 59–69.
Rock, A. J., & Krippner, S. (2012). States of Consciousness or States of Phenomenology? In A. E. Cavanna & A. Nani (Eds.), *Consciousness: States, Mechanisms and Disorders* (pp. 55–65). Hauppauge: Nova Science Publishers.
Rodríguez Pereyra, G. (2007). Descartes's Substance Dualism and His Independence Conception of Substance. *Journal of the History of Philosophy, 46*(1), 69–89.
Rogo, D. S. (1976). Research on Psi-Conducive States: Some Complicating Factors. *Journal of Parapsychology, 40*(1), 34–45.
Rogo, D. S. (1979). *The Poltergeist Experience: Investigations into Ghostly Phenomena.* New York: Penguin.
Roll, W. G., & Joines, W. T. (2013). RSPK and Consciousness. *Journal of Parapsychology, 77*(2), 192–211.
Roney-Dougal, S. M., Ryan, A., & Luke, D. (2013). The Relationship Between Local Geomagnetic Activity, Meditation and Psi Part 1: Literature Review and Theoretical Model. *Journal of the Society for Psychical Research, 77*(911), 72–88.
Rosen, D. H., Smith, S. M., & Huston, H. L. (1991). Empirical Study of Associations Between Symbols and Their Meanings: Evidence of Collective Unconscious (Archetypal) Memory. *Journal of Analytical Psychology, 36*(2), 211–228.
Rosenthal, O., & Behrmann, M. (2006). Acquiring Long-Term Representations of Visual Classes Following Extensive Extrastriate Damage. *Neuropsychologia, 44*(5), 799–815.
Roudinesco, E. (2016). *Freud: In His Time and Ours* (C. Porter, Trans.). Cambridge: Harvard University Press.
Sabom, M. B., & Kreutziger, S. (1977). The Experience of Near Death. *Death Education, 1*(2), 195–203.
Saethre, E. (2007). UFOs, Otherness, and Belonging: Identity in Remote Aboriginal Australia. *Social Identities, 13*(2), 217–233.
Sayin, H. U. (2014a). Does the Nervous System Have an Intrinsic Archaic Language? Entoptic Images and Phosphenes. *Neuroquantology, 12*(3), 427–445.
Sayin, H. U. (2014b). The Consumption of Psychoactive Plants During Religious Rituals: The Roots of Common Symbols and Figures in Religions and Myths. *Neuroquantology, 12*(2), 276–296.
Schacter, D. L., Chiao Chiao, J. Y., & Mitchell, J. P. (2003). The Seven Sins of Memory: Implications for Self. *Annals of the New York Academy of Sciences, 1001*, 226–239.

Schäfer, L. (2006). A Response to Ervin Laszlo: Quantum and Consciousness. *Zygon: Journal of Religion and Science, 41*(3), 573–582.

Schmidt, A., Müller, F., Lenz, C., Dolder, P. C., Schmid, Y., Zanchi, D., et al. (2018). Acute LSD Effects on Response Inhibition Neural Networks. *Psychological Medicine, 48*(9), 1464–1473.

Schnaper, N., & Panitz, H. L. (1990). Near-Death Experiences: Perception Is Reality. *Journal of Near-Death Studies, 9*(2), 97–104.

Schwartz, G. E. (2010). Possible Application of Silicon Photomultiplier Technology to Detect the Presence of Spirit and Intention: Three Proof-of-Concept Experiments. *Explore, 6*(3), 166–171.

Scroppo, J. C., Drob, S. L., Weinberger, J. L., & Eagle, P. (1998). Identifying Dissociative Identity Disorder: A Self-Report and Projective Study. *Journal of Abnormal Psychology, 107*(2), 272–284.

Segal, T. L., & Talbott, V. (2016). Close Encounters of the Familiar Kind. *Journal for Spiritual and Consciousness Studies, 39*(2), 120–129.

Self, W. (1997). The Edison of the Toolshed Orgasm. *Spy Magazine, 11*(5), 34.

Sellers, E. M., Romach, M. K., & Leiderman, D. B. (2018). Studies with Psychedelic Drugs in Human Volunteers. *Neuropharmacology, 142*, 116–134.

Serdahely, W. J. (1990). Pediatric Near-Death Experiences. *Journal of Near-Death Studies, 9*(1), 33–39.

Shaw, D. W. (2014). Reconsidering Maxwell's Aether. *Physics Essays, 27*(4), 601–607.

Sheldrake, R. (2006). Morphic Fields. *World Futures: The Journal of General Evolution, 62*(1/2), 31–41.

Shermer, M. (2005). Abducted. *Scientific American, 292*(2), 34.

Sherwood, S. J. (2002). Relationship Between the Hypnogogic/Hypnopompic State and Reports of Anomalous Experiences. *Journal of Parapsychology, 66*(2), 127–150.

Shiah, Y.-J. (2012). A Possible Mechanism for ESP at the Initial Perceptual Stage. *Journal of Parapsychology, 76*(1), 147–159.

Slade, P. D., & Bentall, R. P. (1988). *Sensory Deception: A Scientific Analysis of Hallucination*. Baltimore: Johns Hopkins University Press.

Sloan, M. C. (2010). Aristotle's Nicomachean Ethics as the Original Locus for the Septem Circumstantiae. *Classical Philology, 105*, 236–251.

Smith, D. E. (2007). Lysergic Acid Diethylamide: An Historical Perspective. *Journal of Psychoactive Drugs, 1*(1), 1–5.

Smythies, J. (2011). Ketamine, Bergson and NDEs. *Journal of the Society for Psychical Research, 75*(904), 148–150.

Solms, M., & Turnbull, O. (2002). *The Brain and the Inner World. An Introduction to the Neurosciences of Subjective Experiences.* New York: Other Press.
Solomonova, E., & Wei, S. X. (2016). Exploring the Depth of Dream Experience. *Constructivist Foundations, 11*(2), 407–416.
Song, D. (2017). Comment on Information and Dualism. *Neuroquantology, 15*(2), 141–144.
Song, D. (2018). Visualization of Information. *Neuroquantology, 16*(1), 32–36.
Spanos, N. P. (1996). Hypnosis: Mythology Versus Reality. In N. P. Spanos (Ed.), *Multiple Identities & False Memories: A Sociocognitive Perspective* (pp. 17–28). Washington, DC: American Psychological Association.
Sparks, G., Pellechia, M., & Irvine, C. (1998). Does Television About UFOs Affect Viewers' UFO Beliefs? An Experimental Investigation. *Communication Quarterly, 46*(3), 19.
Spatt, J. (2002). Déjà Vu: Possible Parahippocampal Mechanisms. *Journal of Neuropsychiatry and Clinical Neurosciences, 4*(1), 6–10.
Stafford Betty, L. (2004). Mind, Paranormal Experience, and the Inadequacy of Materialism. *International Philosophical Quarterly, 44*(3), 373–392.
Staniloiu, A., Markowitsch, H. J., & Brand, M. (2010). Psychogenic Amnesia—A Malady of the Constricted Self. *Consciousness and Cognition, 19*, 778–801.
Stefula, J. J., Butler, R. D., & Hansen, G. P. (1993). *A Critique of Budd Hopkins' Case of the UFO Abduction of Linda Napolitano.* Retrieved November 20, 2018, from http://tricksterbook.com/ArticlesOnline/LindaCase.pdf.
Stevens, A. (1998). Response to P. Pietikainen. *Journal of Analytical Psychology, 43*, 345–355.
St John, G. (2016). The DMT Gland: The Pineal, the Spirit Molecule, and Popular Culture. *International Journal for the Study of New Religions, 7*(2), 153–174.
Strassman, R. (2001). *DMT: The Spirit Molecule: A Doctor's Revolutionary Research into the Biology of Near Death and Mystical Experiences.* Rochester: Park Street Press.
Strassman, R. (2005). Hallucinogens. In M. Earleywine (Ed.), *Mind-Altering Drugs: The Science of Subjective Experience* (pp. 49–85). Oxford: Oxford University Press.
Stuber, G. D., Britt, J. P., & Bonci, A. (2012). Optogenetic Modulation of Neural Circuits That Underlie Reward Seeking. *Biological Psychiatry, 71*(12), 1061–1067.
Studerus, E., Kometer, M., Hasler, F., & Vollenweider, F. X. (2011). Acute, Subacute and Long-Term Subjective Effects of Psilocybin in Healthy Humans: A Pooled Analysis of Experimental Studies. *Journal of Psychopharmacology, 25*(11), 1434–1452.

Stumbrys, T., & Erlacher, D. (2017). Inner Ghosts: Encounters with Threatening Dream Characters in Lucid Dreams. *Dreaming, 27*(1), 40–48.

Sturma, M. (2002). Aliens and Indians: A Comparison of Abduction and Captivity Narratives. *Journal of Popular Culture, 36*(2), 318–334.

Sudduth, M. (2009). Super-Psi and the Survivalist Interpretation of Mediumship. *Journal of Scientific Exploration, 23*(2), 167–193.

Sully, J. (1927). Illusions and Hallucinations. In W. S. Taylor (Ed.), *Readings in Abnormal Psychology and Mental Hygiene* (pp. 546–550). New York: D. Appleton & Company (Original work published 1897).

Szara, S., Rockland, L., Rosenthal, D., & Handlon, J. H. (1966). Psychological Effects and Metabolism of N, N-Diethyltryptamine. *Archives of General Psychiatry, 15*(3), 320–329.

Talarico, J. M., & Rubin, D. C. (2007). Flashbulb Memories Are Special After All; in Phenomenology, Not Accuracy. *Applied Cognitive Psychology, 21*(5), 557–578.

Tammet, D. (2006). *Born on a Blue Day; Inside the Extraordinary Mind of an Autistic Savant.* New York: Free Press.

Tart, C. T. (1972). States of Consciousness and State-Specific Sciences. *Science, 176*(4040), 1203–1210.

Tart, C. T. (1986). Consciousness, Altered States, and Worlds of Experience. *Journal of Transpersonal Psychology, 18*(2), 159–170.

Tart, C. T. (1994). *Living the Mindful Life.* Boston, MA: Shambhala.

Tart, C. T. (1998). Six Studies of Out-of-Body Experiences. *Journal of Near-Death Studies, 17*(2), 73–99.

Tart, C. T. (1999). Moving Psychology Toward the 21st Century. *Contemporary Psychology, 44*(4), 275–276.

Tart, C. T. (2008). Accessing State-Specific Transpersonal Knowledge: Inducing Altered States. *Journal of Transpersonal Psychology, 40*(2), 137–154.

Taylor, A. K. (2013). Leveling and Sharpening. In A. K. Taylor (Ed.), *Encyclopedia of Human Memory* (Vol. 1, p. 618). Santa Barbara, CA: ABC-CLIO LLC.

Taylor, K. A. (1994). How Not to Refute Eliminative Materialism. *Philosophical Psychology, 7*(1), 101–125.

Taylor, W. D. (2011). Yes, There Is a "Happiness" Gene … but It's Not That Simple; the 5-HTT Gene That Transports Serotonin in the Brain Can Lead to Feelings of Satisfaction in Life. *Duke Medicine Health News, 17*(7), 7.

Thalbourne, M. A. (2009). Reports of Paranormal Experiences: Can Transliminality Tell Us Anything About Them. *Archive for the Psychology of Religion, 31*(3), 375–386.

Thalbourne, M. A., & Houran, J. (2000). Transliminality, the Mental Experiences Inventory, and Tolerance of Ambiguity. *Personality and Individual Differences, 28,* 853–863.
Thalbourne, M. A., & Maltby, J. (2008). Transliminality, Thin Boundaries, Unusual Experiences, and Temporal Lobe Lability. *Personality and Individual Differences, 44*(7), 1617–1623.
Tillett, G. (2012). Modern Western Magic and Theosophy. *Aries: Journal for the Study of Western Esotericism, 12*(1), 17–51.
Titchener, E. B. (1899). Idea and the Association of Ideas. In E. B. Titchener (Ed.), *A Primer of Psychology*. London: Macmillan.
Tong, F. (2003). Out-of-Body Experiences: From Penfield to Present. *Trends in Cognitive Science, 7*(3), 104–106.
Totton, N. (2003). Introduction. In N. Totton (Ed.), *Psychoanalysis and the Paranormal: Lands of Darkness* (pp. 1–14). London: Karnac.
Tschigguel, W., & Hunter, M. (2008). Awakening in Hypnosis. *Contemporary Hypnosis, 25*(1), 39–45.
Tucker, G. J., & Neppe, V. M. (1988). Neurology and Psychiatry. *General Hospital Psychiatry, 10*(1), 24–33.
Tuske, J. (1999). Being in Two Minds: The Divided Mind in the Nyayasutras. *Asian Philosophy, 9*(3), 229–239.
Tyrell, G. N. M. (1953/1963). *Apparitions*. New York: Collier Books.
Urban, E. (2005). Fordham, Jung and the Self: A Re-examination of Fordham's Contribution to Jung's Conceptualization of the Self. *Journal of Analytical Psychology, 50*(5), 571–594.
Valle, M., Maqueda, A. E., Rabella, M., Rodríguez-Pujadas, A., Antonijoan, R. M., Romero, S., et al. (2016). Inhibition of Alpha Oscillations Through Serotonin-2A Receptor Activation Underlies the Visual Effects of Ayahuasca in Humans. *European Neuropsychopharmacology, 26*(7), 1161–1175.
van der Sluijs, M. A., & Peratt, A. L. (2009). The Ourobóros as an Auroral Phenomenon. *Journal of Folklore Research, 46*(1), 3–41.
van Erkelens, H. (1991). Wolfgang Pauli's Dialogue with the Spirit of Matter. *Psychological Perspectives: A Semi-Annual Journal of Jungian Thought, 24*(1), 34–53.
Vollenweider, F. X., & Geyer, M. A. (2001). A Systems Model of Altered Consciousness: Integrating Natural and Drug-Induced Psychoses. *Brain Research Bulletin, 56,* 495–507.
Vollenweider, F. X., & Kometer, M. (2010). The Neurobiology of Psychedelic Drugs: Implications for the Treatment of Mood Disorders. *Nature Reviews Neuroscience, 11,* 642–651.

von Lucadou, W., & Wald, F. (2014). Extraordinary Experiences in Its Cultural and Theoretical Context. *International Review of Psychiatry, 26*(3), 324–334.
Wagstaff, G. F., Cole, J., Wheatcroft, J., Anderton, A., & Madden, H. (2008). Reducing and Reversing Pseudomemories with Hypnosis. *Contemporary Hypnosis, 25*(3/4), 178–191.
Wahbeh, H., Sagher, A., Back, W., Pundhir, P., & Travis, F. (2018). A Systematic Review of Transcendent States Across Meditation and Contemplative Traditions Explore. *The Journal of Science and Healing, 14*(1), 19–35.
Weitz, L. J. (1976). Jung's and Freud's Contributions to Dream Interpretation: A Comparison. *American Journal of Psychotherapy, 30*(2), 289–293.
Welman, M., & Faber, P. A. (1992). The Dream in Terminal Illness: A Jungian Formulation. *Journal of Analytical Psychology, 37*(1), 61–81.
Wickens, A. (2009). *Introduction to Biopsychology* (3rd ed.). Harlow, UK: Pearson Education Ltd.
Wilber, K. (1984). Sheldrake's Theory of Morphogenesis. *Journal of Humanistic Psychology, 24*(2), 107–116.
Williams, C. (2014). All-Seeing Eyes. *New Scientist, 221*(2961), 40–41.
Williams, R. J. (2005). *The Symbol of the Hybrid Human/Alien Child in the Abduction Phenomenon: Rebirthing Within the Psyche and the Psychosomatic Imagination.* Proceedings of the Eighth Australian and International Religion, Literature and the Arts Conference 2004. http://hdl.handle.net/2123/1263.
Wilson, C. (1971/1983). *The Occult.* St Albans: Grenada.
Winkelman, M. J. (1990). Shamans and Other 'Magico-Religious' Healers: A Cross-Cultural Study of Their Origins, Nature, and Social Transformations. *Ethos, 18*(3), 308–352.
Wright, C. (1993). Eliminative Materialism: Going—Concern or Passing Fancy? *Mind and Language, 8*(2), 316–326.
Wright, P. A. (1994). A Psychobiological Approach to Shamanic Altered States of Consciousness. *ReVision, 16*(4), 164–172.
Wright, S. H. (2012). An Interview with Anabela Cardoso: Anomalous Electronic Voices Received via Instrumental Transcommunication. *Neuroquantology, 10*(3), 515–518.
Yong, A. (2008). Science and the Akashic Field: An Integral Theory of Everything (Book Review). *Perspectives on Science and Christian Faith, 60*(1), 53.
Yoshimi, J. (2012). Supervenience, Dynamical Systems Theory, and Non-reductive Physicalism. *British Journal for the Philosophy of Science, 63*(2), 373–398.
Zakirov, U. N. (2017). Model of Decaying Dark Energy Based on the Kaluza-Klein Theory. *Russian Physics Journal, 60*(4), 678–684.

Zegans, L. S., Pollard, J. C., & Brown, D. (1967). The Effects of LSD-25 on Creativity and Tolerance to Regression. *Archives of General Psychiatry, 16*(6), 740–749.

Zeligman, M., Greene, J. H., Hundley, G., Graham, J. M., Spann, S., Bickley, E., et al. (2017). Lived Experiences of Men With Dissociative Identity Disorder. *Adultspan Journal, 16*(2), 65–79.

Zhao, B., & Sala, S. D. (2018). Different Representations and Strategies in Mental Rotation. *Quarterly Journal of Experimental Psychology, 71*(7), 1574–1583.

Zhu, C. (2013). Jung on the Nature and Interpretation of Dreams: A Developmental Delineation with Cognitive Neuroscientific Responses. *Behavioral Sciences, 3*(4), 662–675.

Index

T. Jinks, *Psychological Perspectives on Reality, Consciousness and Paranormal Experience*,
https://doi.org/10.1007/978-3-030-28902-7

FSC
www.fsc.org
MIX
Papier | Fördert gute Waldnutzung
FSC® C083411

Kolibri 360 GmbH
Ferdinand-Jühlke-Straße 7
99095 Erfurt, Deutschland
produktsicherheit@kolibri360.de